T0336263

VOLUME ONE HUNDRED AND TWENTY ONE

Advances in COMPUTERS

The Blockchain Technology for Secure and Smart Applications across Industry Verticals

VOLUME ONE HUNDRED AND TWENTY ONE

ADVANCES IN COMPUTERS

The Blockchain Technology for Secure and Smart Applications across Industry Verticals

SHUBHANI AGGARWAL

Thapar Institute of Engineering & Technology, Patiala, Punjab, India

NEERAJ KUMAR

Thapar Institute of Engineering & Technology, Patiala, Punjab, India

PETHURU RAJ

Site Reliability Engineering (SRE) Division, Reliance Jio Platforms Ltd. (JPL), Bangalore, India

Academic Press is an imprint of Elsevier
50 Hampshire Street, 5th Floor, Cambridge, MA 02139, United States
525 B Street, Suite 1650, San Diego, CA 92101, United States
The Boulevard, Langford Lane, Kidlington, Oxford OX5 1GB, United Kingdom
125 London Wall, London, EC2Y 5AS, United Kingdom

First edition 2021

ISBN: 978-0-12-821991-1
ISSN: 0065-2458

For information on all Academic Press publications
visit our website at https://www.elsevier.com/books-and-journals

Publisher: Zoe Kruze
Editorial Project Manager: Leticia Lima
Production Project Manager: James Selvam
Cover Designer: Alan Studholme

Typeset by SPi Global, India

Contents

Contributors

Shubhani Aggarwal
Thapar Institute of Engineering & Technology, Patiala, Punjab, India

Neeraj Kumar
Thapar Institute of Engineering & Technology, Patiala, Punjab, India

Pethuru Raj
Site Reliability Engineering (SRE) Division, Reliance Jio Platforms Ltd. (JPL), Bangalore, India

Preface

Blockchain technology is being passionately presented as one of the breakthrough technological solutions for empowering the real digital transformation across industry verticals. Most of the business domains including supply chain, logistics, healthcare, real estate, telecommunication, banking, and retail have meticulously embraced this powerful technology to visualize and realize a bevy of next-generation mission-critical business applications. A number of current and classic business and IT operations challenges are being identified and addressed through the smart leverage of the blockchain techniques and tools. In short, a dazzling array of automation and acceleration requirements of globally diversified enterprises are getting fulfilled. This technology-enabled transition is meticulously enabling worldwide business behemoths and houses to gain heightened revenue and reputation. Especially blockchain technology is being positioned as the unbreakable and impenetrable data security mechanism for the ensuing digital era. It is also being promoted as the pioneering method for facilitating and proclaiming blockchain ledger/database as the single source of truth for all kinds of corporate, customer, and confidential data. Precisely speaking, it is to redefine and revive the business relationships through trust, transparency, and collaboration for the betterment of the human society. Blockchain bestows legacy processes to be modernized and optimized for the deeply and decisively connected world.

Through this book, we would like to dig deeper and deal with a few prime aspects for our esteemed readers.

1. Firstly, it is to give a solid and succulent introduction to the blockchain paradigm. The first few chapters are allocated to clearly articulate the various foundational and fundamental technical concepts such as cryptography, data structures, hashing techniques, digital signatures, accumulators, consensus algorithms, blockchain architecture including the core components.
2. The second part of the book is primarily for explaining the recent developments in the blockchain space. New blockchain use cases such as smart contracts, decentralized applications (DApps), decentralized autonomous organizations (DAOs), the state-of-the-art industrial internet of things (IIoT) systems, voting systems, and secured digital twins are described in detail.

3. The final portion of the book is to accentuate the various technical challenges and concerns of the blockchain technology and to illustrate how data analytics methods and artificial intelligence (AI) algorithms come handy in minimizing the shortcomings of blockchain. Predominantly the performance, scalability, and security issues are widely expounded.

Blockchain is definitely one of the pioneering technologies for the future of businesses. It had a humble beginning of bitcoin cryptocurrencies. Having understood the special significance of the blockchain technology, which is behind the overwhelming success of bitcoins, other industry verticals have explored, experimented, and expounded the distinct contributions of the blockchain technology. Today several business domains are benefiting immensely out of the unique automation capabilities of blockchain. Along its journey, it also faces a number of loopholes and performance issues. Researchers started to combine AI tools in order to overcome the key vulnerabilities of blockchain. Thereby, there are new buzzwords such as "Blockchain 2.0 and 3.0 and 4.0". The purpose of this book is to expound how enterprises, governments, institutions, and communities can deeply and decisively understand the various technological advancements and accomplishments in the blockchain space in order to build and deploy sophisticated applications.

Pethuru Raj PhD
Site Reliability Engineering (SRE) Division,
Reliance Jio Platforms Ltd. (JPL),
Bangalore, India
E-mail: peterindia@gmail.com
https://sweetypeterdarren.wixsite.com/pethuru-raj-books

CHAPTER ONE

Demystifying the blockchain technology

Pethuru Raj
Site Reliability Engineering (SRE) Division, Reliance Jio Platforms Ltd. (JPL), Bangalore, India

Contents

Abstract

The blockchain paradigm is being widely touted by many as the innovative and disruptive one capable of bringing in a few exemplary and elegant transformations in the IT space. As business operations and offerings are substantially enabled through

Advances in Computers, Volume 121
ISSN 0065-2458
https://doi.org/10.1016/bs.adcom.2020.08.001

the various crucial accomplishments and advancements in the IT field, business executives across the globe are equally keen to experiment with and embrace this new and futuristic technology to reap a slew of business benefits. Interestingly, blockchain has the inherent potential and promise to bring forth a bevy of strategically sound implications for various industry verticals. Cryptocurrency is one of the finest and foremost applications of the blockchain technology. The supply chain domain is exploring this new phenomenon for realizing some crucial advantages. The IoT discipline is another one capable of attaining a number of distinct benefits out of all the trendsetting improvisations being realized in the blockchain space. This chapter is specially crafted to tell what, why, how, and where the indispensable blockchain paradigm is being used toward real digital transformations.

1. Introduction

Newer technologies are consistently emerging and evolving fast in the hot and happening IT space with the overwhelming support of IT professionals and professors. Businesses are typically handicapped in some of the important areas due to the unavailability of cutting-edge technologies and state-of-the-art tools. These limitations are being steadily surmounted through the smart analysis and application of newly incorporated technological paradigms. On the other hand, fresh possibilities and opportunities are bound to emerge and get noticed with the conscious adoption and adaption of new technologies. Some of the tough-to-achieve business requirements such as customer delight, higher productivity, new business models, and revenue-generation windows, etc. can be easily and quickly realized and delivered through newly introduced technologies and their methodical leverage. Thus, worldwide business enterprises, in order to be ahead of their competitors, readily and rewardingly embrace promising and potential technologies with all the alacrity and clarity.

In the recent past, there came a few pioneering technologies. Also, previously discarded technologies get a fresh life because of a dazzling array of distinct improvisations in the IT infrastructures domain and also due to the steady emergence of integrated and insightful platforms. Artificial intelligence (AI) with a bombardment of heavily improved machine and deep learning (ML/DL) algorithms are garnering a lot of attraction and attention these days. With a host of digitization and edge technologies gaining uninhibited prominence and dominance, the aspects of extreme connectivity and deeper integration among digitized entities and devices are seeing the reality. The Internet of things (IoT) and cyber physical systems (CPS) concepts are traversing through a surging popularity these days. Another highly visible trend is to club together multiple proven technologies to present a new and clustered technology. The blockchain technology is a neat combination of several pioneering technologies and this blended paradigm is being pronounced as one of the most significant technologies capable of making solid and spectacular impacts. This chapter is all about explaining the blockchain technology and how it is going to be hugely impactful for several industry verticals.

2. The key motivations for the blockchain technology

Many technical, business, and user cases are emerging for experimenting with and embracing the blockchain technology, which is seen largely as a disruptive one by many. Today banks play a very vital role in facilitating financial transactions. Banks provide the much-needed trust when we exchange money with our friends and colleagues across the world. But this model is very expensive and slow. These intermediary-based financial transactions stifle the innovation culture. The centralized computing model being adopted by banks comes in the way of exploiting the distinct benefits of distributed and decentralized computing. If the centralized database gets broken, then all kinds of customer, confidential, and corporate data gets stolen. This model is vulnerable to distributed denial of service (DDoS) attacks. Hence, the brewing trend is that we need trust but at the same time, we want to discard banks and other intermediaries forthwith. Employing banks makes the whole process a bit complicated. For legal transactions, lawyers and judges play that intermediary role. Contract documents are being made and used for eliminating any kind of confusions and cheatings. Precisely speaking, trust and formal contracts have been the predominant ways and means for all kinds of transactions. Unfortunately, both trust and contract are not optimal solutions. The blockchain technology offers a third option, which is affordable, inventive, fast, and secure.

Secondly, we are heading toward the connected era. We are being bombarded with a dazzling array of slim and sleek, trendy, and handy devices. These embedded and networked devices come handy in producing next-generation services and applications not only for business houses but also for common people. That is, we are at the verge of entering into the era of the IoT and cyber physical systems (CPS). The well-known challenge and concern here is the security, safety, and privacy of IoT devices and data. The traditional security solution approaches are found incompetent in guaranteeing the utmost security due to the participation of resource-constrained devices. Any everyday environment stuffed and equipped with intelligent sensors and actuators is typically dynamic and liable for instability. Herein, the arrival of the blockchain paradigm is being applauded. That is, futuristic environments (personal, industrial, social, and professional) are going to be not only smart but also secure and safe.

Finally, suppose you are a retailer sourcing from over 100 different fast-moving consumer goods (FMCG) companies, including over 10,000 stock keeping units, in one store, and your B2B transactions amount to over 1000 high transactions a day. Now, these 1000 transactions are just from one store. You may have 10 or more stores. Each of these transactions gets recorded on a combination of paper and digital methods. Now the problem here is that your purchase department can underreport transactions and bill you for the ordered quantity. So, while your stock shows the right amount, your store shelves are never stacked up after a sale. This presents a problem because it hurts top line and bottom line revenues. Further on, there can also be price changes very often being made by the FMCG company. That details cannot be recorded by the retailer.

Now enter blockchain, assuming that the retailer has moved all transactions digitally, which creates blocks of each transaction like Vendor A supplied X, Vendor B supplied Y and Vendor C supplied YY and in the process the system ropes in the finance and the purchase team together with each receipt generated. Now each transaction becomes part of a block and gets recorded in the ledger of all parties involved. Any changes made to a price or purchase agreement, another block gets added, which has to be verified by all parties. The algorithm secures each block of transactions and it cannot be tampered with because all parties have the keys to make the transactions legitimate. The technology brings transparency and traceability.

Newer technologies, architectural patterns, and tools are forthcoming in order to assist deeper and decisive automation in business operations. Also, disruptive applications can be realized through fresh technology-inspired approaches and algorithms. Decentralized applications (DApps) and smart contracts are being demanded these days. Blockchain has showed the initial exuberance and the innate strength to setup and sustain DApps across multiple industry verticals. There are a number of distinct advantages of distributed databases and decentralized applications. IT, business, and device services can do certain actions automatically through smart contracts. That is, contract logic gets embedded inside the application and service code in order to automate the contractual obligations. Thus, eliminating third-party involvement through automation is the core strength of this new technological paradigm. Several other business verticals are keenly experimenting and exploring to gain something new out of this concept.

3. Delineating the blockchain technology

Governments and corporates all over world are slowly and steadily realizing the value in blockchain, which is a new kid on the block. It originated from the financial circles and is permeating into the business world. Its ardent supporters sincerely applaud it for its unbreakable and impenetrable security features. The shortest definition for blockchain is it is a distributed ledger. That is, it stores any list of transactions in a peer-to-peer (P2P) network. Data in a blockchain are stored in fixed structures called "blocks."

- *Cryptocurrency* is a digital currency in which encryption techniques are used to regulate the generation of units of currency, and to verify the transfer of funds operating independently of a central bank.
- *Distributed ledger* is a database that is consensually shared and synchronized across the network spread across multiple sites, institutions, or geographies. It allows transactions to have public "witnesses," thereby making a cyberattack more difficult.
- *Smart contracts* are agreements that are encoded in a computer program and automatically executed upon certain criteria being met. Advantages of smart contracts include improved quality, reduced contract execution costs, and increased speed. Smart contracts can be stored on the blockchain.
- *Miners* are the people keeping the blockchain running by providing a huge amount of computing resources competing to solve a cryptographic puzzleand upon solving the puzzle, they generate a block and they also get rewarded. Miners compete with each other to generate a valid block of transactions.

  Miners collect all pending transactions from the decentralized network then they guess a random number (nonce) to solve cryptographic puzzle, on successfully solving the puzzle they generate a block then they push that block into the network for verification from other nodes so that other nodes after verification can add that block in their copy of blockchain.

- *Nonce*—The cryptographic puzzle that miners solve is to identify the value of nonce. A nonce is a random number which can be used only one time. Mostly it is a random number with combination of some data. Blockchain adds a value called nonce in each block. This nonce is like a salt added to the contents of a block. By adding nonce, the hash output of the contents of the block will change.
- *Hash*—Hashing is a cryptographic technique which maps input data to data of a fixed size output. Bitcoin uses the SHA-256 algorithm for it. SHA-256 output a fixed length number. A slight change in input would change the complete output but the output would always of the same length.

Consensus algorithm is a process in computer science used to achieve agreement on a single data value among distributed systems. Consensus algorithms are designed to achieve reliability in a network involving multiple nodes. Consensus algorithms are capable of doing two things: ensuring that the next block in a blockchain is the one and only version of the truth, and keeping powerful adversaries from derailing the system and successfully forking the chain.

A blockchain can be defined as an anonymous online ledger that uses the data structure to simplify the way we transact. Blockchain allows the users to manipulate the ledger in a secure way and without the help of any third party. A blockchain is anonymous, thus protecting the identities of the users. This makes blockchain a more secure means to carry out financial transactions in the extremely and deeply connected world. The algorithm used in blockchain reduces the dependence on the people to verify the transactions. A blockchain is a kind of transparent, independent, and permanent database coexisting in multiple locations and shared by a common community.

Bitcoin is the first and foremost application of the blockchain technology. We would like to explain the blockchain nitty-gritty through the famous bitcoin application. As we all know, bitcoin is the well-known and increasingly used application of the blockchain technology. We often hear and read about blockchain-compliant cryptocurrencies. Bitcoin is a digital currency that can be used to exchange products and services over, just like we use our paper currencies such as Indian Rupees (INR), United States Dollar (USD), Euro (EUR), etc.

4. Briefing of the blockchain system elements

We know that any blockchain system is simply a distributed and decentralized ledger. The ledger is a chain of blocks. Let us have a look at what a block comprises of. Every block in a blockchain comprises of three core elements.

1. Data inside the block
2. Hash or the unique identity of the block.
3. Hash of the previous block.

Data inside the block—The data inside each block changes as per the technology used. For example, in a bitcoin blockchain, the blocks contain the sender's ID, the receiver's ID as well as the amount of bitcoins being transferred.

The blocks on the bitcoin blockchain are 1 MB of data each. At this point of time, there are around 525,000 blocks. The total size of the bitcoin blockchain is around 525,000 MB. The primary data on the bitcoin blockchain is transaction data of all the bitcoin transactions.

Block 1 (1 MB)	Block 2 (1 MB)	Block 3 (1 MB)
Transaction data 1 Transaction data 2 Et cetera...	Transaction data 13 Transaction data 14 Et cetera...	Transaction data 28 Transaction data 29 Et cetera...

Block 1 chronologically describes the first transactions that have occurred up to 1 MB. The subsequent transactions weighing 1 MB would be stored in block 2 and so on. These blocks are now being chained together. To chain them, every block gets a unique (digital) signature that corresponds to exactly the string of data in that block. If there is any change inside a block, the block will get a new digital signature. This is done through the activity called hashing, which is explained below in detail.

Let us say block 1 registers two transactions, transaction 1 and transaction 2. Assume that these transactions make up a total of 1 MB. This block of data now gets a signature for this specific string of data. Let's say the signature is "X32." This is pictorially displayed below.

A single digit change to the data in block 1 will produce a new signature. The data in block 1 is now linked to block 2 by adding the signature of block 1 to the data of block 2. With this arrangement, the signature of block 2 depends on the signature of block 1. The linkage between block 1 and 2 is vividly illustrated in this diagram.

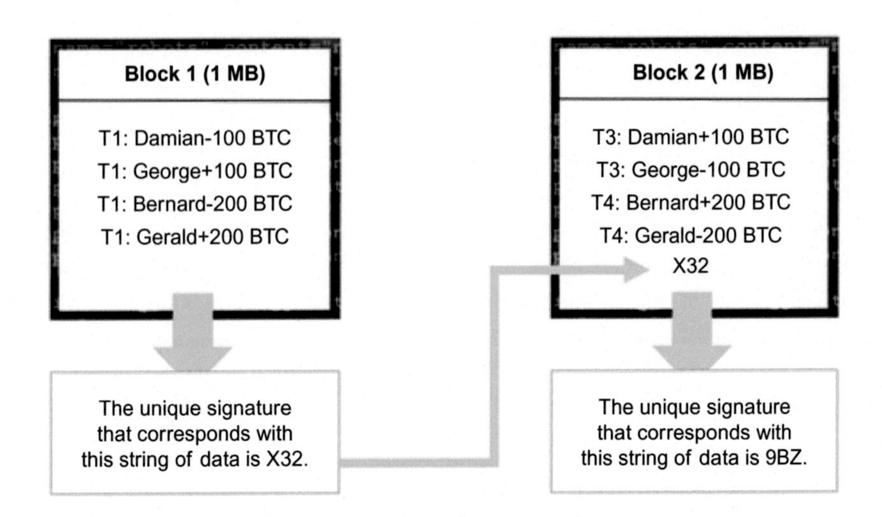

The signatures link the blocks together and help in realizing a chain of blocks, which are a collection of multiple transactions. The blockchain is displayed below.

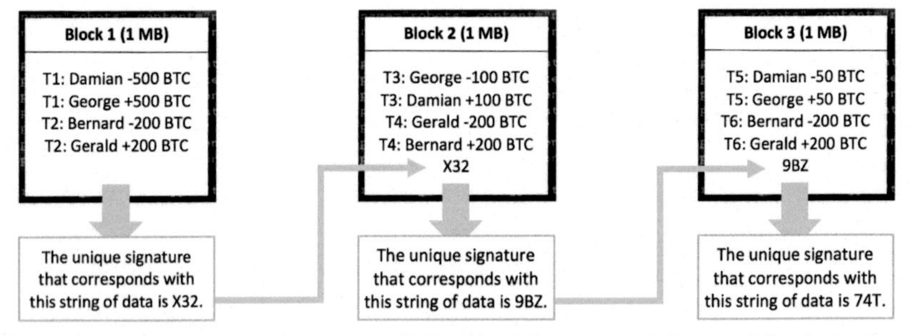

As indicated above, if there is a slight modification of data in block 1, then the block 1 gets a new digital signature and hence block 1 and 2 cannot be joined to each other as illustrated in the below diagram.

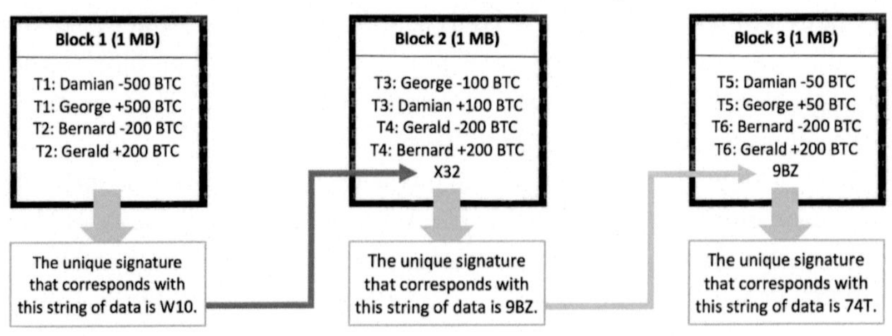

This gives an indication to other users of this blockchain that some data gets changed in block 1. They reject this unauthorized change by shifting back to a previous record of the blockchain where all the blocks are still chained together. The only way for an alteration to be accepted or undetected is if all the blocks stay chained together.

The hash of a block—As mentioned above, the hash of a block is its digital signature and the hash technique plays out a pivotal role in shaping up the concept of blockchain. Each block in a blockchain contains a unique hash and can be generally compared to a fingerprint. If a hash of a block is changed, then the block loses its original form. The hash of the previous block contributes for the formation of the chain between the blocks. It acts as a valid proof of concept for the further blocks to act upon and is the foremost security measure of a blockchain against tampering.

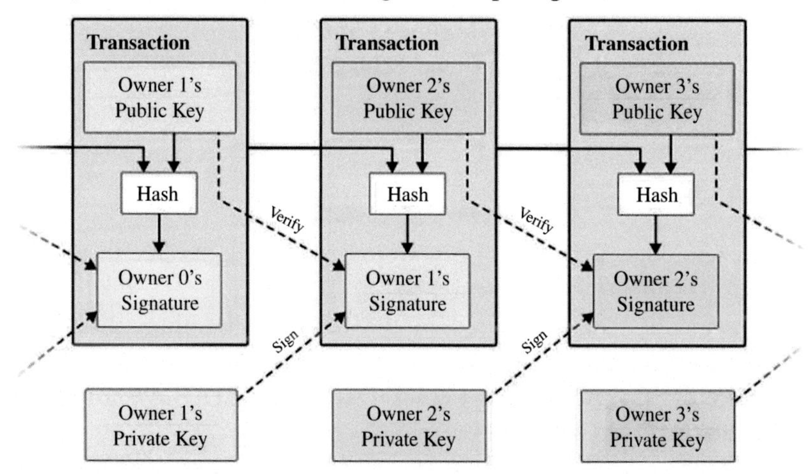

We now understand what forms a block in a blockchain and how blocks get limed up together to form a chain. This link is established and sustained through the hash concept. Now if in case, the hash of any block gets changed in a blockchain, the consecutive blocks no longer recognize it as a valid block. Thus, any kind of tampering makes the block invalid. This is a prominent security mechanism provided natively by blockchain.

How the signature (hash) is created—Block 1 is a record of only one transaction. A sends 100 bitcoin to B. This specific string of data now requires a signature. In blockchain, this signature is created by a cryptographic hash function, which is a very complicated formula that takes any string of input and turns it into a unique 64-digit string of output. For example, if we insert the word "Jinglebells" into this hash function, the output for this specific string of data is:

761A7DD9CAFE34C7CDE6C1270E17F773025A61E511A56F 700D415F0D3E199868

If a single digit of the input changes, including a space, changing a capital letter or adding a period, for example, the output will be totally different. A cryptographic hash function always gives the same output for the same input, but always a different output for different input. This cryptographic hash function is used by the bitcoin blockchain to give the blocks their signatures. The input of the cryptographic hash function in this case is the data in the block, and the output is the signature that relates to that. Let us have a look at block 1 again. Thomas sends 100 Bitcoin to David.

Now imagine that the string of data from this block looks like this:

Block 1 Thomas −100 David +100

If this string of data is inserted in the hashing algorithm, the output (signature) will be this:

BAB5924FC47BBA57F4615230DDBC5675A81AB29E2E0FF85 D0C0AD1C1ACA05BFF

This signature is now added to the data of block 2. Let's say that David now transfers 100 bitcoin to Jimi. The blockchain now looks like this:

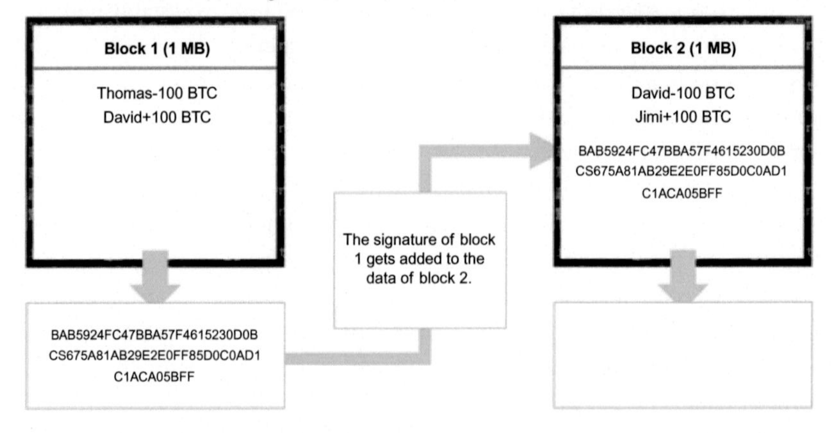

The string of data of block 2 now looks like:

Block 2 David −100 Jimi +100
BAB5924FC47BBA57F4615230DDBC5675A81AB29E2E0FF85D0
C0AD1C1ACA05BFF

If this string of data is inserted in the hashing algorithm, the output (signature) will be this:

25D8BE2650D7BC095D3712B14136608E096F060E32CEC 7322D22E82EA526A3E5

And so, this is the signature of block 2.

There are several hash algorithms. For bitcoin application, SHA-256 hashing algorithm is preferred to generate digital signature for each of the blocks in any blockchain.

The question here is how the signatures stop someone from simply inserting a new signature for each block after altering one. As mentioned above, any change goes unnoticed if all the participating blocks are properly linked. The answer is that only hashes (signatures) that meet certain requirements are accepted on the blockchain.

When does the signature qualify and who signs a block?—A signature does not always qualify. A block will only be accepted on the blockchain if its digital signature starts with a consecutive number of zeroes. For example; only blocks with a signature starting with at least 10 consecutive zeroes qualify to be added to the blockchain. If the signature (hash) of a block does not start with 10 zeroes, the string of data of a block needs to be changed repeatedly until a specific string of data is found that leads to a signature starting with 10 zeroes. Because the transaction data and metadata (block number, timestamp, etc.) need to stay the way they are, a small specific piece of data is added to every block that has no purpose except for being changed repeatedly in order to find an eligible signature. This piece of data is called the nonce of a block. The nonce is completely random and could literally form any set of digits, ranging from spaces to question marks to numbers, periods, capital letters, and other digits.

To summarize, a block now contains: (1) transaction data, (2) the signature of the previous block, and (3) a nonce. The process of repeatedly changing the nonce to find an eligible signature is called mining and is what miners do. Miners spend electricity in the form of computational power in order to constantly try different nonces. The more computational power they have, the faster they can insert random nonces and the more likely they are to find an eligible signature faster. It is a form of trial and error as illustrated in the below diagram.

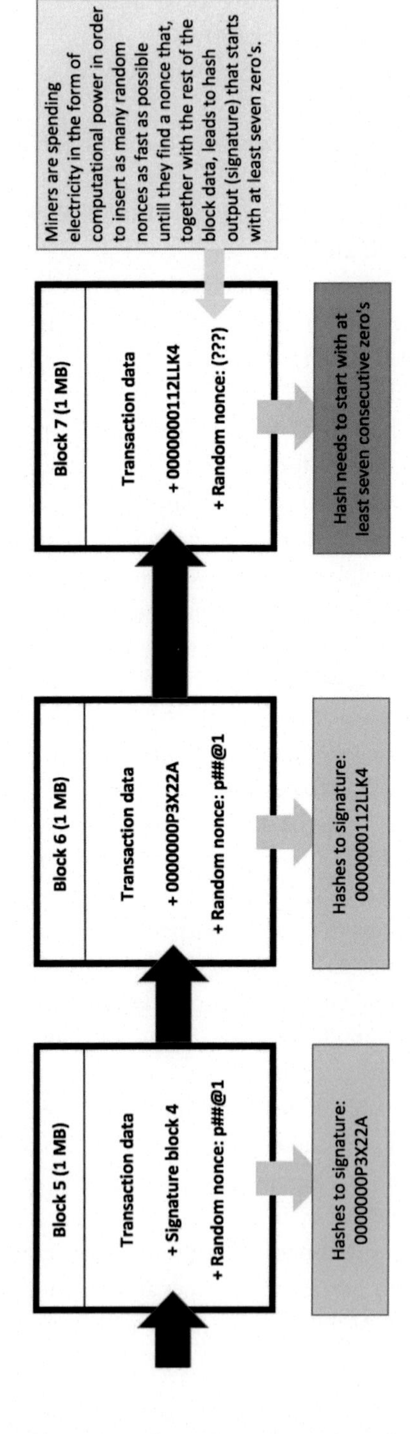
Block 5 (1 MB)
Transaction data
+ Signature block 4
+ Random nonce: p##@1
Hashes to signature: 0000000P3X22A
Block 6 (1 MB)
Transaction data
+ 0000000P3X22A
+ Random nonce: p##@1
Hashes to signature: 0000000112LLK4
Block 7 (1 MB)
Transaction data
+ 0000000112LLK4
+ Random nonce: (???)
Hash needs to start with at least seven consecutive zero's
Miners are spending electricity in the form of computational power in order to insert as many random nonces as fast as possible untill they find a nonce that, together with the rest of the block data, leads to hash output (signature) that starts with at least seven zero's.

Thus, the blockchain technology is a combination of multiple proven technologies in order to bring in utmost security for all kinds of digital data and assets in the extremely online world.

5. Blockchain consensus algorithms

A consensus in a blockchain is a set of rules to be followed to maintain a blockchain system. Consensus is to achieve an agreement on adding new data value in distributed servers or node in the network. Blockchain generally consists of several blocks of information and it is distributed over the network. Each node in the network has the same copy of blockchain and they follow same consensus rules to validate and generate a new block. Each block in blockchain has information of many transactions.

Without consensus, blockchain is simply a way of storing encrypted/unencrypted data. Consensus allows it to decentralized because all the nodes in network follow the same rules and it will maintain uniformity in all the copies of a blockchain. So each and every change in a single blockchain is verified and adopted by another blockchain node in the network.

A fundamental problem in distributed computing and multiagent systems is to achieve overall system reliability in the presence of a number of faulty processes. This often requires processes to agree on some data value that is needed during computation. These processes are described as consensus.

- What happens when an actor decides to not follow the rules and to tamper with the state of his ledger?
- What happens when these actors are a large part of the network, but not the majority?

In order to create a secure consensus protocol, it must be fault tolerant. Blockchain has a distributed nature. That is, the blockchain network has several distributed nodes and each node has to have a copy of the same blockchain ledger. This setup makes blockchain immutable but at the same time to share the same kind of information across the nodes, all the nodes must comply to the same agreement. When a new block gets added in a chain, the same block has to be added in all the copies of the chain. In case of proof of work (PoW) to add a new block, all the nodes in the network try to solve complex cryptographic puzzles and the one which solves it first will gain the authority to add the new block and other nodes have to update their blockchain to adopt this change.

6. Power of work

PoW is a protocol and its main goal is for deterring cyber-attacks such as a distributed denial-of-service attack (DDoS). As we all know, the purpose of DDoS is to exhaust the precious resources of a blockchain node by sending multiple fake requests from geographically distributed systems.

This agreement is maintained by implementing the consensus algorithm. It enables all the node in the network for a common agreement. PoW is one of the consensus algorithms. As the name suggests for validating a transaction, a node has to publicly prove that it has done a certain amount of work. Node has to show the proof for its work and they do it by solving a complex cryptographic puzzle.

In PoW blockchain, all nodes validate transactions to create next block and only the first node that successfully computes it will add the new block. As soon as the other nodes in the network receive it, they will stop trying to build it and move on to the next block. PoW is expensive and time-consuming to produce but easy for others to verify. The hash of the last recorded block in nonce is not known to anyone. The miner needs to find it, and that is the cryptographic puzzle. The miner needs to try one number after another. The cryptographic puzzle is asymmetric that means puzzle is hard for the miners to solve but easy for the nodes to review.

Each block that is added to the blockchain must follow consensus rules if any block does not follow those rules will be rejected by the network nodes. Bitcoin uses PoW to ensure miner has the correct copy of the blockchain and validated all transaction to be in sync with all other nodes in the network. Once a miner solves their PoW puzzle, the results including the previous block's address, the collection of transactions in the block, and the nonce are published to the network for verification. The other nodes from the network automatically check whether the results are valid. If the results are valid, the other nodes simply add the verified block to their copies of the blockchain.

The difficulty of cryptographic puzzle increases by adding leading zeroes. A single change in the string of nonce completely change its output hash value. So adding leading zeroes increases the difficulty of the puzzle. This difficulty is decided by the Blockchain network itself. The higher the difficulty setting, the more difficult it will be to evaluate the nonce. There is no formal formula to calculate nonce. It is just a series of hit and trials. This evaluation of nonce to achieve a hash value of the block data is the "work" that the miners perform and the right output of the puzzle is the "proof."

The proof of work (PoW) algorithm mandates network participants to verify transactions through a computationally complex mathematics puzzle. This algorithm recommends to give appropriate incentives to verify transactions. That is, by providing a reward to the node that successfully completes the puzzle. This idea of computational complexity acts as a huge deterrent. In other words, this guarantees an additional security to the system because it requires for any hacker and evildoer to spend a tremendous amount of time, treasure, and talent to break the chain and to make any retroactive changes after consensus has been reached.

Let us say that a number of bitcoin transactions have occurred and they ought to be verified and validated. All of the nodes that are connected to the network (known as miners) now have to compete to mine the identified block of transactions. They can do this by taking a hash of a few different inputs including (i) all of the transactions in the block, (ii) the timestamp of the time at which the block got hashed, (iii) the difficulty of the block, (iv) the hash of the previous block of transactions, and (v) an extra number called a "nonce."

The problem to be solved here is that miners have to take all of those hash inputs and find a specific hash output. The only way to do this is to guess a nonce and hash the inputs. If those inputs do not lead to the correct output, the miner has to guess a different nonce and rehash the inputs. The miners have to repeat this until they arrive at the correct output. In other words, the math problem is a guess and check problem where the nonce is the variable that can be changed to arrive at correct output. The first miner, who is able to quickly guess the right nonce wins the competition handsomely. He gets a preset number of bitcoins. After this occurs, all of the other nodes on the network check the transactions and the nonce to make sure that the transactions are valid. Then, miners move onto the next block and start the process all over again.

6.1 Advantages of PoW

- A complete decentralization is being achieved by it.
- It is quite easy to implement.
- A huge cost and computing power are required to destroy the system.

6.2 Disadvantages of PoW

- Waste of energy and resource as miners perform work on the creation of blocks that consume a lot of computing power and only one miner will get reward who resolve the puzzle first.

- For resolving PoW problem, miner must have a computer with a lot of processing power and electricity with that other peripherals. For cooling of the system and the environment, miners need to invest more money temperature.

6.3 Where POW used?

PoW is used in many blockchain applications. Some of them are given below:

1. *Bitcoins*—Bitcoin is the first cryptocurrency and therein, each transaction gets recorded in a public ledger.
2. *Ethereum* is an open-source blockchain platform. It is a distributed blockchain network. Smart contract is one of the most popular features of Ethereum.
3. *Bitcoin cash* is also a kind of cryptocurrency. In Bitcoin cash, a block size is increased from 1 to 8 MB.
4. *Litecoin* is a decentralized online cryptocurrency. It can be used to purchase services and products. It is a peer-to-peer (P2P) cryptocurrency.
5. *Zcash*—This is one of the new privacy cryptocurrencies. Its main goal is to provide enhanced privacy.

For mining miner required powerful hardware such as Graphics Processing Units (GPUs). The mining processes required a high amount of electricity. Rewards for mining is to ensure that a blockchain network keeps increase with tough competition to generate new blocks.

7. Proof of stake

In a lottery, probabilistically speaking, if Bob has more tickets than Alice, he is more likely to win. In PoW, if Bob has more computational power and energy than Alice (thus can output more work), he is more likely to win to mine the next block. In proof of stake (PoS), if Bob has more stake than Alice, he is more likely to win ("mine" the next block). PoS takes away the energy and computational power requirement of PoW and replaces it with stake. Stake is referred to as an amount of currency that an actor is willing to lock up for a certain amount of time. In return, they get a chance proportional to their stake to be the next leader and select the next block. PoS replaces miners with "validators," who are given the chance to validate (mine) the next block with a probability proportional to their stake. There are various existing coins which use pure PoS, such as Nxt and Blockchain.

The main issue with PoS is the so-called nothing-at-stake problem. Essentially, in the case of a fork, stakers are not disincentivized from staking in both chains, and the danger of double-spending problems increase. In order to avoid that, hybrids consensus algorithms such as the PoW-PoS combination are recommended.

8. Delegated proof of stake

Another consensus algorithm is delegated proof of stake (DPoS), which is a variant of PoS that provides a high level of scalability at the cost of limiting the number of validators on the network. DPoS is a system in which a fixed number of elected entities (called block producers or witnesses) are selected to create blocks in a round-robin order. Block producers are voted into power by the users of the network, who each get a number of votes proportional to the number of tokens they own on the network (their stake). Alternatively, voters can choose to delegate their stake to another voter, who will vote in the block producer election on their behalf.

- *Block producers* are those responsible for creating and signing new blocks. They are limited in number, and are elected by the voters.
- *Block validators* in DPoS refer to full nodes who verify that the blocks created by block producers follow the consensus rules. Any user is able run a block validator and verify the network.

The number of block producers in a DPoS network is up to the consensus rules of that chain. A round in a DPoS blockchain with N block producers is as follows:

1. N block producers get elected from the pool of block producer candidates.
2. The ith block producer signs the ith block, until $i = N$.

A block is finalized (i.e., cannot be reversed) when it is voted on by $(2/3 + 1)$ of the block producers. Otherwise, the longest chain rule is followed. Finality is an extremely important property when aiming for interoperability between blockchains.

The block reward and inflation mechanism for each DPoS implementation depend on each project's reward model. Voters can also "fire" a block producer if they are found to be malicious (i.e., try to censor transactions or double-spend) by not voting for them in the next round. As a result of the limited number of block producers, DPoS is able to handle transaction throughput that is multiple orders of magnitude greater than today's PoW.

9. Proof of burn

The miners of the proof of burn (PoB) coins will send coins to an unspendable address (otherwise known as an "eater address") in order to take them forever out of circulation or burning them. These transactions are recorded on the blockchain, ensuring that there is a necessary proof that the coins cannot be spent again, and the user who burned the coins is issued a reward.

The entire idea behind PoB consensus is that the user burning the cryptocurrency is showing long-term commitment to the coin by burning it. This is because they are taking a short-term loss in exchange for a long-term gain. Additionally, burning coins is also viewed as less resource intensive by some since the main resource being used is the person's willingness to delay their profits. As long as someone is okay with being patient for their gains, then this may be the consensus algorithm from him or her.

As time progresses, the user of a PoB coin continues receiving rewards, either increasing their stake of alternative coins or earning greater privileges for mining on the network. This way, if a user burns more coins, they will have a greater chance of successfully mining the next block and further increasing their overall rewards. Moreover, there are actually a few different ways to implement a PoB consensus mechanism. In some cases, an existing PoW coin can be burned in exchange for the PoB coin. In other cases, the actual PoB coin is burned in order to gain increased mining privileges. In much the same way that the cost of mining Bitcoin increases over time in the form of hardware and electricity costs, the cost of mining a PoB coin also increases over time as more coins need to be burned to maintain the same odds of being selected to mine the next block.

10. Practical Byzantine fault tolerance

This is a consensus algorithm. It aims to improve upon original Byzantine fault tolerance (BFT) consensus mechanisms and has been implemented and enhanced in several modern distributed computer systems, including a few blockchain platforms.

Byzantine fault tolerance (*BFT*)—In the context of distributed systems, BFT is the ability of a distributed computer network to function as desired and correctly reach a sufficient consensus despite malicious components (nodes) of the system failing or propagating incorrect information to other peers. The objective is to defend against catastrophic system failures by mitigating the influence these malicious nodes have on the correct function of the network and the right consensus that is reached by the honest nodes in the system.

The practical Byzantine fault tolerance (pBFT) model primarily focuses on providing a practical Byzantine state machine replication that tolerates Byzantine faults (malicious nodes) through an assumption that there are independent node failures and manipulated messages propagated by specific and independent nodes. The algorithm is designed to work in asynchronous systems and is optimized to be high-performance with an impressive overhead runtime and only a slight increase in latency.

Essentially, all of the nodes in the pBFT model are ordered in a sequence with one node being the primary node (leader) and the others referred to as the backup nodes. All of the nodes within the system communicate with each other and the goal is for all of the honest nodes to come to an agreement of the state of the system through a majority. Nodes communicate with each other heavily, and not only have to prove that messages came from a specific peer node, but also need to verify that the message was not modified during transmission.

For the pBFT model to work, the assumption is that the amount of malicious nodes in the network cannot simultaneously equal or exceed ⅓ of the overall nodes in the system in a given window of vulnerability. The more nodes in the system, then the more mathematically unlikely it is for a number approaching ⅓ of the overall nodes to be malicious. The algorithm effectively provides both liveness and safety as long as at most (n – ⅓), where n represents total nodes, are malicious or faulty at the same time. The subsequent result is that eventually, the replies received by clients from their requests are correct due to linearizability.

Each round of pBFT consensus (called views) comes down to four phases. This model follows more of a "Commander and Lieutenant" format than a pure Byzantine Generals' Problem, where all generals are equal, due to the presence of a leader node. The phases are given below.

1. A client sends a request to the leader node to invoke a service operation.
2. The leader node multicasts the request to the backup nodes.
3. The nodes execute the request and then send a reply to the client.
4. The client awaits f+1 (f represents the maximum number of nodes that may be faulty) replies from different nodes with the same result. This result is the result of the operation.

The requirements for the nodes are that they are deterministic and start in the same state. The final result is that all honest nodes come to an agreement on the order of the record and they either accept it or reject it. The leader node is changed in a round-robin type format during every view and can even be replaced with a protocol called view change if a specific amount of time has passed without the leader node multicasting the request. A supermajority of honest nodes can also decide whether a leader is faulty and remove them with the next leader in line as the replacement.

The pBFT model has the capability to provide transaction finality without the need for confirmations like in PoW, if a proposed block is agreed upon by the nodes in a pBFT system, then that block is final. This is enabled by the fact that all honest nodes are agreeing on the state of the system at that specific time as a result of their communication with each other. The pBFT model is energy-efficient unlike PoW. A PoW round is required for every block and this has led to higher electrical consumption. With pBFT not being computationally intensive, a substantial reduction in electrical energy is inevitable as miners are not solving a PoW computationally intensive hashing algorithm every block.

On the other hand, this model only works well in its classical form with small consensus group sizes due to the cumbersome amount of communication that is required between the nodes. This uses digital signatures and MACs (method authentication codes) as the format for authenticating messages. Using MACs is extremely inefficient with the amount of communication needed between the nodes in large consensus groups such as cryptocurrency networks, and with MACs, there is an inherent inability to prove the authenticity of messages to a third party.

The pBFT model is also susceptible to sybil attacks where a single party can create or manipulate a large number of identities (nodes in the network), thus compromising the network. This is mitigated against with larger network sizes, but scalability and the high-throughput ability of the pBFT model is reduced with larger sizes and thus needs to be optimized or used in combination with another consensus mechanism.

Platforms implementing optimized versions of pBFT—Today, there are a handful of blockchain platforms that use optimized or hybrid versions of the pBFT algorithm as their consensus model or at least part of it, in combination with another consensus mechanism.

- *Zilliqa* employs a highly optimized version of classical pBFT in combination with a PoW consensus round every ~100 blocks. They use multisignatures to reduce the communication overhead of classical pBFT and in their own testing environments, they have reached a TPS of a few thousand with hopes to scale to even more as more nodes are added. This is also a direct result of their implementation of pBFT within their sharding architecture so that pBFT consensus groups remain smaller within specific shards, thus retaining the high-throughput nature of the mechanism while limiting consensus group size.

- *Hyperledger* is an open-source collaborative environment for blockchain projects and technologies that is hosted by the Linux Foundation and uses a permissioned version of the pBFT algorithm for its platform. Since permissioned chains use small consensus groups and do not need to achieve the decentralization of open and public blockchains such as Ethereum, pBFT is an effective consensus protocol for providing high-throughput transactions without needing to worry about optimizing the platform to scale to large consensus groups. Additionally, permissioned blockchains are private and by invite with known identities, so trust between the parties already exists, mitigating the inherent need for a trustless environment since it is assumed less than ⅓ of the known parties would intentionally compromise the system.

BFT is a well-studied concept in distributed systems and its integration through the PBFT algorithm into real-world systems and platforms, whether through an optimized version or hybrid form, remains a key infrastructure component of cryptocurrencies today. As platforms continue to develop and innovate in the field of consensus models for large-scale public blockchain systems, providing advanced BFT mechanisms will be crucial to maintaining various systems' integrity and their trustless nature.

Some of the criticisms of consensus algorithms are they require enormous amounts of computational energy, that it does not scale well (transaction confirmation typically takes about 10–60 min) and that the majority of mining is centralized in areas of the world where electricity is cheap.

Consensus algorithms ensure that the next block in a blockchain is fully validated and secured. There are multiple kinds of consensus algorithms which currently exist, each with different fundamental processes. We have discussed the leading ones. There are other methods such as proof of identity, proof of importance, proof of elapsed time, etc. The simplicity, modifiability, and usability of consensus algorithms enable building and sustaining decentralized systems. Readers can find more information on consensus algorithms at this page (https://blockgeeks.com/guides/blockchain-consensus/) and (https://101blockchains.com/consensus-algorithms-blockchain/).

11. Blockchain types

Due to its unique advantages and applications, the blockchain paradigm is on fast track. The adoption and adaptation are on the higher side. Enterprises are rolling out newer blockchain use cases in order to make it more visible and viable. Resultantly there are a few blockchain types. There are certain use cases mandating for a couple of different types: public/permissionless and private/permissioned. With a public blockchain, anyone can join and leverage blockchain functionalities whereas private blockchain can be accessed only by authorized users. Within a public blockchain, the much-required trust is being realized through game-theory and cryptography concepts. In case of public blockchain, anyone can simply download the appropriate application and join the decentralized blockchain to involve in transactions. It is not mandatory to have a previous relationship with the ledger, and there is no approval needed also to join public blockchain.

On the other hand, private/permissioned blockchains do not require such artificial incentives since all actors in the network are known to each other. New actors that want to join the network have to be approved by existing actors in the network. This enables more flexibility and efficiency of validating transactions. Private blockchains are typically used by a consortium of organizations that like to keep a shared ledger for settlement of transactions. Predominantly, financial services providers such as banks form a kind of a consortium to securely do transactions. The members of the organizations of the consortium can see transactions.

Who controls blockchain?—It depends on the type of blockchain. It starts from permissioned blockchain, where the verification is done by a central authority or consortium and tends toward permissionless, where anyone can participate in the verification process.

Blockchain's potential goes beyond cryptocurrencies. Blockchain ledgers can include land titles, loans, identities, logistics, and IoT device data to bring transparency to any multiparty transaction. Blockchain is a file system that creates blocks of transactions, called the distributed ledger in a multiparty system, where all parties have access to the ledger and know the nature of transactions in their chain. Blockchain works when there is vast amount of data and uses (AI) to determine the authenticity of complexities in multiparty transactions. The technology is such that there is no way a human can tamper with the system unless all parties agree to manipulate the system across the business chain.

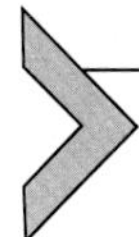

12. About the bitcoin application of the blockchain technology

Bitcoin gives a way for one Internet user to transfer a unique piece of digital property to another Internet user. The real beauty here is that the transfer is guaranteed to be safe and secure. Everyone knows that the transfer has taken place, and nobody can challenge the legitimacy of the transfer. To keep track of the number of bitcoins, each of us uses a ledger, which is a kind of digital file. The ledger file is not stored in central servers. Centralized storage is liable for security attacks. Today's banks use one or more centralized servers to store any ledger. But the blockchain paradigm is for distributing ledger files across servers, which are situated at geographically distributed data centers. That is, multiple servers are being used to store ledger files so that any kind of security attacks on ledgers can be foiled. These servers not only store ledgers but also contribute in data processing and transactions fulfillment. Each of these servers acts as a "node" of the blockchain network and has a copy of the ledger file.

If A wants to send bitcoins to B, A has to broadcast a message to the bitcoin network. The message is the number of bitcoins in A's account has to go down by 5 BTC and the amount of B's account should go up by the same quantity. Each node in the network would receive the message and apply the requested transaction to its copy of the ledger. The fact that the ledger is maintained by a group of networked computers rather than by a centralized entity brings forth a number of distinct advantages.

In our banking system, we can get to know our own transactions and account balances. But as per the blockchain paradigm, everyone can see others' transactions. The blockchain system is designed in such a way that no trust is needed. The prime requirements such as security and reliability are being guaranteed via specialized mathematical functions and code. Precisely speaking, blockchain is a new-generation security-enablement solution that allows a group of distributed computers to maintain a single updated and secure ledger.

In order do transactions on the blockchain network, we need a wallet. This is a software solution that allows us to store and exchange our bitcoins. Since only we should be able to spend our bitcoins, each wallet has to be protected by a cryptographic method that intrinsically uses a unique pair of different but connected keys: a private and a public key.

If a message is encrypted with a public key, then only the owner of the paired private key will be able to decrypt and read the message. On the other way, if we encrypt a message with our private key, only the paired public key can be used to decrypt it. When A wants to send bitcoins, he needs to broadcast a message encrypted with the private key of his wallet. Each node in the network can cross check that the transaction request is coming from A by decrypting the transaction request message with the public key of his wallet.

When encrypting a transaction request with our wallet's private key, what is happening is that a digital signature is generated and it is used by blockchain computers to double check the source and the authenticity of the transaction. The digital signature is typically a string of text and it is being obtained by combining our transaction request and the private key. This signature cannot be used for other transactions. If we change a single character in the transaction request message, the digital signature also get altered. This prevents any potential attacker modifying our transaction requests or altering the amount of bitcoins we are sending. As we broadcast the message only after it has been encrypted, no one can get our private key by any means. As indicated above, multiple computers are being leveraged to keep a copy of the ledger in a safe manner.

How to get know our account balance?

This is an often-repeated question "How does a node know what's our account balance?" The bitter truth is that the blockchain system does not keep track of account balances at all. It only records each and every transaction that is requested. To know our wallet balance, we need to analyze and verify all the transactions that ever took place on the whole network connected to our wallet.

This "balance" verification is easily done thanks to links to previous transactions. In order to send 10 bitcoins to Y, X has to generate a transaction request that includes links to previous incoming transactions whose total balance equals or exceeds 10 bitcoins. These links are called inputs. Nodes in the network will verify that the total amount of these transactions equals or exceeds 10 bitcoins. It also checks that these inputs were not yet spent. In fact, each time we reference inputs in a transaction, those are considered not valid in any future transaction. This is performed automatically in X's wallet and double checked by the bitcoin network nodes, X only sends a 10 BTC transaction to Y's wallet using Y's public key.

The blockchain system checks all the previous transactions that are correlated to the wallet we use to send bitcoins via the references that each one has as inputs. To simplify and speed up the verification process, a special record of unspent transactions is kept by the network nodes. Thanks to this security check, it is not possible to double-spend received bitcoins. Owning Bitcoins means that there are transactions written in the ledger that point to our wallet address and have not been used as inputs yet.

Anyone can access the bitcoin network via an anonymous connection and submit or receive transactions revealing nothing more than his public key. However, if someone uses the same public key over and over, it is possible to connect all the transactions to the same owner. The bitcoin network allows to generate as many wallets as we like, each one with its own private and public keys. This allows to receive payments on different wallets that cannot be linked together. There is no way to know that we own all these wallets private keys unless we send all the received Bitcoins to a single wallet.

The total number of possible Bitcoin addresses is 2^{160}. To be precise, the number arrived is 146150163733090291820368483271628301965593254 2976. This large number protects the network from possible attacks while allowing anyone to own a wallet.

13. Enterprise blockchain use cases

Enterprising businesses all over the world are consciously working on a number of strategically sound initiatives using proven and potential digital transformation technologies and tools. Blockchain is being positioned and pronounced as one of the core digital transformation technologies capable of bringing in scores of innovations and inventions for corporates. Digital transformation projects result in greater amount of digital data. Digital data have to be captured, cleansed, and crunched to extricate actionable insights. That is, the domain of digital intelligence is picking up steadily so that insights-driven businesses are to see the reality. The safety, privacy, and security of digital data and assets have to be ensured through technological solutions. Blockchain is being projected as the prime technology here.

Digital data comes in blocks. These blocks are then chained together and this makes the data immutable and tamper-proof. When a block of data is chained to the other blocks, its data can never be modified again. The data are made available for anyone to see it again and again. The chain of blocks allows us to keep track any digital data and records. For example, bitcoin data, land records, property rights, user identities, medical records, etc. can be secured without getting tampered. If I buy a house and add a photo of the property rights to a blockchain, I will always and forever be able to prove that I owned those rights at that point. Nobody can change that information if it is put on the blockchain. So, blockchain is an efficient and easy way to save confidential, customer, and corporate data and make it immutable.

The leverage of blockchain paradigm is being expedited these days. Many enterprises have started to experiment with this quite new technology. The awareness and understanding of several unique disruptions and innovations of the blockchain technology have resulted in many industry verticals to embrace and explore toward fresh opportunities and possibilities. Cryptocurrencies are the widely used application of blockchain. As of today, blockchain is being considered as the next big step for a range of businesses such as banking, healthcare, transport, etc. Governments are increasingly leveraging this new paradigm in order to be in the good books of their citizens. Land records are being made tamper-proof. The security of IoT devices and their data is being strengthened with the usage of the blockchain concept.

Reliance Jio, the largest telecommunication service provider in India is working on a couple of value-adding project implementations using the blockchain technology. Even the Indian government has made an announcement that the IIT Degrees henceforth will be handed out through blockchain technology. Saudi Arabia recently tied up with Ripple to use blockchain in its central bank. With a greater and deeper understanding of the power of blockchain paradigm, national governments across the globe, IT organizations, research labs, and business establishments are keen in drawing highly impacting blockchain strategy. One thing is clear. Blockchain can be one of the trailblazing technologies for the future. It is to bring in a series of revolutions in producing and presenting premium and path-breaking services to people. Here are a few more examples, where blockchain is being seen as an enabler.

Modum (https://modum.io/)—Blockchain can address challenges in today's supply chains. Business executives argue that the logistics industry can substantially benefit and profit from this technology. Modum is at the forefront by applying the blockchain technology in commercially available environmental condition monitoring solution. It also is able to provide actionable insights on how digital supply chain ecosystems could work.

Modum is dealing with supply chain validation specific to pharmaceutical supply chain. Validating if the package of tablets has actually been manufactured by who it said it is manufactured by. It can also help in validating the characteristic of the shipment conditions such as temperature. It checks the supply chain information such as the origin, the destination, the number of hubs through which the package traverses through, etc. Blockchain is proving to be a game-changer for self-governing, and decentralized supply chain ecosystem.

Slock (https://slock.it/)—This enables devices of any size to access blockchain data securely, take payments autonomously and interact with humans, machines, and anything in between.

Chronicled (https://www.chronicled.com/)—Companies in the blockchain industry expect their solutions to be truly decentralized and distributed. This distinction helps to avoid vendor lock-in. The second advantage is data security. Typically, the security of data is at risk when software solutions are distributed and decentralized. Businesses, their competitors and partners are in the network. The need is to ensure the fool-proof security for data of these participants. But for enforcing business rules for enabling smart contracts to fruition, all the transactions have to be on the blockchain instead of on private channels. This company exclusively focuses on blockchain implementation aimed at product provenance and product validation to authenticate secure identities.

Lighting network (https://lightning.network/)—For the forthcoming era of IoT devices, there is a need for lightning-fast blockchain payments without worrying about block confirmation times. Security is being enforced by blockchain smart contracts without creating an on-blockchain transaction for individual payments. Payment speed is being measured in milliseconds to seconds. Secondly, the blockchain system has to be capable of millions to billions of transactions per second across the network. This company is focused on making bitcoin more suitable for IoT with high speed and high-volume transactions.

Car leasing—Blockchain theoretically enables the car to become a smart asset because it can auto-manage services like auto insurance, lease payments, tolls, parking even our coffee at Starbucks. Car leasing is a big business these days. With millions of leased vehicles on the roads, tracking the status of them can tax even the most robust systems and processes. Leasing a car invokes many systems across a chain of processes including verifying the driver's financial status, driving eligibility, available vehicle inventory, and available car features. Then, leasing companies and financiers need to know what is happening with vehicles after the lease is signed. There is a partnership between DocuSign and Visa, which have a blockchain pilot designed to introduce greater speed and automation to the car-leasing process.

The proof of concept employs blockchain to bring together smart contracts and secure payments to create a connected and IoT-enabled car-leasing experience. Smart contracts are the most important component of blockchain, where any agreement or contract that is highly sensitive and involves multiple parties, which can benefit from the blockchain's independence and distributed system of record. With smart contracts, we are able to create self-executing contractual states, stored on the blockchain as a public ledger, which can act based on information or events from other systems, while keeping the data durable and reliable. The shared ledger can record signatures and agreements that can be validated on the blockchain. Smart contracts like this could be used in financial services, the public sector, healthcare, or legal proceedings, for example. Anywhere where a public ledger account of a contract or agreement adds value.

We have discussed about various enterprise use cases getting easily and quickly implemented using this new technological paradigm. These developments ultimately bring in the much-needed clarity and confidence.

14. The blockchain technology benefits

Having seen the benefits of a shared ledger, many banks are teaming up together to create and sustain private blockchains. Enterprise blockchain solutions are gaining higher acceptance and importance. The financial services industry is the first and foremost to embrace the blockchain paradigm. Other industries are showing exemplary interest in leveraging the blockchain technology. Several proof of concepts (PoCs) and pilots are being initiated across industry verticals in order to deeply and decisively understand its distinct competencies. On the other hand, there are some limitations and risks with it. Researchers and enthusiasts are working in unison to bring forth technological solutions to surmount those identified issues. Futuristically speaking, there are industry case studies in support of this disruptive paradigm. The widely exclaimed advantages are

- We have complete control of the value we own. There is no intermediary and third-party organizations to hold our value. There is no one to limit our access to it.
- The cost to perform a value transaction from and to anywhere in the planet earth is very low and hence this blockchain solution allows micropayments.
- Value can be transferred in a few minutes and the transaction can be considered secure in a few hours, not days or weeks.

- Since anyone at any point of time can verify every transaction made on the blockchain, the much-needed transparency is granted.
- The blockchain technology contributes immensely to build decentralized applications that simplifies information management and value transfer in a secure and faster manner.
- The concept of smart contracts is fast emerging and with the maturity and stability of right technologies, a number of activities can be easily automated and accelerated programmatically.

The bitcoin cryptocurrency can reduce or eliminate the need for certain intermediaries and automate manual tasks. Distributed ledger stores the entire ownership history of an asset. The other use cases are:

1. Know Your Customer (KYC),
2. Antimoney Laundering (AML),
3. Trade surveillance,
4. Collateral management,
5. Settlement and clearing,
6. The ability to capture historical and current ownership of high-value items.

Blockchain already has a significant impact on industries on how they react, adapt, and change. It has the inherent potential to revolutionize how businesses operate.

- Blockchain is a true peer-to-peer network that will reduce reliance on some types of third-party intermediaries like banks, lawyers, and brokers.
- Blockchain can speed up process execution in multiparty scenarios and allow for faster transactions that are not limited by office hours.
- Information in blockchains is viewable by all participants and cannot be altered. This will reduce risk and fraud and create trust.
- Distributed ledgers will provide quick ROI by helping businesses create leaner, more efficient and more profitable processes.
- The distributed and encrypted nature of blockchain means it will be difficult to hack. This shows promise for business and IoT security.
- Blockchain is programmable, which will make it possible to automatically trigger actions, events, and payments once conditions are met.

The blockchain technology is still evolving and there are many product and platform vendors contributing immensely for making it penetrative, persuasive, and pervasive. Business houses are experimenting with this new entrant and exploring the possibility of leveraging this unique idea to optimize business processes to realize bigger and better applications. With cloud centers are being recognized as the highly optimized and organized IT environments to deploy and deliver blockchain platforms and services, this nascent technology is finding a solid, strong, and stimulating base to grow faster and glow in the days to unfurl.

15. The blockchain challenges and concerns

However, everything is not rosy with blockchain. There are a few critical challenges and concerns to be surmounted to boost the confidence of business decision-makers. This section lists out the important drawbacks and in the subsequent sections, we discuss how blockchain experts analyze those issues and put forth their views on overcoming them. There are articles and blogs accentuating and articulating several consensus algorithms, performance-increment measures, remedial approaches, best practices, specific solutions, etc. The distributed and decentralized computing insisted by blockchain demands constant computational power at multiple locations. This ultimately results wastage of a large amount of electrical power. Though blockchain brings down cost and inefficiencies to a larger extent. The excitement and elegance being associated with blockchain has inspired many security experts and IT professionals to stretch further to unearth competent methods as a solace for the various limitations. Blockchain can be a strategically sound technological paradigm if these drawbacks are moderated.

- We know that transactions can be sent and received anonymously. On one side, this preserves the users' privacy but on the other side, it allows illegal activity on the network as institutions cannot track user's identity.
- Even if there are many exchange platforms, it is still not that easy to trade bitcoins for goods and services.
- Bitcoin, like many other cryptocurrencies, is very volatile. Firstly, there are not many bitcoins available in the market and the demand is also changing rapidly. Bitcoin price is hugely impacted by large events or announcements in the cryptocurrency industry.
- On some occasions, performance issues crop up. Typically it takes a few minutes to process and complete a transaction. This There are new initiatives and implementations for enhancing the performance and throughput of blockchain networks.

The blockchain technology is still in its infancy. New tools are being developed and released to the market to substantially improve the blockchain stability while offering a range of fresh features, tools and services. Even blockchain systems and networks are being attacked. There are many pundits and pioneers recommending various ways and means of suppressing security loopholes in the blockchain paradigm. The next section has the details of understanding the kind of security attacks and how they can be overcome. The various features and functionalities of the shared ledger method can be used to proactively and preemotively surmount various attacks on blockchain systems.

16. How blocks and their chains nullify security attacks on blockchain networks

As discussed above, there are some crucial challenges being expounded through various media. This section explains the various problems and how they are being addressed through various techniques and tips, which are part and parcel of this futuristic and flexible technology.

The order of transactions—If there is any mismatch in the order of transactions, there is a possibility for misuse. There is a major security hole that could be exploited to recall bitcoin after spending them. Transactions are typically passed from node to node within the network, so the order in which two transactions reach each node can be different. An attacker could send a transaction, wait for the counterpart to ship a product, and then send a reverse transaction back to his own account. In this case, some nodes could receive the second transaction before the first one and therefore consider they first payment transaction invalid as the transaction inputs result already spent. How do we know which transaction has been requested first? It is not secure to order the transactions by timestamp because it could easily be counterfeited. Therefore, there is no way to tell if a transaction happened before another, and this opens up the potential for fraud. If this happens, there will be disagreements between the network nodes regarding the order of transactions each of them received. Therefore, any blockchain system has to be designed to use nodes agreement to order transactions and prevent the fraud described above.

As indicated above, the bitcoin network orders transaction by putting them together into groups called blocks, each block contains a definite amount of transactions and a link to the previous block. This is what puts one block after the other in time. Blocks are therefore organized into a time-related chain that gives the name to the whole system: blockchain. Transactions in the same block are considered to have happened at the same time and transactions not yet in a block are considered unconfirmed.

The Sequence of blocks—Each node can group transactions together into a block and broadcast it to the network as a suggestion for what block should be the next. Since any node can suggest a new block, how does the system agree on what block should be the next?

In order to be added to the blockchain, each block must contain the answer to a complex mathematical problem, which is typically created using an irreversible cryptographic hash function. The only way to solve such mathematical problem is to guess random numbers. The numbers get combined with the previous block content to generate a defined result, which is usually a number below a certain value. A normal computing takes an year time to arrive at the right number and to solve the mathematical problem. However, there are many computers participating in the bitcoin network and therefore guessing random numbers is quite fast and hence a block can be solved every 10 min on an average. The compute node that solves such mathematical problem gains the right to place the next block on the blockchain and it can broadcast newly incorporated block to the whole network.

Double-spending attack—Another potential issue is as follows. If two nodes solve the problem at the same time and spread their blocks to the network simultaneously, then both blocks are broadcasted and each node builds on the block that it received first. The blockchain system mandates each node to build immediately on the longest block chain available. So if there is ambiguity about which is the last block, as soon as the following block gets solved each node will adopt the longest chain as the only option.

Due to the low probability of solving blocks simultaneously, it is almost impossible that multiple blocks are solved at the same time over and over again building different "tails," so the whole blockchain stabilizes quickly to one single string of blocks that every node agrees on. The disagreement about which block represent the end of the chain "tail," opens up the potential for fraud again. If a transaction happens to be in a block that belongs to a shorter tail, once the next block is solved such transaction will go back to the unconfirmed transactions as all the others included in block B.

Let us see how Mary could leverage the end-of-chain ambiguity to perform a double-spending attack. Mary sends money to John, John then ships the product to Mary, now since nodes always adopt the longer tail as the confirmed transactions, if Mary could generate a longer tail that contains a reverse transaction with the same input references, John would be out of both his money and his product.

Additional security mechanisms—The blockchain paradigm use a "PoW" concept to increase the security. A PoW, which is explained in detail in the beginning of this chapter, can be equated to the processing time taken by a computer. This essentially slows down the creation of a new block in a blockchain. In a bitcoin blockchain, it generally takes around 10 min to create a new block. So to alter an entire blockchain, one has to spend several hours. These measures, however, do not guarantee that the value of a block cannot be changed. So there is one final security measure that blockchains use. As explained above, blockchains use a "distributed ledger," (a peer-to-peer network). This means that even if one manages to alter one particular copy of a blockchain by adding a defected block, the altered block will still not be accepted into the system. This is simply because the altered block will not match with the copies that others have in a system and hence will be discarded. In order for a block to be accepted into the system, it has to pass a Consensus, meaning, it has to be validated by more than 50% of the network. This makes the alteration of any blockchain almost impossible.

As each block contains a reference to the previous block and that reference is a part of the mathematical problem. As indicated above, the mathematical problem has to be solved in order to spread the following block to the network. So it is hard to precompute a series of blocks due to the high number of random guesses needed to solve a block and place the same on the blockchain. Here, Mary is in a race against the rest of the network to solve the mathematical problem that allows her to place the next block on the chain. And even if she solves it before anyone else, it is very unlikely she could solve two, three, or more blocks in a row, since every time she is competing against a set of competitors. So, it is very unlikely Mary could solve several blocks in a row at the exact time needed to perform a double-spending attack. She would need control of 50% of the computing power of the whole network to have a 50% chance to solve a block before some other node does, and even in this case, she has a 25% chance to solve two blocks in a row. The more blocks to be solved in a row, the lower the probability that Mary can succeed.

Therefore, transactions get more and more secure with time. Those included in blocks that have been confirmed in the past are more secure than those included in the last block. Since a block is added to the chain every 10 min on average, waiting for about 1 h from when the transaction is included in a block for the first time gives a quite high probability that the transaction has been processed and is nonreversible.

In order to be able to send bitcoins, we need to reference a transaction that sends bitcoin to our wallet, and this applies to all the transactions that ever took place in the network.

Mining and mining pools—To balance the deflationary nature of bitcoin due to software errors and wallets password loss, a reward is given to those that solve the mathematical problem of each block. The activity of running the bitcoin blockchain software in order to obtain these bitcoin rewards is called "mining." This is the main incentive that pushes private people to operate the nodes. They provide the necessary computing power needed to process bitcoin transactions and to stabilize the blockchain network. As a single computer takes more time, the idea of clubbing together many computers to divide the number of guesses each one has to try in order to solve the next block. In this way, it is faster for the group to guess the right number and get the reward that is then shared among the group members. Such groups are called mining pools.

Some of these mining pools are pretty large and represent more than 20% of the total network computing power. This has clear implications for the network security as seen in the double-spend attack example. Even if one of these pools could potentially gain 50% of the network computing power, the further back along the chain a block gets, the more secure are the transactions included in it. Since the overall network computing power is likely to increase over time due to technological innovation and the increasing number of nodes, the blockchain system recalibrates the mathematical problem difficulty to solve the next block in order to target 10 min on average for the whole network. This ensures the network stability and overall security.

Moreover, every 4 years the block reward is cut in half, so mining bitcoin (=running the network) gets less interesting over time. To prevent nodes from stopping running the network, small reward fees can be attached to each transaction. These rewards are collected by the node that successfully include such transactions in a block and solves its mathematical problem. Due to this mechanism, transactions associated with a higher reward are usually processed faster than those associated with a low reward. This means that when sending a transaction you can decide if you would like to process it faster (=more expensive) or cheaper (=takes more time). Transactions fees in the bitcoin network are currently very small if compared with what banks charges and are not associated with the transactions amount.

Precisely speaking, the blockchain technology has the disruptive and transformative potential to revolutionize several industries from advertising to energy distribution. Its main power lies in its ability of not requiring trust. Also, it is following the proven decentralized architecture. Not only decentralized applications (DApps) but also the concept of smart contracts is flourishing with the faster maturity and stability of the blockchain technology. Integrated and insightful platforms, acceleration engines, and frameworks are being built in order to produce and sustain blockchain applications.

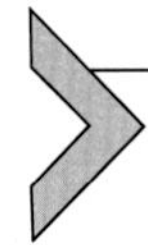

17. An efficient cross-organizational information transfer method through blockchain

By design, a blockchain business network securely shares information between different organizations by distributing ledger transactions to peer nodes located throughout the business network, including nodes physically located within a competing organization's security environment and domain. The same blockchain security design features that enable these secure cross-organizational information transfers are also ideally suited to ensure the safe, efficient, and cost-effective transfer of information across different government and military network security domains.

The security controls and assured sharing inherent to open-source blockchain Hyperledger Fabric hosted on high assurance off-the-shelf hardware infrastructures, can provide secure, timely, and consistent end-to-end sharing of information within and across disparate security domains. A blockchain-based cross-domain solution is likely to be less complex, more effective, and less expensive than traditional, special-purpose cross-domain guards when mitigating the high stakes security risks of cross-domain information transfer.

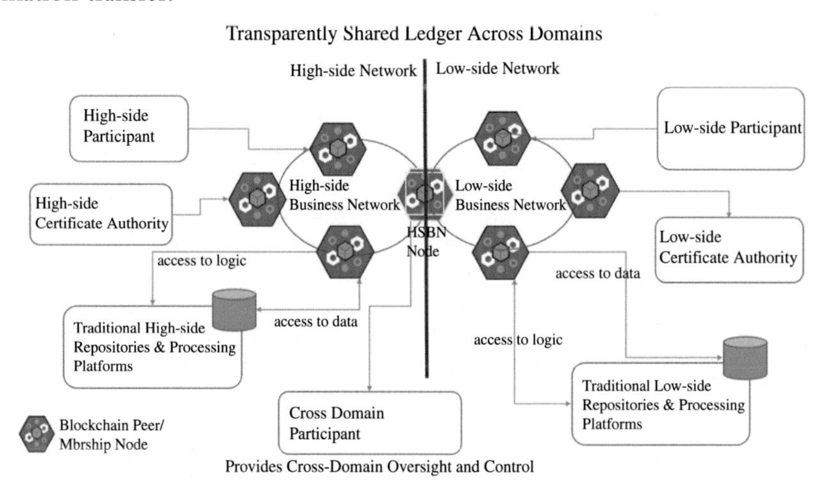

Public, private, government, and military organizations classify, label, and protect information commensurate with the security risk it poses if disclosed without authorization. For example, unclassified information poses no risk if disclosed while inadvertent disclosure of top-secret information could be expected to cause exceptionally grave damage. The most rudimentary form of protection is segregating information of different security levels on completely separate IT infrastructures and communication networks. These separate networks, however, hinder day-to-day work because typically real-world work and business processes require accessing and exchanging information at varying classifications.

Traditionally, highly specialized IT systems known as cross-domain guards sit at the boundary of each security domain reading files from a designated source location in one security domain and applying a preapproved filter for that specific source to each file. These filters and controls verify the integrity and contents of the file to satisfy security policies then write the file to a designated target location in the other security network. These cross-domain data flows are either from lower security classifications to higher ones or vice-versa.

The concern for flowing from lower security classifications to high security classifications guards is preventing the introduction of dangerous computer files like viruses or malware and preventing a reverse flow of data. The chief concern for flowing from higher security classifications to lower classifications guards is preventing unauthorized leakage of higher classification data to the lower classification network. Also, for the higher-to-lower scenario, the data transfer rates and volume are typically much slower and smaller than for the lower-to-higher scenario because the data must be manually verified as appropriately classified prior to sending to the lower classification network.

Traditional cross-domain solutions (CDS) are expensive due to lengthy and complex approval and implementation processes, and specialized equipment and skill sets. Any changes to a data flow also require extensive, expensive, and lengthy approval and implementation processes. These frictions inherent to traditional CDS's hinder effective and timely information sharing. The blockchain technology is an appropriate one here.

In a blockchain network, the role of cross-domain guard would be performed by a blockchain network peer installed on a high assurance platform, referred to as a high security business node (HSBN). Data movement within the distributed cross-domain blockchain business network takes the form of cryptographically secured transactional updates to a shared ledger held by each node in a specific members-only channel of the business network.

The other peer nodes of the blockchain business network reside in one or the other security domains, either high or low, receiving the protected blocks of endorsed ledger transactions. Endorsed ledger updates to peers located on the opposite security domain occur via the border HSBN to provide assurance that the blockchain-provided security controls cannot be bypassed or overridden.

A blockchain cross-domain solution reduces frictions in the information exchange process and improves accessibility, accountability, and traceability of information exchange. Specifically, it provides a single shared view of each asset throughout its life cycle regardless of the network domain. Further on, the blockchain solution gives an auditable control and oversight of asset information throughout the life cycle. Also, this approach provides information sharing rather than merely moving data. More relevant details on this important use case can be found on this topic at this page https://www.ibm.com/blogs/blockchain/2018/01/blockchain-as-a-cross-domain-security-solution/.

18. The marriage between blockchain, AI, and the IoT

We have been using physical machines (bare metal (BM) servers) as the nodes for storing and processing transaction data. There is a recent and tectonic shift toward devices instead of compute servers (commoditized).

Our everyday devices are getting increasingly connected. It is being projected that there will be billions of connected devices in the years ahead. When connected instrument, equipment, appliances, and machines abound, context-aware applications can be realized and used. Experts are of the opinion that we can realize a greater value when machines start using blockchain as the engine of innovation.

As we all know, with the greater usage of the IoT concept, we can aspire to have smart systems, solutions, and services. A variety of smart environments such as smart homes, hotels, hospitals, etc. can be set up and sustained through the application of various advancements in the IoT discipline. That is, business applications and IT services become smart. With device-to-device (D2D) and device-to-cloud (D2C) integration capabilities, the optimism of building and running connected cars has gone up significantly. The concepts of IoT and cyber physical systems (CPS) have gained momentum in order to produce complicated yet sophisticated systems.

For example, we hear and read about smart cars. There are other names such as connected cars, self-driving or autonomous cars. Smart cars can innately interact with smart parking applications, car chargers, car manufacturers and mechanics, can talk to infotainment systems, home electronics, fuel stations, etc. Thus, a number of tasks get automated through seamless and spontaneous integration. Ultimately the long-standing goal of smart cities can be fulfilled through smart cars, drones, robots, and applications like garbage and drainage management. There are multiple participants in any connected system. Further on, the participants do not fully trust each other.

But they ought to share information, money, and virtual currency that cannot be forged. There are no centralized authorities in such kinds of new-generation smart systems. Herein, the value of the blockchain technology is being felt. Not only this technology ensures the tightest security for information and money exchange but also it accomplishes every activity in a transparent, faster, and safer manner. Blockchain contributes in sharing trusted information such as medical records shared between patient, doctor and insurance company.

19. Conclusion

Blockchain, the technology that had laid a scintillating foundation for the hugely popular cryptocurrency bitcoin, has been a revolutionary paradigm in many ways. It intrinsically addresses many challenges associated with digital transactions such as double-spending and currency reproduction. Blockchain also reduces the cost of online transactions while ensuring highest security for digital assets and data. The game-changing phenomenon is that the need for payment processors, custodians, and reconciliation bodies is eliminated. The merits of the blockchain technology are not limited to securing digital transactions alone. Cyber-attacks on mission-critical applications and cloud infrastructures can be proactively stopped. Cybercriminals cannot exploit digital data blocks because the blocks are anonymously stored with other stakeholders within a network. All transactional data is systematically verified with every relevant stakeholder. The distributed and decentralized architectural arrangement helps in minimizing data manipulation tricks.

Appendix

There are blockchain application development languages and platforms to speed up the process of building blockchain applications. These tools help in moderating developmental and operational complexities. They are listed below with the basic details.

Solidity is an object-oriented, high-level language for implementing smart contracts. Smart contracts are programs which govern the behavior of accounts within the Ethereum state. Solidity was influenced by C++, Python, and JavaScript and is designed to target the Ethereum Virtual Machine (EVM). Solidity is statically typed, supports inheritance, libraries, and complex user-defined types among other features. With Solidity, we can create contracts for uses such as voting, crowdfunding, blind auctions, and multisignature wallets.

Cakeshop—Cakeshop lets you manage a local blockchain node. It comes with APIs and tools that we can use to set up the cluster node, work with contracts, and explore the chain. This is for working with an Ethereum-based blockchain and packaged as a Java web application archive (WAR) that can be dropped into any application server. This can get up and running in under 60 s.

Hyperledger caliper—This is a tool for benchmarking blockchain performance indicators such as success rate, throughput, latency, and resource consumption.

Geth is Ethereum node implementation. It is created using the Go programming language. Geth is used in a variety of tasks on the Ethereum blockchain. It can be used to transfer tokens, mine Ether tokens, and create smart contracts. Furthermore, it can also be used to explore block history. Geth is available Windows, Mac, and Linux systems. Once we install Geth, we need to either connect to an existing blockchain or create our own. To simplify things, Geth automatically connects to the Ethereum mainnet.

Blockchain testnet—This is an essential tool, as it lets test our dApps before making them live. Each blockchain solution has its testnet, and we need to use the respective testnet. Testnets are especially useful, as it lets us test without spending real resources. Ethereum, for example, uses gas as the fuel for carrying out different actions. Developers cannot spend gas every time they do a test run. This means spending thousands of dollars to test. It is not feasible. A testnet lets blockchain developer iron out bugs without spending large amounts of cash. The choice of testnet depends on our dApp. We can use public test, private test, or GanachiCLI—a customizable blockchain emulator.

Blockchain-as-a-service (*BaaS*)—This is a recent offering that allows customers to leverage cloud-based solutions to build, host and use their own blockchain apps, smart contracts, and functions on the blockchain while the cloud-based service provider manages all the necessary tasks and activities to keep the infrastructure agile and operational.

This web page (https://builtin.com/blockchain/blockchain-as-a-service-companies) has listed out the company names along with the relevant details about the hugely popular BaaS offerings. All the major cloud service provides such as AWS, Microsoft Azure, Alibaba cloud, etc. are making BaaS available to worldwide users.

Truffle—This is an Ethereum blockchain framework. It offers an asset pipeline and development environment for Ethereum development.

With Truffle, we can develop complex Ethereum dApps and smart contracts. It has a vast library that lets tackling challenging requirements. The key features offered by Truffle include the following:
- Automate contract testing using Chai and Mocha.
- Do full smart contract development including linking, compilation, and deployment.
- Do custom build procedures with the configurable build pipeline.

Ether.js is a handy tool when it comes to developing client-side JavaScript wallets. It lets interacting with Ethereum blockchain. Initially, it was only used to work with ethers.io, but now, it is a full-fledged general-purpose library.

Remix IDE is a popular integrated development environment (IDE) that runs from the browser. It lets developing Solidity contracts from the browser. It is developed using JavaScript and hence it runs on any modern browser. It comes with module support that brings more functionality to the IDE. For example, we can use a file explorer module to save or load files from our computer. Other useful modules include plugin manager, solidity editor, terminal, and settings.

dAppBoard is an analytical platform used for Ethereum smart contracts. It also comes with Ethereum blockchain explorer. dAppBoard is web-based and lets us to monitor smart contracts running on Ethereum networks. It can give information, such as the total number of users of a particular dApp or an overview of the whole Ethereum network.

A.1 The blockchain frameworks

To better articulate a blockchain framework, it is helpful to divide it into two portions: the infrastructure and the application. This division helps us understand which concepts belong to which portion, and also understand how and why blockchain applications can be developed independently on top of the infrastructure.

Infrastructure, or sometimes it is called *network*, is composed of the compute nodes and the software running on them. The compute nodes can be physical machines/bare metal (BM) servers, virtual machines (VMs), or containers. Software provides features and capability such as users (identity or accounts), transaction formation and processing, consensus protocol, and something unique to specific frameworks, like native currency and mining for permissionless blockchains, channel, and identity management for permissioned blockchains, etc.

Application is composed of the code running inside the infrastructure (known as smart contract in general) and the client application interacting with the infrastructure. The latter serves as the accessing point from external world. A good blockchain framework should allow application development outside the actual deployment of the network. We do not need to deploy the network before developing the application. If our application can run on a small network (even of one node), it should run well in a bigger infrastructure. These frameworks assist in enterprise Blockchain implementation.

1. Hyperledger; Supported by Linux Foundation and IBM.
2. Ethereum; a private blockchain framework from Ethereum.
3. Multichain; an open platform for building blockchains.
4. Eris Industries; a low cost blockchain implementation framework.
5. R3 Corda; Corda is a blockchain framework designed specifically for BFSI industry.
6. Openblockchain; an open blockchain fabric code framework.

About the author

Pethuru Raj is working as the chief architect and vice-president in the Site Reliability Engineering (SRE) division of Reliance Jio Platforms Ltd., Bangalore. His previous stints are in IBM global Cloud center of Excellence (CoE), Wipro consulting services (WCS), and Robert Bosch Corporate Research (CR). In total, he has gained more than 19 years of IT industry experience and 8 years of research experience.

Finished the CSIR-sponsored Ph.D. degree at Anna University, Chennai and continued with the UGC-sponsored postdoctoral research in the Department of Computer Science and Automation, Indian Institute of Science (IISc), Bangalore. Thereafter, he was granted a couple of international research fellowships (JSPS and JST) to work as a research scientist for 3.5 years in two leading Japanese universities. Published more than 30 research papers in peer-reviewed journals such as IEEE, ACM, Springer-Verlag, Interscience, etc. Has authored and edited 20 books thus far and contributed 35 book chapters thus far for various technology books edited by highly acclaimed and accomplished professors and professionals.

Focuses on some of the emerging technologies such as the Internet of Things (IoT), Artificial Intelligence (AI), Big and Fast Data Analytics, Blockchain, Digital Twins, Cloud-native computing, Edge/Fog Clouds, Reliability Engineering, Microservices architecture (MSA), Event-driven Architecture (EDA), etc. His personal web site is at www.tinyurl.com/peterindia.

CHAPTER TWO

Data structures☆

Shubhani Aggarwal and Neeraj Kumar
Thapar Institute of Engineering & Technology, Patiala, Punjab, India

Contents

Abstract

Data and its information is an essential part, and various implementations are being made to store in different ways. Data are just a collection of facts and figures that are stored in a particular format. A data item refers to a single set of values, which are categorized into subitems or the group of items. In computer terms, a data structure is a specific way to store and organize data in a computer's memory so that the data can be used efficiently and effectively. Data may be arranged in many different ways, such as

☆Cryptographic primitives used in blockchain.

Advances in Computers, Volume 121
ISSN 0065-2458
https://doi.org/10.1016/bs.adcom.2020.08.002

the logical or mathematical model for a particular organization of data is termed as a data structure. First, the data must be loaded in structure to show the actual relationships of the data with the real-world object. Second, the formation should be simple so that anyone can efficiently process the data. In this chapter, we represent the different forms of data structure to ease access of the data.

Chapter points

- In this chapter, we discuss the organization of data in a logical or mathematical model form as data structures.
- Here, we discuss several data structures, which can be used in a blockchain to organize and manage a large amount of data.

Data structures can be defined as the organization of data in a logical or mathematical model. It must be simple enough that one can effectively process the data when necessary [1]. The data present in the data structure are processed using several operations, which are described as follows.

- *Traversing*: In this operation, access to each element present in the data structure exactly once.
- *Searching*: In this operation, finding the location of a particular element with a key attribute.
- *Inserting*: In this operation, adding new elements to the data structures.
- *Deleting*: In this operation, removing old elements from the data structures.

There are two more operations in the data structures, which are used in special situations that are described as follows.

- *Sorting*: In this operation, arrange the elements in some logical order, i.e., ascending order, descending order, alphabetical order with some numeric key value.
- *Merging*: In this operation, combine two elements or lists of two different sorted files into one sorted file.

Data structures are normally classified into two broad categories, which are shown in Fig. 1 and their description is described in following sections.

1. Primitive data structure

Primitive data structures are basic in nature and directly operated by the machine instructions. These data structures are divided into four categories, which are described as follows.

1.1 Integers

It is defined as a whole number, not a fraction number, the integer number can be positive, negative, or zero. For example, 0, 2, 3, −78, and 5467. It cannot have decimal numbers.

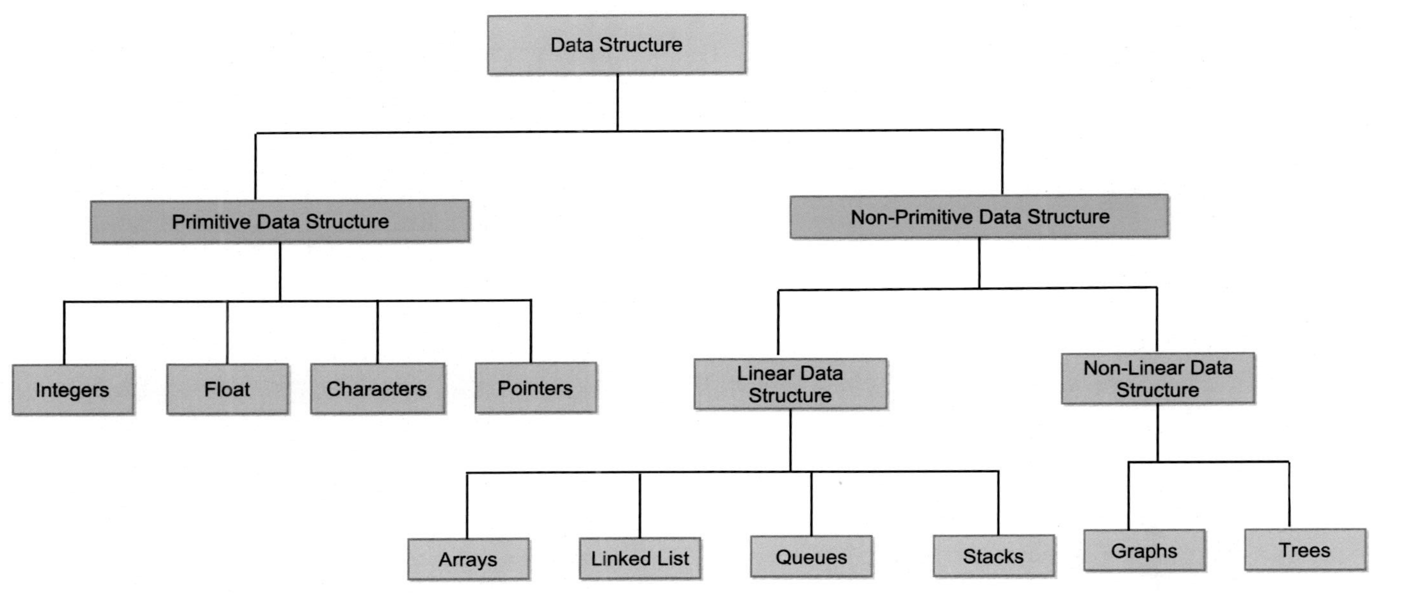

Fig. 1 Classification of data structures.

1.2 Float

A variable used to define with a fractional or decimal point. It can also be positive or negative. For example, 0.5, −0.987, and 455.70 are floating-point numbers.

1.3 Characters

It is defined as letters or symbols used in the data structures. For example, upper and lower case alphabetic characters such as F, S, a, etc., special symbols like backspace, whitespace, hyphen, etc.

1.4 Pointers

It is a variable used to find the location of another variable because it stores the address of that variable, which means the direct address of memory location. The pictorial representation of pointers is as shown in Fig. 2.

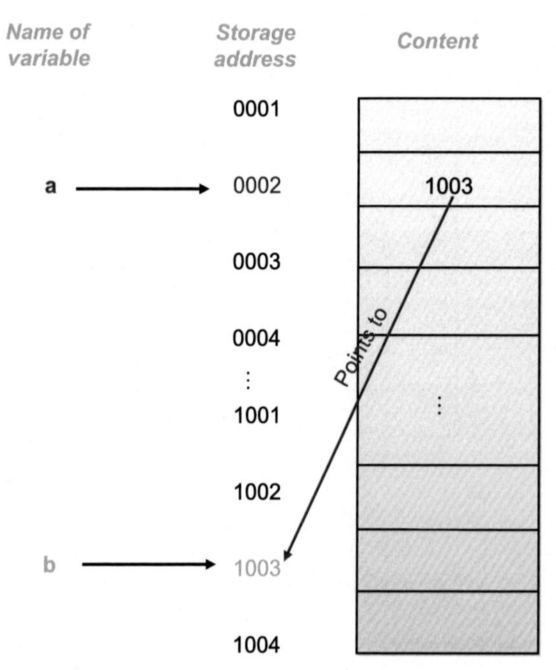

Fig. 2 Pointers.

2. Nonprimitive data structure

Nonprimitive data structure can be classified as linear and nonlinear. If the elements of the data structures make a sequence then, it is called a linear data structure that includes stacks, queues, linked lists, arrays whereas nonlinear data structures consist of graphs and trees.

2.1 Arrays

A linear relationship between the elements, which are described utilizing sequential memory locations is called arrays. It is a collection of elements stored at the contiguous location, which means a set of homogeneous data elements stored in random access memory (RAM). The data may be all floating numbers or all integer numbers or all characters as shown in Fig. 3.

The computer finds the address of any element by simply adding an offset to a base value as defined in the following Eq. (1).

$$\mathrm{LOC}(\mathrm{LA}[K]) = \mathrm{Base}(\mathrm{LA}) + w(K - \mathrm{LB}) \tag{1}$$

where as, Base(LA) represents a base address of the linear array, LB represents the lower bound, w is the number of words per memory call for linear array, and LOC(LA[K]) finds the location of linear array for any value of K.

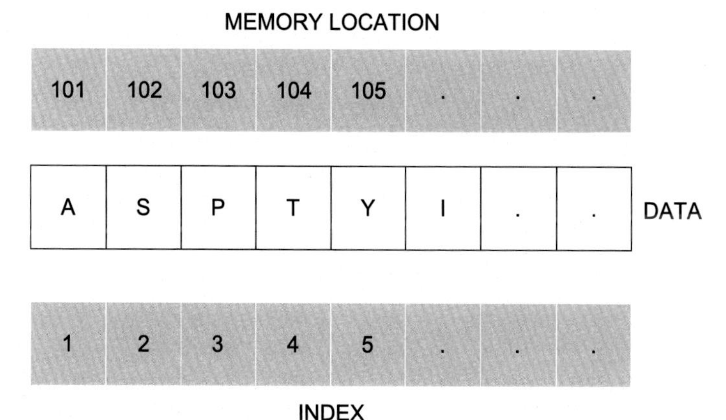

Fig. 3 Array representation.

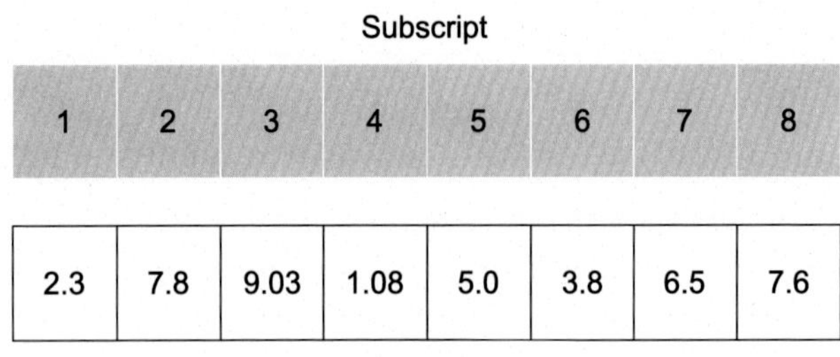

Fig. 4 One-dimensional array.

Array Name: X

Column Subscript

Row Subscript	0	1	2	3
0	2	78	90	10
1	-2	7	3	100
2	20	8	9	1
3	3	-8	3	190

Fig. 5 Two-dimensional array.

Similarly, the number of data elements presents in the linear array can be represented by the following Eq. (2).

$$\text{Length} = \text{UB} - \text{LB} + 1 \tag{2}$$

where as, UB represents the upper bound and LB represents the lower bound of the linear array.

Therefore, single subscripted values are called as one-dimensional array or linear array and two-subscripted values are called as two-dimensional array as shown in Figs. 4 and 5.

2.1.1 Creating an array

In python, to create an array by importing array (data type, value_list) module, which is used for the data type and value list specified in its arguments. The coding lines of creating an array are defined in Listing 1.

LISTING 1 Creation of array

```
# Python program to demonstrate
#Creation of Array
# importing "array" for array creations
import array as arr

# creating an array with integer type
a = arr.array('i', [1, 2, 3])

# printing original array
print ("The new created array is : ", end =" ")
for i in range (0, 3):
print (a[i], end =" ")
print()

# creating an array with float type
b = arr.array('d', [2.5, 3.2, 3.3])

# printing original array
print ("The new created array is : ", end =" ")
for i in range (0, 3):
print (b[i], end =" ")
```

2.1.2 Adding elements to an array

Data elements can be added to an array using insert() in-built function. By this, a new element can be added at the front, end, or any given index of an array. An append() function is also used to add the data elements at the end of the array. The coding lines of adding elements to an array are defined in Listing 2.

2.1.3 Accessing elements from the array

To access elements from the array by using index operator[] and it must be a integer number. The coding lines for accessing elements from an array are defined in Listing 3.

2.1.4 Removing elements from the array

The elements of an array can be removed by using remove() in-built function. Instead of this, the pop() function can also be used. But pop() function

LISTING 2 Adding elements to a array

```
# Python program to demonstrate
#Adding Elements to a Array
# importing "array" for array creations
import array as arr

# array with int type
a = arr.array('i', [1, 2, 3])


print ("Array before insertion : ", end =" ")
for i in range (0, 3):
print (a[i], end =" ")
print()

# inserting array using
# insert() function
a.insert(1, 4)

print ("Array after insertion : ", end =" ")
for i in (a):
print (i, end =" ")
print()

# array with float type
b = arr.array('d', [2.5, 3.2, 3.3])

print ("Array before insertion : ", end =" ")
for i in range (0, 3):
print (b[i], end =" ")
print()

# adding an element using append()
b.append(4.4)

print ("Array after insertion : ", end =" ")
for i in (b):
print (i, end =" ")
print()

```

LISTING 3 Accessing of element from list

```
# Python program to demonstrate
# Accessing of element from list

# importing array module
import array as arr

# array with int type
a = arr.array('i', [1, 2, 3, 4, 5, 6])

# accessing element of array
print("Access element is: ", a[0])

# accessing element of array
print("Access element is: ", a[3])

# array with float type
b = arr.array('d', [2.5, 3.2, 3.3])

# accessing element of array
print("Access element is: ", b[1])

# accessing element of array
print("Access element is: ", b[2])
```

returns an element from an array and by-default it removes the last element of an array. The coding lines for removing elements from an array are defined in Listing 4.

2.1.5 Searching element in an array

To search an element in an array, use the in-built index() method. This function returns the index of the first occurrence of value defined in arguments. The coding lines for searching elements from an array are defined in Listing 5.

2.1.6 Updating elements in an array

To update elements in an array, simply reassign a new value to the desired index where it wants to update. The coding lines for updating elements in an array are defined in Listing 6.

LISTING 4 Removal of elements in a array

```
1 # Python program to demonstrate
2 #Removal of elements in a Array
3 # importing "array" for array operations
4 import array
5 # initializing array with array values
6 # initializes array with signed integers
7 arr = array.array('i', [1, 2, 3, 1, 5])
8
9 # printing original array
10 print ("The new created array is : ", end ="")
11 for i in range (0, 5):
12 print (arr[i], end =" ")
13
14 print ("\r")
15
16 # using pop() to remove element at 2nd position
17 print ("The popped element is : ", end ="")
18 print (arr.pop(2))
19
20 # printing array after popping
21 print ("The array after popping is : ", end ="")
22 for i in range (0, 4):
23 print (arr[i], end =" ")
24
25 print("\r")
26
27 # using remove() to remove 1st occurrence of 1
28 arr.remove(1)
29
30 # printing array after removing
31 print ("The array after removing is : ", end ="")
32 for i in range (0, 3):
33 print (arr[i], end =" ")
```

2.2 Linked list

In a linked list, the elements are linked using pointers instead of the elements are stored at contiguous memory locations. It consists of nodes where each node contains an information field and a pointer field to the next node in the list as shown in Fig. 6.

LISTING 5 Searching an element in array

```
1  # Python code to demonstrate
2  # searching an element in array
3
4
5  # importing array module
6  import array
7
8  # initializing array with array values
9  # initializes array with signed integers
10 arr = array.array('i', [1, 2, 3, 1, 2, 5])
11
12 # printing original array
13 print ("The new created array is : ", end ="")
14 for i in range (0, 6):
15 print (arr[i], end =" ")
16
17 print ("\r")
18
19 # using index() to print index of 1st occurrenece of 2
20 print ("The index of 1st occurrence of 2 is : ", end ="")
21 print (arr.index(2))
22
23 # using index() to print index of 1st occurrenece of 1
24 print ("The index of 1st occurrence of 1 is : ", end ="")
25 print (arr.index(1))
```

2.2.1 Linked list insertion

A new node in a linked list can be added in three ways, i.e., at the front of the linked list, after a given node, and at the end of the linked list. The code lines to insert elements in a linked list are defined in Listing 7.

2.2.2 Deletion of an element in a linked list

Follow the three steps to delete a node from the linked list.

- First, find the previous node of the node to be deleted.
- Second, change the next of the previous node.
- Third, free memory for the node to be deleted.

The code lines to delete elements in a linked list are defined in Listing 8.

LISTING 6 Update an element in array

```
1 # Python code to demonstrate
2 # how to update an element in array
3
4 # importing array module
5 import array
6
7 # initializing array with array values
8 # initializes array with signed integers
9 arr = array.array('i', [1, 2, 3, 1, 2, 5])
10
11 # printing original array
12 print ("Array before updation : ", end ="")
13 for i in range (0, 6):
14 print (arr[i], end =" ")
15
16 print ("\r")
17
18 # updating a element in a array
19 arr[2] = 6
20 print("Array after updation : ", end ="")
21 for i in range (0, 6):
22 print (arr[i], end =" ")
23 print()
24
25 # updating a element in a array
26 arr[4] = 8
27 print("Array after updation : ", end ="")
28 for i in range (0, 6):
29 print (arr[i], end =" ")
```

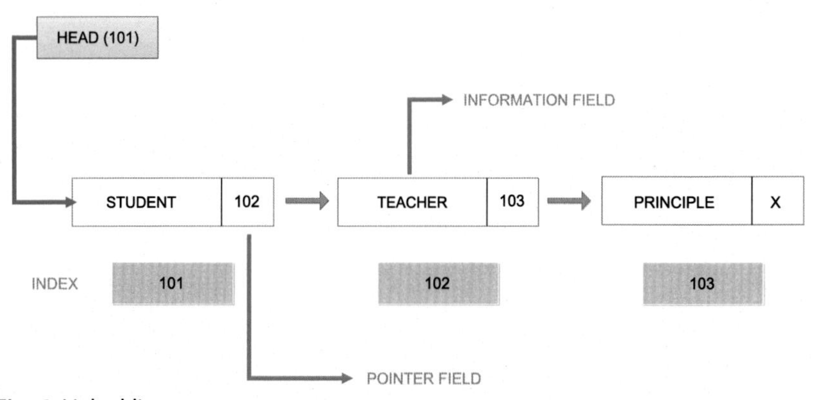

Fig. 6 Linked list.

LISTING 7 Linked list insertion

```
1 # A complete working Python program to demonstrate all
2 # insertion methods of linked list
3
4 # Node class
5 class Node:
6
7 # Function to initialize the node object
8 def __init__(self, data):
9 self.data = data # Assign data
10 self.next = None # Initialize next as null
11
12
13 # Linked List class contains a Node object
14 class LinkedList:
15
16 # Function to initialize head
17 def __init__(self):
18 self.head = None
19
20
21 # Functio to insert a new node at the beginning
22 def push(self, new_data):
23
24 # 1 & 2: Allocate the Node &
25 #  Put in the data
26 new_node = Node(new_data)
27
28 # 3. Make next of new Node as head
29 new_node.next = self.head
30
31 # 4. Move the head to point to new Node
32 self.head = new_node
33
34
35 # This function is in LinkedList class. Inserts a
36 # new node after the given prev_node. This method is
37 # defined inside LinkedList class shown above */
38 def insertAfter(self, prev_node, new_data):
39
40 # 1. check if the given prev_node exists
41 if prev_node is None:
42 print "The given previous node must inLinkedList."
43 return
```

Continued

LISTING 7 Linked list insertion—cont'd

```
44
45 # 2. create new node &
46 #  Put in the data
47 new_node = Node(new_data)
48
49 # 4. Make next of new Node as next of prev_node
50 new_node.next = prev_node.next
51
52 # 5. make next of prev_node as new_node
53 prev_node.next = new_node
54
55
56 # This function is defined in Linked List class
57 # Appends a new node at the end. This method is
58 # defined inside LinkedList class shown above */
59 def append(self, new_data):
60
61 # 1. Create a new node
62 # 2. Put in the data
63 # 3. Set next as None
64 new_node = Node(new_data)
65
66 # 4. If the Linked List is empty, then make the
67 # new node as head
68 if self.head is None:
69 self.head = new_node
70 return
71
72 # 5. Else traverse till the last node
73 last = self.head
74 while (last.next):
75 last = last.next
76
77 # 6. Change the next of last node
78 last.next = new_node
79
80
81 # Utility function to print the linked list
82 def printList(self):
83 temp = self.head
84 while (temp):
85 print temp.data,
86 temp = temp.next
87
88
89
90 # Code execution starts here
91 if __name__=='__main__':
```

LISTING 7 Linked list insertion—cont'd

```
92
93 # Start with the empty list
94 llist = LinkedList()
95
96 # Insert 6. So linked list becomes 6->None
97 llist.append(6)
98
99 # Insert 7 at the beginning. So linked list becomes 7->6->
     None
100 llist.push(7);
101
102 # Insert 1 at the beginning. So linked list becomes
     1->7->6->None
103 llist.push(1);
104
105 # Insert 4 at the end. So linked list becomes 1->7->6->4->
     None
106 llist.append(4)
107
108 # Insert 8, after 7. So linked list becomes 1 -> 7-> 8->
     6-> 4-> None
109 llist.insertAfter(llist.head.next, 8)
110
111 print 'Created linked list is:',
112 llist.printList()
```

LISTING 8 Deletion of an element in a linked list

```
1 # Python program to delete a node from linked list
2
3 # Node class
4 class Node:
5
6 # Constructor to initialize the node object
7 def __init__(self, data):
8 self.data = data
9 self.next = None
10
11 class LinkedList:
```

Continued

LISTING 8 Deletion of an element in a linked list—cont'd

```
12
13 # Function to initialize head
14 def __init__(self):
15 self.head = None
16
17 # Function to insert a new node at the beginning
18 def push(self, new_data):
19 new_node = Node(new_data)
20 new_node.next = self.head
21 self.head = new_node
22
23 # Given a reference to the head of a list and a key,
24 # delete the first occurrence of key in linked list
25 def deleteNode(self, key):
26
27 # Store head node
28 temp = self.head
29
30 # If head node itself holds the key to be deleted
31 if (temp is not None):
32 if (temp.data == key):
33 self.head = temp.next
34 temp = None
35 return
36
37 # Search for the key to be deleted, keep track of the
38 # previous node as we need to change 'prev.next'
39 while(temp is not None):
40 if temp.data == key:
41 break
42 prev = temp
43 temp = temp.next
44
45 # if key was not present in linked list
46 if(temp == None):
47 return
48
49 # Unlink the node from linked list
50 prev.next = temp.next
51
52 temp = None
53
54
55 # Utility function to print the linked LinkedList
56 def printList(self):
57 temp = self.head
58 while(temp):
59 print (" %d" %(temp.data)),
```

LISTING 8 Deletion of an element in a linked list—cont'd

```
60 temp = temp.next
61
62
63 # Driver program
64 llist = LinkedList()
65 llist.push(9)
66 llist.push(1)
67 llist.push(5)
68 llist.push(3)
69
70 print ("Created Linked List: ")
71 llist.printList()
72 llist.deleteNode(1)
73 print ("\nLinked List after Deletion of 1:")
74 llist.printList()
```

2.3 Queue

A queue can be defined as the first in first out (FIFO) type of data structure. In this data structure, insert an element in a queue from the REAR end and delete an element in a queue from the FRONT end as shown in Fig. 7. There are four operations performed on queue.

- *Enqueue*: This operation is used to add an element to the queue. If the queue is full then, an overflow condition occurs.
- *Dequeue*: This operation is used to remove an element from the queue. The items are extracted in the same order in which they are inserted in a queue. If the queue is empty then, an Underflow condition occurs.
- *Front*: This operation is used to obtain the front item from the queue.
- *Rear*: This operation is used to obtain the last item from the queue.

To implement queues, we can use two ways, which are described as follows.

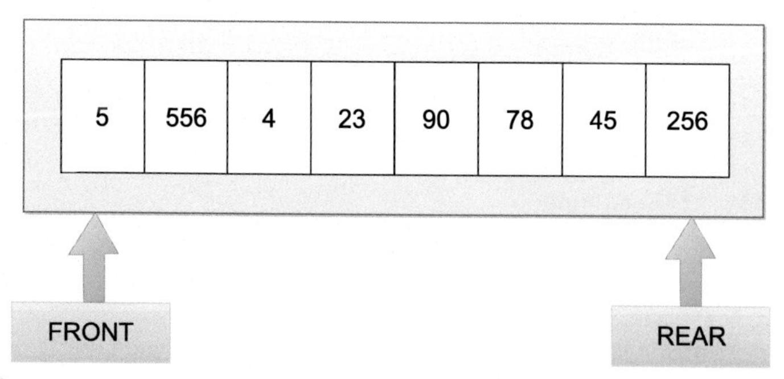

Fig. 7 Queue.

2.3.1 Implement queues using linked list

For linked list implementation in a queue, there are two operations used.

- *enQueue()*: This operation adds a new node after rear and moves rear to the next node.
- *deQueue()*: This operation removes the front node and moves front to the next node.

The coding lines to implement linked list in a queue are defined in Listing 9.

LISTING 9 Linked list implementation of queue

```
1 # Python3 program to demonstrate linked list
2 # based implementation of queue
3
4 # A linked list (LL) node
5 # to store a queue entry
6 class Node:
7
8 def __init__(self, data):
9 self.data = data
10 self.next = None
11
12 # A class to represent a queue
13
14 # The queue, front stores the front node
15 # of LL and rear stores the last node of LL
16 class Queue:
17
18 def __init__(self):
19 self.front = self.rear = None
20
21 def isEmpty(self):
22 return self.front == None
23
24 # Method to add an item to the queue
25 def EnQueue(self, item):
26 temp = Node(item)
27
28 if self.rear == None:
29 self.front = self.rear = temp
30 return
31 self.rear.next = temp
32 self.rear = temp
33
34 # Method to remove an item from queue
```

LISTING 9 Linked list implementation of queue—cont'd

```
35 def DeQueue(self):
36
37 if self.isEmpty():
38 return
39 temp = self.front
40 self.front = temp.next
41
42 if(self.front == None):
43 self.rear = None
44
45 # Driver Code
46 if __name__== '__main__':
47 q = Queue()
48 q.EnQueue(10)
49 q.EnQueue(20)
50 q.DeQueue()
51 q.DeQueue()
52 q.EnQueue(30)
53 q.EnQueue(40)
54 q.EnQueue(50)
55 print("Queue Front " + str(q.front.data))
56 print("Queue Rear " + str(q.rear.data))
```

2.3.2 Implement queues using arrays

For implementing queue using an array, we insert an element at the rear and extract an element from the front. If we simply increment front and rear indices, then there may be problems, the front may reach the end of the array. The coding lines of implementing the queue using an array are defined in Listing 10.

LISTING 10 Array implementation of queue

```
1 # Python3 program for array implementation of queue
2
3 # Class Queue to represent a queue
4 class Queue:
5
6 # __init__ function
7 def __init__(self, capacity):
8 self.front = self.size = 0
9 self.rear = capacity -1
10 self.Q = [None]*capacity
11 self.capacity = capacity
```

Continued

LISTING 10 Array implementation of queue—cont'd

```
12
13 # Queue is full when size becomes
14 # equal to the capacity
15 def isFull(self):
16 return self.size == self.capacity
17
18 # Queue is empty when size is 0
19 def isEmpty(self):
20 return self.size == 0
21
22 # Function to add an item to the queue.
23 # It changes rear and size
24 def EnQueue(self, item):
25 if self.isFull():
26 print("Full")
27 return
28 self.rear = (self.rear + 1) % (self.capacity)
29 self.Q[self.rear] = item
30 self.size = self.size + 1
31 print("%s enqueued to queue" %str(item))
32
33 # Function to remove an item from queue.
34 # It changes front and size
35 def DeQueue(self):
36 if self.isEmpty():
37 print("Empty")
38 return
39
40 print("%s dequeued from queue" %str(self.Q[self.front]))
41 self.front = (self.front + 1) % (self.capacity)
42 self.size = self.size -1
43
44 # Function to get front of queue
45 def que_front(self):
46 if self.isEmpty():
47 print("Queue is empty")
48
49 print("Front item is", self.Q[self.front])
50
51 # Function to get rear of queue
52 def que_rear(self):
53 if self.isEmpty():
54 print("Queue is empty")
55 print("Rear item is", self.Q[self.rear])
```

LISTING 10 Array implementation of queue—cont'd

```
56
57
58 # Driver Code
59 if __name__ == '__main__':
60
61 queue = Queue(30)
62 queue.EnQueue(10)
63 queue.EnQueue(20)
64 queue.EnQueue(30)
65 queue.EnQueue(40)
66 queue.DeQueue()
67 queue.que_front()
68 queue.que_rear()
```

2.4 Stack

A stack is a linear data structure that stores items in a last-in-first-out (LIFO) or first-in-last-out (FILO) way. The insert and delete operations in a stack are called push and pop operations as shown in Fig. 8. A new element is added at one end and an old element is removed from the same end. There are several functions associated with stack, which are described as follows.

- *empty()*: It returns the stack is empty.
- *size()*: It returns the size of the stack.
- *top()*: It returns a reference to the topmost element of the stack.
- *push(s)*: It adds an element "s" at the top of the stack.
- *pop()*: It deletes the topmost element of the stack.

2.4.1 Stack implementation using list

The items in a list are stored next to each other in memory, if the stack grows bigger than the block of memory that currently hold it, then Python needs to do some memory allocations using the append() function. The coding lines of stack implementation using list are defined in Listing 11.

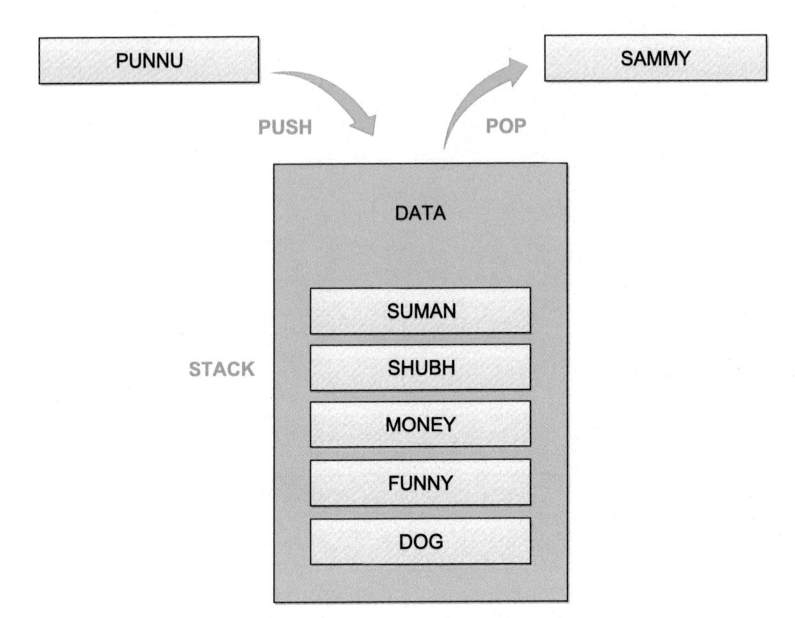

Fig. 8 Stack.

LISTING 11 Stack implementation using list

```
# Python program

stack = []
# append() function to push
# element in the stack
stack.append('a')
stack.append('b')
stack.append('c')

print('Initial stack')
print(stack)

# pop() function to pop
# element from stack in
# LIFO order
print('\nElements poped from stack:')
print(stack.pop())
print(stack.pop())
print(stack.pop())

print('\nStack after elements are poped:')
print(stack)
```

2.4.2 Stack implementation using queue module

Queue module has a LIFO queue. The data is inserted into Queue using put() function and takes data out from the Queue using get() function. There are several functions used in this module.

- *maxsize*: Total number of items allowed in the queue.
- *empty()*: This function returns True if the queue is empty, otherwise False.
- *full()*: This function returns True if the queue is full.
- *get()*: This function removes and returns an item from the queue.
- *get_nowait()*: This function returns an item if one is immediately available, else raise QueueEmpty.
- *put(item)*: It puts an item into the queue. If the queue is full, wait until a free slot is available before adding the item.
- *put_nowait(item)*: It puts an item into the queue without blocking.
- *qsize()*: This function returns the number of items in the queue.

The coding lines of stack implementation using queue module are defined in Listing 12.

2.5 Graphs

A graph is a set of points termed as vertices, and the links that connect the vertices are called edges. It is a nonlinear data structure. To create a graph and add several data elements are the basic operations that implements on the graph, which are described as follows.

2.5.1 Creating a graph

To represent the vertices as the keys of the dictionary and the connection between the vertices called edges as the values in the dictionary. The coding lines of creating a graph are defined in Listing 13 and the pictorial representation of the created graph is shown in Fig. 9.

2.5.2 Display vertices of graph

To represent the graph vertices, use the keys() method to find the vertices from the dictionary. The coding lines of displaying vertices of graph are defined in Listing 14.

2.5.3 Display edges of graph

To represent the edges of the graph, create an empty list of edges then iterate through the edge values associated with each of the vertices. A list is created

LISTING 12 Stack implementation using queue module

```
# Python program
# Initializing a stack
stack = LifoQueue(maxsize = 3)

# qsize() show the number of elements
# in the stack
print(stack.qsize())

# put() function to push
# element in the stack
stack.put('a')
stack.put('b')
stack.put('c')

print("Full: ", stack.full())
print("Size: ", stack.qsize())

# get() function to pop
# element from stack in
# LIFO order
print('\nElements poped from the stack')
print(stack.get())
print(stack.get())
print(stack.get())

print("\nEmpty: ", stack.empty())
```

LISTING 13 Creating a graph

```
# Create the dictionary with graph elements
graph = { "a" : ["b","c"],
"b" : ["a", "d"],
"c" : ["a", "d"],
"d" : ["e"],
"e" : ["d"]
}

# Print the graph
print(graph)
```

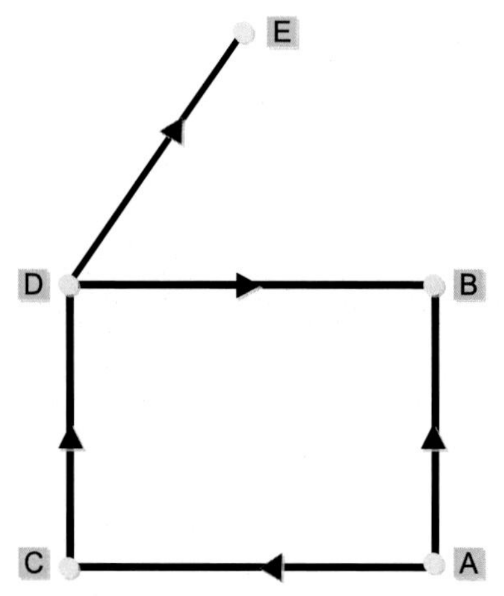

Fig. 9 Created graph representation.

LISTING 14 Display vertices of graph

```
class graph:
def __init__(self,gdict=None):
if gdict is None:
gdict = []
self.gdict = gdict

# Get the keys of the dictionary
def getVertices(self):
return list(self.gdict.keys())

# Create the dictionary with graph elements
graph_elements = { "a" : ["b","c"],
"b" : ["a", "d"],
"c" : ["a", "d"],
"d" : ["e"],
"e" : ["d"]
}

g = graph(graph_elements)

print(g.getVertices())
```

LISTING 15 Display edges of graph

```
1 class graph:
2
3 def __init__(self,gdict=None):
4 if gdict is None:
5 gdict = {}
6 self.gdict = gdict
7
8 def edges(self):
9 return self.findedges()
10 # Find the distinct list of edges
11
12 def findedges(self):
13 edgename = []
14 for vrtx in self.gdict:
15 for nxtvrtx in self.gdict[vrtx]:
16 if {nxtvrtx, vrtx} not in edgename:
17 edgename.append({vrtx, nxtvrtx})
18 return edgename
19
20 # Create the dictionary with graph elements
21 graph_elements = { "a" : ["b","c"],
22 "b" : ["a", "d"],
23 "c" : ["a", "d"],
24 "d" : ["e"],
25 "e" : ["d"]
26 }
27
28 g = graph(graph_elements)
29
30 print(g.edges())
```

that contains distinct edges found from the vertices. The coding lines for displaying edges of the graph are defined in Listing 15.

2.5.4 Adding a vertex to the graph

To add a vertex to the graph, use additional key to the dictionary of the graph. The coding lines of adding a vertex to the graph are defined in Listing 16.

LISTING 16 Adding a vertex to the graph

```
1 class graph:
2
3 def __init__(self,gdict=None):
4 if gdict is None:
5 gdict = {}
6 self.gdict = gdict
7
8 def getVertices(self):
9 return list(self.gdict.keys())
10
11 # Add the vertex as a key
12 def addVertex(self, vrtx):
13 if vrtx not in self.gdict:
14 self.gdict[vrtx] = []
15
16 # Create the dictionary with graph elements
17 graph_elements = { "a" : ["b","c"],
18 "b" : ["a", "d"],
19 "c" : ["a", "d"],
20 "d" : ["e"],
21 "e" : ["d"]
22 }
23
24 g = graph(graph_elements)
25
26 g.addVertex("f")
27
28 print(g.getVertices())
```

2.5.5 Adding an edge to the graph

To add a new edge to the graph, firstly add new vertex as a tuple and validating the edge is present or not. The coding lines of adding an edge to the graph are defined in Listing 17.

2.6 Trees

A tree represents the nodes connected by the edges. It is a nonlinear data structure. It has the following properties.

- One node is marked as a root node.
- Every node other than the root is followed by one parent node.
- Each node can have an arbitrary number of child nodes.

LISTING 17 Adding an edge to the graph

```
1 class graph:
2
3 def __init__(self, gdict=None):
4 if gdict is None:
5 gdict = {}
6 self.gdict = gdict
7
8 def edges(self):
9 return self.findedges()
10 # Add the new edge
11
12 def AddEdge(self, edge):
13 edge = set(edge)
14 (vrtx1, vrtx2) = tuple(edge)
15 if vrtx1 in self.gdict:
16 self.gdict[vrtx1].append(vrtx2)
17 else:
18 self.gdict[vrtx1] = [vrtx2]
19
20 # List the edge names
21 def findedges(self):
22 edgename = []
23 for vrtx in self.gdict:
24 for nxtvrtx in self.gdict[vrtx]:
25 if {nxtvrtx, vrtx} not in edgename:
26 edgename.append({vrtx, nxtvrtx})
27 return edgename
28
29 # Create the dictionary with graph elements
30 graph_elements = { "a" : ["b","c"],
31 "b" : ["a", "d"],
32 "c" : ["a", "d"],
33 "d" : ["e"],
34 "e" : ["d"]
35 }
36
37 g = graph(graph_elements)
38 g.AddEdge({'a','e'})
39 g.AddEdge({'a','c'})
40 print(g.edges())
```

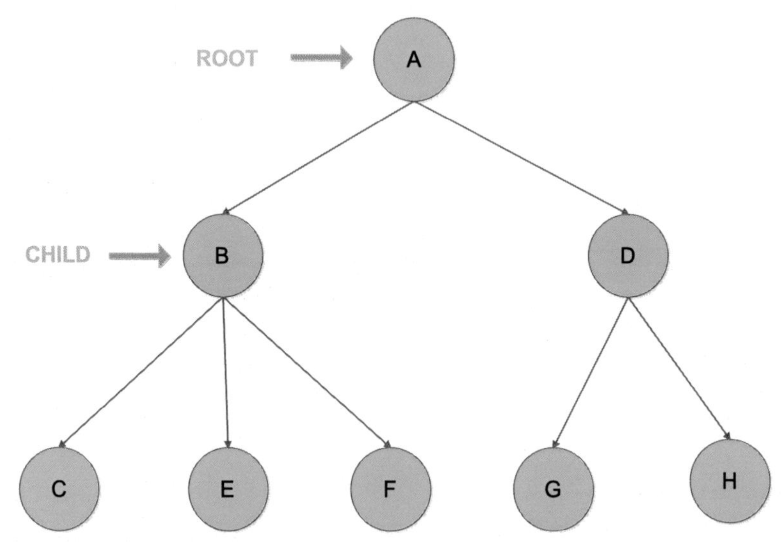

Fig. 10 Representation of a tree.

The pictorial representation of the tree is as shown in Fig. 10.

2.6.1 Creating a root

To create a root of the tree, create a node class and assign some value to the node. The coding lines of creating a root of the tree are defined in Listing 18.

LISTING 18 Creating a root

```
1 class Node:
2
3 def __init__(self, data):
4
5 self.left = None
6 self.right = None
7 self.data = data
8
9
10 def PrintTree(self):
11 print(self.data)
12
13 root = Node(5)
14
15 root.PrintTree()
```

2.6.2 Inserting a node into tree

To insert a node into a tree, use the node class and add an insert class to it. The inserted class compares the value of the node to the parent node and decides to add it as a left node or a right node. The small node adds to the left side of the tree and a large node adds to the right side of the tree after it compares with the parent node. The coding lines of inserting a node into the tree are defined in Listing 19.

LISTING 19 Insert a node into tree

```
1  class Node:
2
3  def __init__(self, data):
4
5  self.left = None
6  self.right = None
7  self.data = data
8
9  def insert(self, data):
10 # Compare the new value with the parent node
11 if self.data:
12 if data < self.data:
13 if self.left is None:
14 self.left = Node(data)
15 else:
16 self.left.insert(data)
17 elif data > self.data:
18 if self.right is None:
19 self.right = Node(data)
20 else:
21 self.right.insert(data)
22 else:
23 self.data = data
24
25 # Print the tree
26 def PrintTree(self):
27 if self.left:
28 self.left.PrintTree()
29 print( self.data),
30 if self.right:
31 self.right.PrintTree()

32
33 # Use the insert method to add nodes
34 root = Node(12)
35 root.insert(7)
36 root.insert(13)
37 root.insert(2)
38
39 root.PrintTree()
```

2.7 Heaps

A heap is a special tree of a nonlinear data structure in which each parent node is less than or equal to its child node is called a min heap. If each parent node is greater than or equal to its child node then, it is called a max heap as shown in Fig. 11. To create a heap, several functions have been used and are described as follows.

- *heapify*: This function converts a regular list to a heap.
- *heappush*: This function adds an element to the heap without changing the current heap.
- *heappop*: This function returns the smallest data element from the heap.

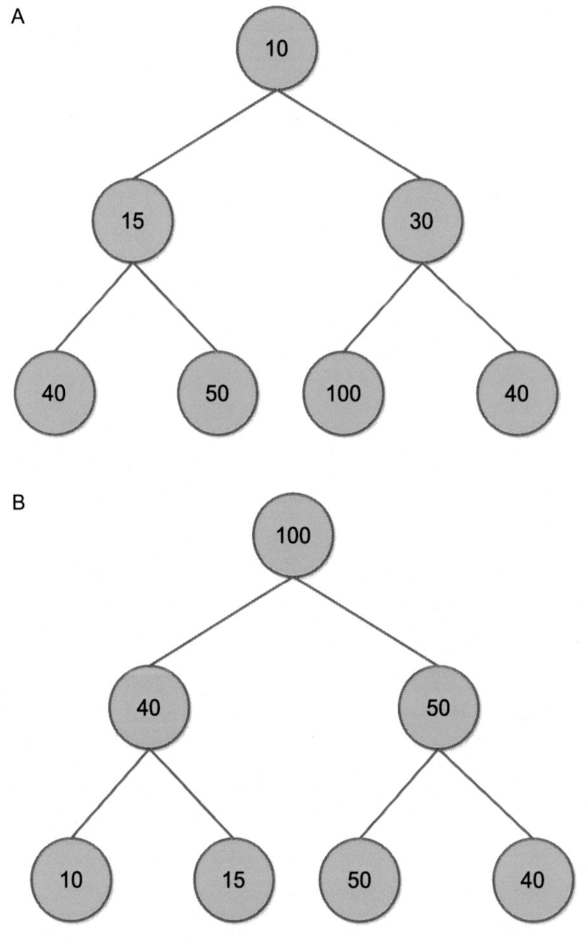

Fig. 11 (A) Min heap and (B) Max heap.

- *heapreplace*: This function replaces the smallest data element with a new value passed in the function.

2.7.1 Creating a heap

To create a heap, use heapify() in-built function. The coding line of creating a heap are defined in Listing 20.

2.7.2 Inserting into heap

To insert an element into a heap always adds an element at the last index. Here, we use heappush() in-built function to insert an element into a heap. The coding lines of inserting an item into a heap are defined in Listing 20.

LISTING 20 Creating a heap

```
1 import heapq
2
3 H = [22,2,46,79,4,6]
4
5 # Use heapify to rearrange the elements
6 heapq.heapify(H)
7 print(H)
```

LISTING 21 Inserting into heap

```
1 import heapq
2 H = [22,2,46,79,4,6]
3 # Covert to a heap
4 heapq.heapify(H)
5 print(H)
6 # Add element
7 heapq.heappush(H,8)
8 print(H)
```

2.7.3 Removing from heap

To remove an item from the heap, use heappop() in-built function. The coding lines of removing an item from the heap are defined in Listing 22.

2.7.4 Replacing in a heap

The heapreplace() in-built function removes the smallest element of the heap and inserts a new element at some place. The coding lines of replacing in a heap are defined in Listing 23.

LISTING 22 Removing from heap

```
1 import heapq
2
3 H = [22,2,46,79,4,6]
4 # Create the heap
5
6 heapq.heapify(H)
7 print(H)
8
9 # Remove element from the heap
10 heapq.heappop(H)
11
12 print(H)
```

LISTING 23 Replacing in a heap

```
1 import heapq
2
3 H = [22,2,46,79,4,6]
4 # Create the heap
5
6 heapq.heapify(H)
7 print(H)
8
9 # Replace an element
10 heapq.heapreplace(H,7)
11 print(H)
```

3. Merkle root

A Merkle tree is also known as a hash tree. It is a data structure used in several applications like blockchain technology to verify and synchronize the data. In this hash tree data structure, each nonleaf node is a hash value of its child node. All the leaf nodes are at the same depth and are as far left as possible. It maintains data integrity, data consistency, and data confidentiality [2].

3.1 Hash function

A hash function maps any length of input to a fixed length of output and this output is called a hash. It is a mathematical function that converts a numerical input value into a compressed another numerical value, which is unique for every input and enables the fingerprinting of data. The pictorial representation oh hash function is shown in Fig. 12. So, a large amount of data can be easily identified through their hash values.

Fig. 13 shows the binary Merkle tree, the root of the tree is a hash of the entire tree.

- The tree data structure allows an efficient mapping of a huge amount of data and small changes made to the data can be easily identified.

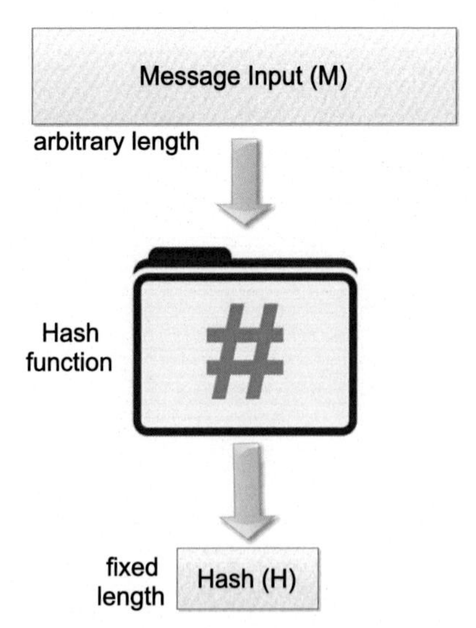

Fig. 12 Hash function.

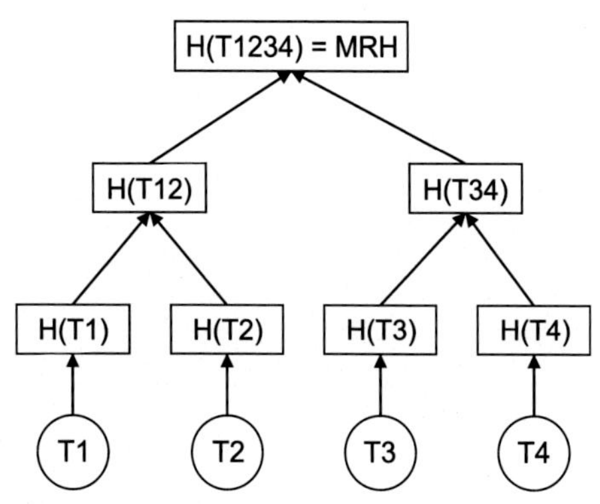

Fig. 13 Merkle tree.

- To know where data change has occurred then, we can traverse to a small part of the tree data structure from the root hash.
- The root hash is used as the fingerprint for the entire data.

3.2 Applications of Merkle tree

Several applications, where the Merkle tree has been used by the researchers to maintain and security of the data.

- Merkle trees are useful in distributed systems where the same data should exist in multiple places and maintains data integrity.
- It can be used to check inconsistencies in the data.
- It is used in bitcoin for proof-of-work (PoW) consensus mechanism in the blockchain technology

4. InterPlanetary file system

InterPlanetary file system (IPFS) is a peer-to-peer (P2P) version-controlled, content-addressed file system. This file system is used for various concepts such as Distributed Hash Table (DHT), BitSwap, and MerkleDAG. IPFS was developed by Juan Bennet at Protocol Labs in 2015. Several applications currently use IPFS for security and privacy of a large amount of data [3]. The pictorial representation of IPFS system is as shown in Fig. 14.

4.1 History of IPFS

The current or traditional way to exchange data across the Internet is HyperText Transfer Protocol (HTTP), but it fails in case of large files.

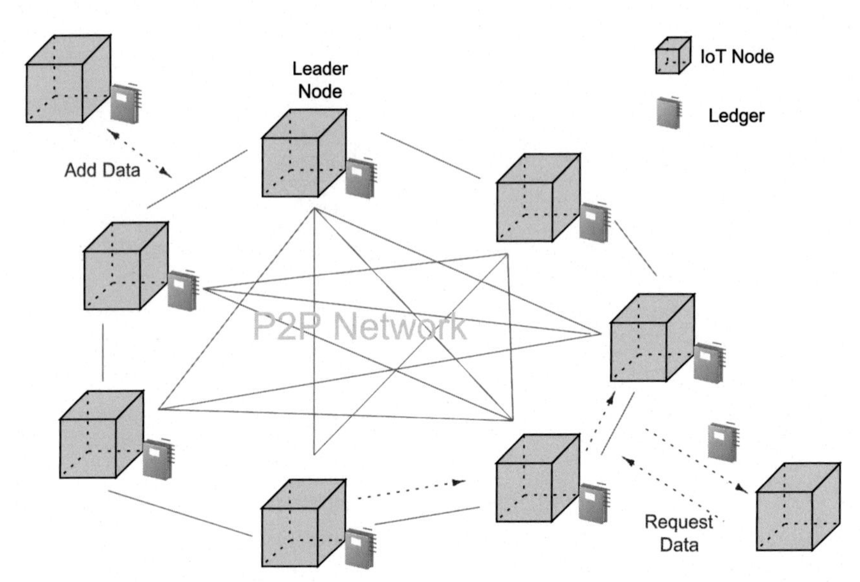

Fig. 14 Architecture of IPFS system.

HTTP mainly uses a client–server protocol that may lead to low latency and makes it difficult to create a P2P connection. Also, real-time media streaming is difficult on HTTP. Because of these failures of HTTP, IPFS prefers to use it in various applications like blockchain.

Unlike HTTP, an IPFS network is content-addressed, where any data is uploaded on an IPFS network and returns in a hash. After that the data is requested to access using that hash value. Anyone can provide storage on the IPFS network and everyone is incentivized with crypto tokens. So, data is distributed and replicated throughout the network which leads to data permanence and high latency. As the data is completely distributed, it has no scope for the centralization of data. The concepts used in IPFS are described as follows.

1. *Distributed hash table*: DHT is used to store and retrieve data across nodes in the network. It is a class that is similar to hash tables. Using a DHT, any node on the network can request the value corresponding to a hash key.
2. *Block exchange*: It is used in the BitTorrent protocol, which is also known as BitSwap to exchange data between nodes. It is a P2P file-sharing protocol that coordinates data exchange between untrusted swarms. It rewards those nodes that contribute to each other and punishes those nodes that only request resources. This helps an IPFS node in retrieving multiple parts of data parallelly.

3. *Merkle directed acyclic graph (DAG)*: It uses a Merkle tree or a Merkle DAG, which is used in the Git Version Control system. It is used to track change to files on the network in a distributed way. Here, the data is stored in the cryptographic hash of the content.

4.2 Nodes of IPFS

Each node on IPFS network is identified by using a NodeID. It is a hash of its public key. They can store files on their local storage and are incentivized to do so. They have maintained a DHT, which is used to find out the NodeIDs of other peers and what data those peers can serve on the network.

4.3 Advantages of IPFS

The advantages of IPFS are described as follows.

- The users can communicate with each other in a local network.
- There is no need for servers anymore, creators can distribute their work without any cost.
- Due to its higher bandwidth, the data load is faster on the network.

4.4 Disadvantages of IPFS

There are some disadvantages of IPFS, which are described as follows.

- IPFS installation is not user friendly and finds a lot of difficulties.
- IPFS consumes a lot of bandwidth which is not acknowledged by Internet users.
- IPFS is used by tech supporter and regular people do not tend to set up their nodes that lead to a shortage of nodes on the network.

References

[1] M. Patel, Data Structure and Algorithm With C, Educreation Publishing, 2018.
[2] C. Liu, R. Ranjan, C. Yang, X. Zhang, L. Wang, J. Chen, MuR-DPA: top-down levelled multi-replica Merkle hash tree based secure public auditing for dynamic big data storage on cloud, IEEE Trans. Comput. 64 (9) (2014) 2609–2622.
[3] J. Benet, Ipfs-content addressed, versioned, p2p file system, arXiv preprint arXiv: 1407.3561 (2014).

About the authors

Shubhani Aggarwal is pursuing PhD from Thapar Institute of Engineering & Technology (Deemed to be University), Patiala, Punjab, India. She received the BTech degree in Computer Science and Engineering from Punjabi University, Patiala, Punjab, India, in 2015, and the ME degree in Computer Science from Panjab University, Chandigarh, India, in 2017. She has many research interests in the area of Blockchain, cryptography, Internet of Drones, and information security.

Neeraj Kumar received his PhD in CSE from Shri Mata Vaishno Devi University, Katra (Jammu and Kashmir), India in 2009, and was a postdoctoral research fellow in Coventry University, Coventry, UK. He is working as a Professor in the Department of Computer Science and Engineering, Thapar Institute of Engineering & Technology (Deemed to be University), Patiala (Punjab), India. He has published more than 400 technical research papers in top-cited journals such as IEEE TKDE, IEEE TIE, IEEE TDSC, IEEE TITS, IEEE TCE, IEEE TII, IEEE TVT, IEEE ITS, IEEE SG, IEEE Netw., IEEE Comm., IEEE WC, IEEE IoTJ, IEEE SJ, Computer Networks, Information sciences, FGCS, JNCA, JPDC, and ComCom. He has guided many research scholars leading to PhD and ME/MTech. His research is supported by funding from UGC, DST, CSIR, and TCS. His research areas are Network Management, IoT, Big Data Analytics, Deep Learning, and Cyber-Security. He is serving as editor of the following journals of repute: ACM Computing Survey, ACM·IEEE Transactions on Sustainable Computing, IEEE·IEEE Systems Journal, IEEE·IEEE Network Magazine, IEE·IEEE Communication Magazine, IEE·Journal of Networks and Computer Applications, Elsevier Computer Communication, Elsevier

International Journal of Communication Systems, Wiley. Also, he has been a guest editor of various international journals of repute such as IEEE Access, IEEE ITS, Elsevier CEE, IEEE Communication Magazine, IEEE Network Magazine, Computer Networks, Elsevier, Future Generation Computer Systems, Elsevier, Journal of Medical Systems, Springer, Computer and Electrical Engineering, Elsevier, Mobile Information Systems, International Journal of Ad Hoc and Ubiquitous Computing, Telecommunication Systems, Springer, and Journal of Supercomputing, Springer. He has also edited/authored 10 books with international/national publishers like IET, Springer, Elsevier, CRC: Security and Privacy of Electronic Healthcare Records: Concepts, Paradigms and Solutions (ISBN-13: 978-1-78561-898-7), Machine Learning for Cognitive IoT, CRC Press, Blockchain, Big Data and IoT, Blockchain Technologies Across Industrial Vertical, Elsevier, Multimedia Big Data Computing for IoT Applications: Concepts, Paradigms and Solutions (ISBN: 978-981-13-8759-3), Proceedings of First International Conference on Computing, Communications, and Cyber-Security (IC4S 2019) (ISBN 978-981-15-3369-3). One of the edited textbook entitled, "Multimedia Big Data Computing for IoT Applications: Concepts, Paradigms, and Solutions" published in Springer in 2019 is having 3.5 million downloads till June 6, 2020. It attracts attention of the researchers across the globe (https://www.springer.com/in/book/9789811387586). He has been a workshop chair at IEEE Globecom 2018 and IEEE ICC 2019 and TPC Chair and member for various international conferences such as IEEE MASS 2020 and IEEE MSN 2020. He is a senior member of the IEEE. He has more than 12,321 citations to his credit with current h-index of 60 (September 2020). He has won the best papers award from IEEE Systems Journal and ICC 2018, Kansas City in 2018. He has been listed in the highly cited researcher of 2019 list of Web of Science (WoS). In India, he is listed in top 10 position among highly cited researchers list. He is an adjunct professor at Asia University, Taiwan, King Abdul Aziz University, Jeddah, Saudi Arabia, and Charles Darwin University, Australia.

CHAPTER THREE

Hashes☆

Shubhani Aggarwal and Neeraj Kumar
Thapar Institute of Engineering & Technology, Patiala, Punjab, India

Contents

Abstract

Hashing is a method to store and retrieve records from a database. Anyone can insert, delete, and search for records based on a search key value. When hashes are implemented properly, the operations like inserting, deleting, etc. can be performed in constant time. Several hashes that have been discussed in this chapter are SHA-2, EtHash, Scrypt, X11, EquiHash, and Ripemd-160. These hash techniques can be used in various technologies like blockchain to secure storage of any type of data and information.

Chapter points

- In this chapter, we discuss different types of hashes that can be used in a blockchain.
- Here, we discuss the functionality of different hashes that can be used in a blockchain.

A cryptographic hash is a kind of signature for a text or a data file. A hash is not encryption and it cannot be decrypted back to the original text. It is a one-way cryptographic function that condenses any size of data to a fixed size. Its applications include hash tables, integrity verification, challenge handshake authentication (a client can send the hash of a password over the internet without transmitting the original password for validation by a

☆ Cryptographic primitives used in blockchain.

Advances in Computers, Volume 121
ISSN 0065-2458
https://doi.org/10.1016/bs.adcom.2020.08.003

server), digital signatures, antitamper (the recipient can rehash the message and compare it to the supplied hash. If they match, the message is unchanged). The pictorial representation of hash from the random text is as shown in Fig. 1.

1. SHA-256 or SHA-2

SHA stands for secure hashing algorithm. SHA-1 and SHA-256 are two different versions of that algorithm. They differ in both constructions and in bit-length of the signature. SHA-1 is a 160-bit hash and SHA-256 generates an almost-unique 256-bit (32-byte) signature for a text. SHA-256 is one of the successor and strongest hash functions to SHA-1. It is not much more complex to code than SHA-1 and has not yet been compromised in any way [1]. The pictorial representation of SHA-256 is as shown in Fig. 2.

2. Ethash

As we know, PoW mining process is used to secure distributed blocks of transactions on to the blockchain [2]. It involves taking data from a block header of block to make an input, and calculated hash the input using a cryptographic hashing algorithm. This will make produces the output hash value of a fixed length. In the case of Ethereum, the algorithm used for this process is Ethash. With Ethash, the output formed in a hash value is below a certain threshold known as difficulty. It involves the Ethereum network

Fig. 1 Cryptographic hash.

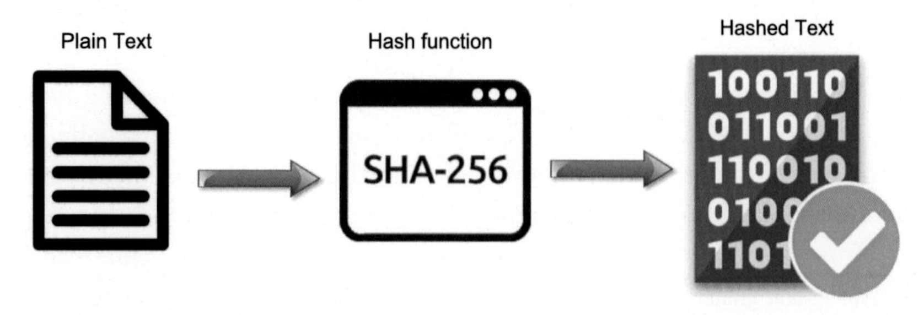

Fig. 2 Hashing algorithm.

that increases and decreases the threshold to control the rate at which the blocks are mined on the blockchain. If the rate to create the blocks are found increases, then the network will automatically increase the difficulty, i.e., it will lower the network threshold so that the number of valid hashes capable of being found decreases. Conversely, if the rate of discovered blocks decreases, the network threshold will increase to make a higher number of correct hash values that could be found. On average, one block is produced after every 12 s in the Ethereum network. If the miner successfully discovers the block that can be added to the blockchain will be represented by following rewards.

- A static block reward consisting of three ether.
- The cost of all the gas spent on the execution of all transactions that are contained within the block will be awarded to the miner's account as part of the PoW mining process.

Ethash is a memory-hard hash algorithm and a capability of ASIC-resistant. As shown in Fig. 3, by combining the header hash value and a nonce (a random value that can change for the attempts we make) to make another hash value for generating the task at the first level. In the mixing level, first

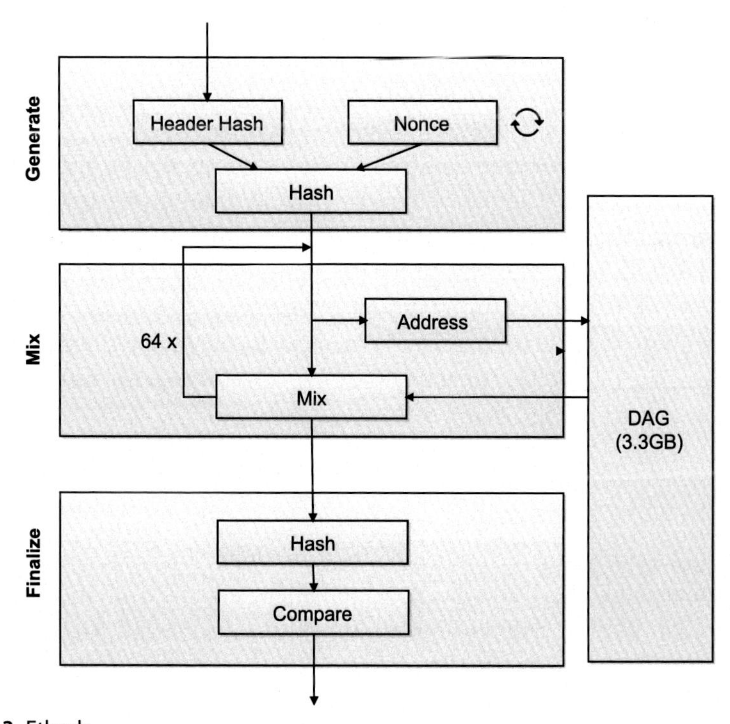

Fig. 3 Ethash.

calculates an address from the current value then, looks up this address in the directed acyclic graph (DAG) (a big block of memory that retrieves corresponding memory line). Further, mix the result of generating hash value from the first level and repeat the whole operation by 64 times. Then, again do some more hashing and compare with the difficulty target. It is very large in size (approx. 3.3 GB) and its growing with time.

3. Scrypt

The scrypt key derivation function was originally developed by Colin Percival and published in 2009 for use in the Tarsnap online backup system and is designed to be far more secure against hardware brute-force attacks than alternative functions such as password-based key derivation function 2 (PBKDF2) or bcrypt. Scrypt is a memory hard key derivation function. It requires a large amount of RAM to be solved. It adjusts and creates a lot of pseudorandom numbers that needs to be stored in a RAM location. The main advantage of scrypt is that there should be the possibility of more miners joining the network and contributing sufficiently to make it worth their while. The other advantage is that there is usage of less energy as the total network power is also less [3]. To define the scrypt, it has some following parameters.

- *cpuCost(N)*: CPU cost of the algorithm and it must be a power of 2 greater than 1 (by default its value is 64 or 2^4)
- *memoryCost(R)*: Memory cost of the algorithm (by default is 8).
- *parallelization(P)*: The parallelization of the algorithm (by default is 1). Note that the implementation does not take advantage of parallelization.
- *keyLength(Length)*: Key length for the algorithm (by default is 32).
- *saltLength(S)*: Salt length (by default is 64).

The parameters should be increased as memory latency and CPU parallelism increases. The pictorial representation of scrypt hash generator and verification is as shown in Figs. 4 and 5

4. X11

X11 is a hashing technique that was developed by Evan Duffield and implemented by the Darkcoin development team in 2014. It works by aggregating the 11 individual hash functions such as BLAKE, BLUE MIDNIGHT WISH (BMW), Grøstl, JH, Keccak, Skein, Luffa, CubeHash, SHAvite-3, SIMD, and ECHO. The algorithm works when

Fig. 4 Scrypt hash generator.

a first value is submitted to the BLAKE function, it produces a hash value which then submitted to the BMW function, which then produces another hash value and submitted to the next function. This process repeats until the very last function in a sequential manner. It is one of the most energy-efficient algorithms and able to decrease the effect of heat by 30% for graphical processing unit miners. In this algorithm, graphic cards use less processing power as compared to mine the coins. So, this algorithm is popular in those areas where electricity cost is high. The icon to represent the X11 is as shown in Fig. 6.

5. Equihash

Equihash is one of the most remarkable hashing algorithms in a blockchain [4]. It was developed by Alex Biryukov and Dmitry Khovratovich at the University of Luxembourg, the Equihash algorithm is an asymmetric memory-orientated, i.e., memory-hard system. This means

scrypt hash generator and verification

Password

Shubhani

Validate Hash

YKNkr4a0vt8GRPjvLHx9cPdkY/eEds9F1QG5vV0jjJc=

Salt

123456789

N 64
R 8
P 1
Length 32

Hash Verification of

YKNkr4a0vt8GRPjvLHx9cPdkY/eEds9F1QG5vV0jjJc=

[hash Verification Sucessfull]

Fig. 5 Scrypt hash verification.

Fig. 6 X11.

a verifier should use the same amount of memory as the prover in order to validate a proof. Effectively, the verifier would have to be almost as powerful as the prover. It plays an important role in the security of blockchain networks and investigates its efficiency and capability at ASIC-resistance. The algorithm prevents ASIC centralization by protecting that the generation of proof is highly intense memory. Because memory is an important and expensive resource in computing and optimizing on an ASIC chip, which will result in a large computational cost to an end-user. The number of cryptocurrencies that make use of the Equihash algorithm such as Zcash, Bitcoin Gold, Bitcoin Private, Komodo, ZenCash, ZClassic, etc.

5.1 Properties of Equihash

Biryukov and Khovratovich assume the algorithm has acquiring the following properties.

- *Progress-free process*: This algorithm must follow a stochastic process for avoiding the centralization of mining. With this, the probability of generating proof at any given instant of time is independent of previous events.
- *Large cost*: To require and optimize high intense memory for the computation of proofs that results in large computational cost.
- *Small proof size and instant verification*: The size of proofs is small which results in faster verification, i.e., less memory required for the verification process.
- *Flexibility:* To further and future improvements and architectural changes in the algorithm, the memory variable must be flexible.

6. RIPEMD-160

RIPEMD-160 [5] is a fast cryptographic hash function and extended version of RIPEMD, which was developed in the framework of the EU project RIPE (Race Integrity Primitives Evaluation). Like the predecessors of message digest (MD4, MD5), RIPEMD, RIPEMD-160 is tuned for 32-bit processors. The algorithm will work on 8-bit and 16-bit microprocessors. It uses the following operations.

- Bitwise Boolean operations.
- Two's complement addition of words.
- Left rotation of words.

It compresses the input by dividing it into blocks each of 64 bytes. The input size must be padded and a multiple of 64 bytes. It must be appended with a single 1 bit followed by a string of 0 bits. Each block is converted into a 16-word block using a Little-endian convention. The output would be contained in five 32-bit words. The main part of the algorithm is compression function. It computes the new state from the old state and the next 16-word block. The pictorial representation of RIPEMD-160 is as shown in Fig. 7. The comparison among all hash functions in blockchain platforms is described in Table 1.

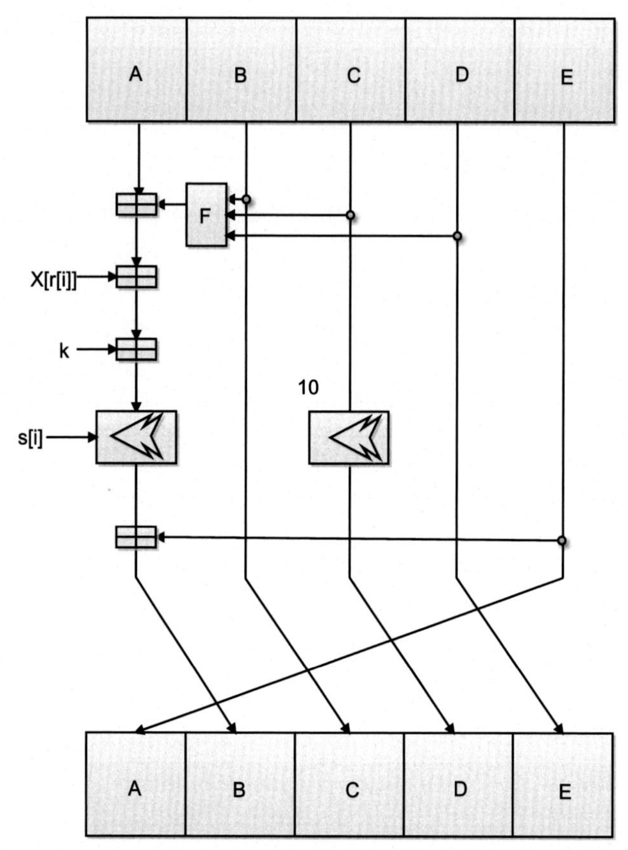

Fig. 7 Compression function of subblock RIPEMD-160 hash algorithm.

Table 1 Hash functions used in blockchain platforms.

Blockchain platforms	SHA256	Ethash	SCrypt	X11	Equihash	RIPEMD160
Bitcoin	Yes	No	No	No	No	Yes
Ethereum	Yes	Yes	No	No	No	Yes
Dash	Yes	No	No	Yes	No	No
Litecoin	Yes	No	Yes	No	No	Yes
Zcash	Yes	No	No	No	Yes	No
Zcoin	Yes	No	No	No	No	No
Ripple	Yes	No	No	No	No	Yes
Blackcoin	Yes	No	Yes	No	No	Yes
Qtum	Yes	Yes	No	No	No	Yes

Table 1 Hash functions used in blockchain platforms.—cont'd

Blockchain platforms	SHA256	Ethash	SCrypt	X11	Equihash	RIPEMD160
BitConnect	No	No	Yes	No	No	No
Komodo	Yes	No	No	No	Yes	No
Dogecoin	Yes	No	Yes	No	No	Yes
Ark	Yes	No	No	No	No	Nov
Byteball	Yes	No	No	No	No	No
Naivecoin	Yes	No	No	No	No	No
RScoin	Yes	No	No	No	No	No

References

[1] H. Gilbert, H. Handschuh, Security analysis of SHA-256 and sisters, in: International Workshop on Selected Areas in Cryptography, Springer, 2003, pp. 175–193.
[2] E. Wiki, Ethash, GitHub Ethereum Wiki (2017). https://github.com/ethereum/wiki/wiki/Ethash.
[3] C. Percival, S. Josefsson, The scrypt password-based key derivation function, IETF Draft (2016). http://tools.ietf.org/html/josefsson-scrypt-kdf-00.txt (accessed: 30.11.2012).
[4] L. Alcock, L. Ren, A Note on the Security of Equihash, in: Proceedings of the 2017 on Cloud Computing Security Workshop, 2017, pp. 51–55.
[5] H. Dobbertin, A. Bosselaers, B. Preneel, RIPEMD-160: a strengthened version of RIPEMD, in: International Workshop on Fast Software Encryption, Springer, 1996, pp. 71–82.

About the authors

Shubhani Aggarwal is pursuing PhD from Thapar Institute of Engineering & Technology (Deemed to be University), Patiala, Punjab, India. She received the BTech degree in Computer Science and Engineering from Punjabi University, Patiala, Punjab, India, in 2015, and the ME degree in Computer Science from Panjab University, Chandigarh, India, in 2017. She has many research interests in the area of Blockchain, cryptography, Internet of Drones, and information security.

Neeraj Kumar received his PhD in CSE from Shri Mata Vaishno Devi University, Katra (Jammu and Kashmir), India in 2009, and was a postdoctoral research fellow in Coventry University, Coventry, UK. He is working as a Professor in the Department of Computer Science and Engineering, Thapar Institute of Engineering & Technology (Deemed to be University), Patiala (Punjab), India. He has published more than 400 technical research papers in top-cited journals such as IEEE TKDE, IEEE TIE, IEEE TDSC, IEEE TITS, IEEE TCE, IEEE TII, IEEE TVT, IEEE ITS, IEEE SG, IEEE Netw., IEEE Comm., IEEE WC, IEEE IoTJ, IEEE SJ, Computer Networks, Information sciences, FGCS, JNCA, JPDC, and ComCom. He has guided many research scholars leading to PhD and ME/MTech. His research is supported by funding from UGC, DST, CSIR, and TCS. His research areas are Network Management, IoT, Big Data Analytics, Deep Learning, and Cybersecurity. He is serving as editor of the following journals of repute: ACM Computing Survey, ACM·IEEE Transactions on Sustainable Computing, IEEE·IEEE Systems Journal, IEEE·IEEE Network Magazine, IEE·IEEE Communication Magazine, IEE·Journal of Networks and Computer Applications, Elsevier Computer Communication, Elsevier International Journal of Communication Systems, Wiley. Also, he has been a guest editor of various international journals of repute such as IEEE Access, IEEE ITS, Elsevier CEE, IEEE Communication Magazine, IEEE Network Magazine, Computer Networks, Elsevier, Future Generation Computer Systems, Elsevier, Journal of Medical Systems, Springer, Computer and Electrical Engineering, Elsevier, Mobile Information Systems, International Journal of Ad Hoc and Ubiquitous Computing, Telecommunication Systems, Springer, and Journal of Supercomputing, Springer. He has also edited/authored 10 books with international/national publishers like IET, Springer, Elsevier, CRC: Security and Privacy of Electronic Healthcare Records: Concepts, Paradigms and Solutions (ISBN-13: 978-1-78561-898-7), Machine Learning for Cognitive IoT, CRC Press, Blockchain, Big Data and IoT, Blockchain Technologies Across Industrial Vertical, Elsevier, Multimedia Big Data Computing for IoT Applications: Concepts, Paradigms and Solutions (ISBN: 978-981-13-8759-3), Proceedings of First International Conference on Computing, Communications, and Cyber-Security (IC4S 2019) (ISBN 978-981-15-3369-3). One of the edited text-book entitled,

"Multimedia Big Data Computing for IoT Applications: Concepts, Paradigms, and Solutions" published in Springer in 2019 is having 3.5 million downloads till June 6, 2020. It attracts attention of the researchers across the globe (https://www.springer.com/in/book/9789811387586). He zhas been a workshop chair at IEEE Globecom 2018 and IEEE ICC 2019 and TPC Chair and member for various international conferences such as IEEE MASS 2020 and IEEE MSN 2020. He is a senior member of the IEEE. He has more than 12,321 citations to his credit with current h-index of 60 (September 2020). He has won the best papers award from IEEE Systems Journal and ICC 2018, Kansas City in 2018. He has been listed in the highly cited researcher of 2019 list of Web of Science (WoS). In India, he is listed in top 10 position among highly cited researchers list. He is an adjunct professor at Asia University, Taiwan, King Abdul Aziz University, Jeddah, Saudi Arabia, and Charles Darwin University, Australia.

CHAPTER FOUR

Digital signatures☆

Shubhani Aggarwal and Neeraj Kumar
Thapar Institute of Engineering & Technology, Patiala, Punjab, India

Contents

Abstract

A digital signature algorithm refers to a standard for digital signatures, which is based on the algebraic properties of discrete logarithm problem and modular exponentiations based on public-key cryptosystems principal. Digital signatures are work on the principle of two mutually authenticating cryptographic keys, i.e., a public key and a private key. In this chapter, we have discussed the digital signature algorithm, elliptic curve digital signature algorithm (ECDSA), and Edward curve digital signature algorithm (EdDSA). With this, the detailed working and cryptographic functionality of these algorithms are described.

Chapter points

- In this chapter, we discuss the digital signatures that can be used in blockchain.
- Here, we discuss the signing and verification process of the digital signature algorithms.

1. Digital signatures

Digital signatures [1] are a cryptographic tool used to sign and verify message signatures to provide message authentication, message integrity, and nonrepudiation of electronic documents, defined as follows.

☆ Cryptographic primitives used in blockchain.

Advances in Computers, Volume 121
ISSN 0065-2458
https://doi.org/10.1016/bs.adcom.2020.08.004

- *Message authentication*: A proof that the message has not been modified during transmission and transmit by a known sender who has created and signed the message.
- *Message Integrity*: A proof that the message was not changed after the signing by the sender.
- *Nonrepudiation*: The sender and the receiver cannot deny the document after the signature and verification are once created.

Digital signatures are widely used in the business and financial industry such as for authorizing bank payments (money transfer), for signing transactions in the public blockchain systems, and in many other scenarios. It cannot identify who is a person that creates a certain signature (Fig. 1). This can be solved in combination with a digital certificate, which binds a public key owner with identity (person, organization, web site, or other). The use of digital signatures in blockchain platforms is as shown in Table 1.

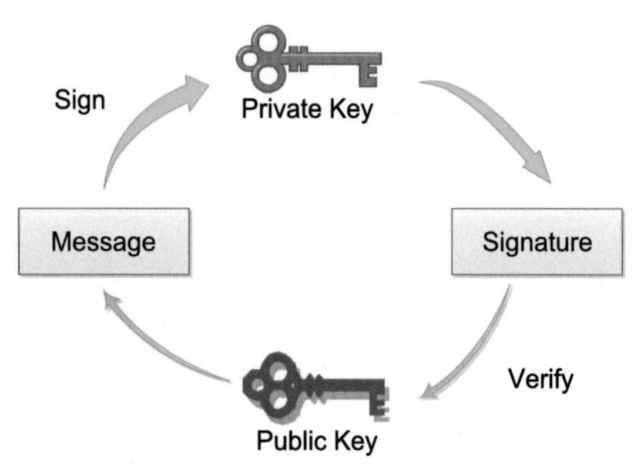

Fig. 1 Digital signatures.

Table 1 Digital signatures used in blockchain platforms.

Blockchain platforms	ECDSA	EdDSA
Bitcoin	Yes	No
Ethereum	Yes	No
Dash	Yes	No
Litecoin	Yes	No
Zcash	Yes	No
Zcoin	Yes	No
Ripple	Yes	No
Blackcoin	Yes	No

Table 1 Digital signatures used in blockchain platforms.—cont'd

Blockchain platforms	ECDSA	EdDSA
Qtum	Yes	No
BitConnect	Yes	No
Komodo	Yes	No
Dogecoin	Yes	No
Ark	Yes	No
Byteball	Yes	No
Naivecoin	No	Yes
RScoin	Yes	No

2. Digital signature algorithm

Digital signatures use a public-key cryptosystem [2] and use a public/private key pairs. A message is signed by a private key and the signature is verified by the corresponding public key is as shown in Fig. 1 [3]. The message is signed by the sender's private key (*PK*). Firstly, the input message is hashed and then the signature is calculated by the signing algorithm as represented in the following Eq. (1).

$$SignMsg = (msg, PK) \tag{1}$$

Further, signature of the message is verified by the corresponding public key (*PU*). Typically the signed message is hashed and some calculation is performed by the signature algorithm using the message hash + the public key. The result from signing is a Boolean value (valid or invalid signature) as represented in the following Eq. (2).

$$VerifyMsg = (msg, signature, PU) = valid/invalid \tag{2}$$

The signed message mathematically guarantees that the message was signed with *PK* by a corresponding *PU*. After signing the message, the message and the signature cannot be modified and thus provides authentication and integrity. Any user who knows the *PU* of the message signer can verify the message and signature at any instant of time. After the signing of the message, the sender cannot deny the document signing known as nonrepudiation.

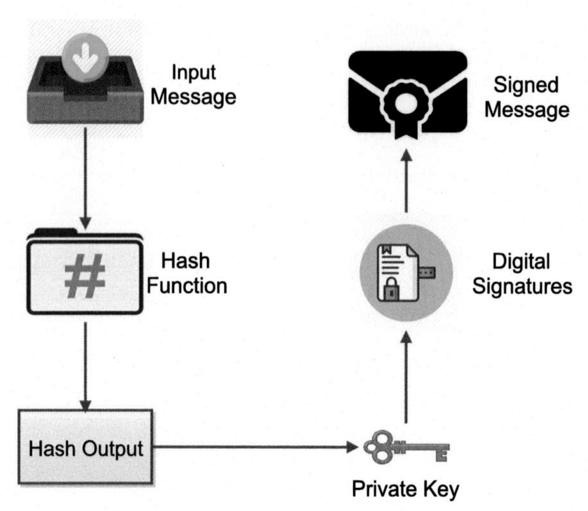

Fig. 2 Signing of the message.

In Fig. 2 during signing the message, the input message is hashed using a hash function then, calculates the digital signature with the private key of the sender and produces a signed message. It consists of the original message and the calculated signature.

In Fig. 3 during signature verification, the message for verification is hashed and some computations are performed between the message hash, the digital signature and the public key, and finally, a comparison decides whether the signature is valid or not.

The most popular digital signature schemes are elliptical curve digital signature algorithm (ECDSA) and Edward curve digital signature algorithm (EdDSA). Modern developers often use EdDSA signatures instead of ECDSA signatures, because the EdDSA signature scheme uses public and private keys of 32 bytes, signatures of 64 bytes, signing and verification are comparatively faster. But in blockchain for Bitcoin and Ethereum platforms, ECDSA signatures are preferably used in comparison to EdDSA signatures. The detail description of these digital signature algorithms is described in the next sections.

3. Elliptic curve digital signature algorithm

ECDSA [4] is used for creating a signature of data to verify its authenticity without compromising its security. It does not encrypt the data but only protects the data and make sure that the data has not tampered

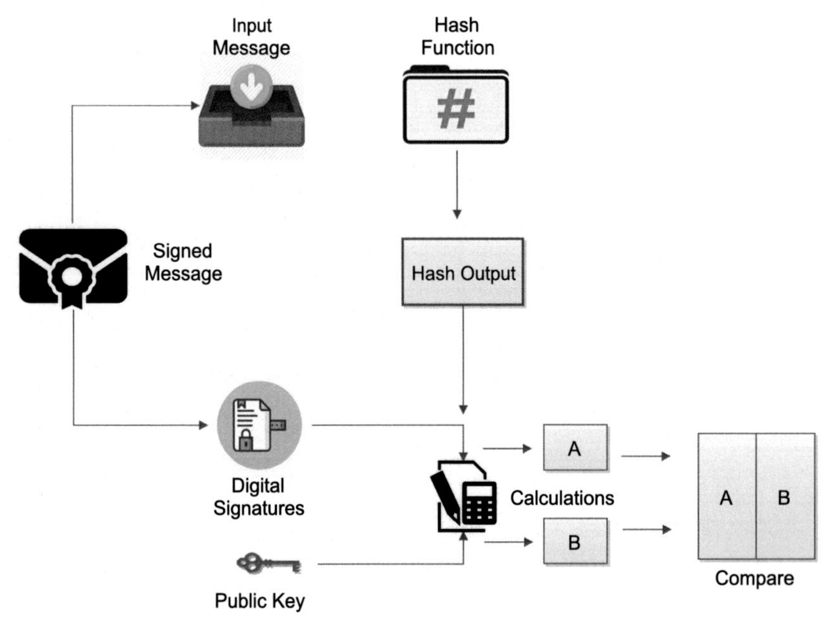

Fig. 3 Verification of the message.

[5, 6]. It is used SHA-1 cryptographic hash to sign the message. The working of ECDSA is described in the following steps.

1. The given Eq. (3) as the curve which is symmetric on the *x-axis*. So, at every x, there are two values of y.

$$y^2 = (x^3 + a * x + b)modq \tag{3}$$

2. The modulo q is a prime number and make sure all the values lie in the range of *160* bits. Therefore, the values of y^2 are between *0* and *q-1*.
3. Assume N possible points on the given curve that generates the square value of two integers, where $N¡q$ (perfect square between *0* and q).
4. As we know the curve is symmetric on the *x-axis* that will results in two values or points (one is on negative axis and other is on positive axis), this means only$N/2$ possible coordinates which are valid and give points on the curve.
5. From the aforementioned information, we conclude that there is a number of finite points on the curve.
6. *Point addition*: Point addition is defined as adding one point M to another point QN that results $S = (M + N)$. Further, these points will intersect the curve on a third point O that is the negative value

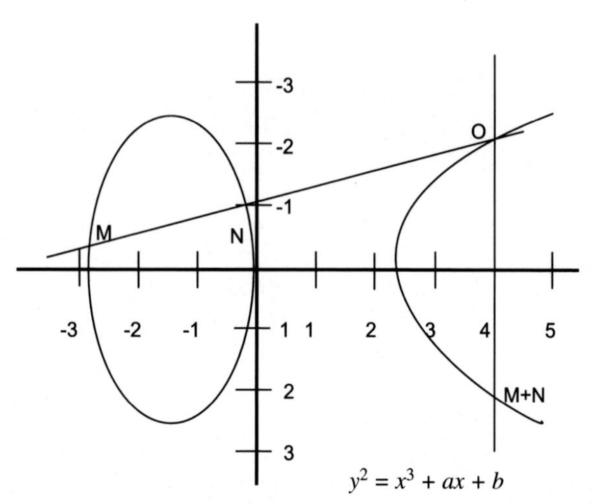

Fig. 4 Point addition.

of S. So, define $O = -S$ that represents the symmetrical point of O on the *x-axis* as shown in Fig. 4.

7. *Point multiplication*: In the same way, $M + M$ will be the symmetric point of O that intersects the line which is the tangent on O point. Similarly, $M + M + M$ is the addition and the resulting point of $M + M$ and point M. Since $M + M + M$ can also be written as $(M + M) + M$. This information defines the point multiplication where $k * M$ is the addition of the point M to itself by k times as shown in Fig. 5.
8. In a given Eq. (3), a, b, p, N, and G need to know. Where, a, b are the parameters of the curve function, p is the prime modulus, and N is the number of points of the curve. Here, there is a need for finding the only G point that represents a *reference point*.
9. Firstly, calculate the private and public key. A private key is a random number that is generated, and the public key is a point on the curve generated from the point multiplication of G with the private key. Assume dA as the private key and Qa as the public key. So, calculate the following Eq. (4).

$$Qa = dA * G \tag{4}$$

10. *Creating a signature:* A signature is of 40 bytes and is represented by O and S values having 20 bytes each. Then, first we generate a random value k is of 20 bytes and calculate the following Eq. (5).

$$M = k * G \tag{5}$$

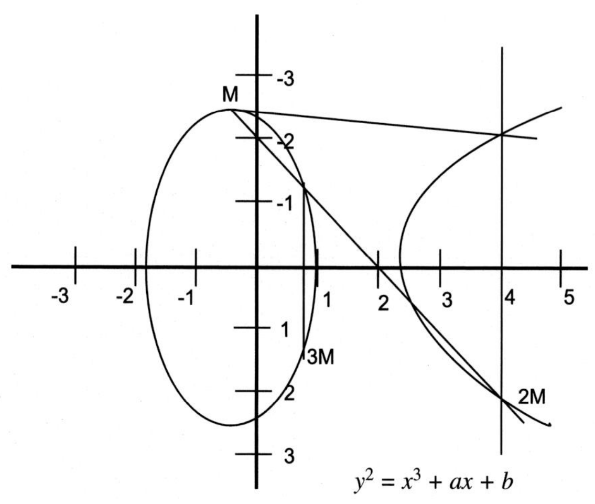

Fig. 5 Point multiplication.

where, M is a point on the curve having coordinates (x, y), x value is represented as O and y is represented as S, k is a random value and G is a reference point or origin of curve.

Now, calculate the S as in given Eq. (6), i.e., y coordinate of the M point.

$$S = k^{-1}(z + dA + O)modq \tag{6}$$

where z is the hash value of the message, dA is the private key, O is the x coordinate of the M point.

11. *Verifying the signature*: To verify the signature, there is a need for public key Qa by calculating the following Eq. (7).

$$M = S^{-1} * z * G + S^{-1} * O * Qa \tag{7}$$

If x coordinate is equal to the O then, validate otherwise not. Check Eq. (4) and put it into Eq. (7) in following Eq. (8) as under.

$$M = S^{-1} * z * G + S^{-1} * O * dA * G \tag{8}$$

Find common S^{-1} and G from Eq. (8) and get Eq. (9).

$$M = S^{-1} * (z + dA * O) * G \tag{9}$$

where, M should watch O and O is the coordinate of $k * G$. So, get the following Eq. (10).

$$k * G = S^{-1} * E(z + dA * O) * G \tag{10}$$

Removing G from both sides as it is common and simplify Eq. (10) and get Eq. (11).

$$k = S^{-1} * (z + dA * O) \tag{11}$$

Now, inverting k and S and we get Eq. (12).

$$S = k^{-1} * (z + dA * O) \tag{12}$$

Eq. (12) is equal to Eq. (6) that proves the validity of the signature.

12. *Security*: To calculate S, there is a need for k and dA, but for verifying the signature, there is a need for $O$and Qa. Since $O = k * G$ and $Qa = dA * G$ and because of the trap door function in the ECDSA point multiplication as discussed in step 7, we cannot calculate dA or k from knowing Qa and O, this makes the ECDSA algorithm secure. There is no other way to find the private keys, and there is no way of faking a signature without knowing the private key.

4. Edward curve digital signature algorithm

EdDSA [7] is a modern and secure algorithm based on perforance-optimized elliptic curves. The signature of EdDSA uses the two forms of edwards such as edwards25519 (255-bit curve) and edwards448 (448-bit curve).

1. *Key generation*: Both forms of edward use small private keys, public keys and small signature with high security level as shown in Table 2. The EdDSA key-pair consists of.
   - $Privatekey = PK(integer)$
   - $Publickey = PU = PK * G$ (Elliptic curve point)
     where, G is the generator.

Table 2 Parameters difference between Ed25519 and Ed448.

Parameters	Ed25519	Ed448
Private key length	32 bytes	57 bytes
Public key length	32 bytes	57 bytes
Signature size	64 bytes	114 bytes
High security level	128 bytes	224 bytes

The private key (*PK*) is generated from a random integer, known as a seed. The public key (*PU*) is a point on the elliptic curve (EC), calculated by the elliptic curve point multiplication as described in the following Eq. (13). The private key, multiplied by the generator point *G* for the curve. The public key is encoded as a compressed EC point, i.e., the *y*-coordinate, combined with the lowest bit of the *x*-coordinate.

$$PU = PK * G \tag{13}$$

2. *Creating the signature*: The EdDSA algorithm takes as input a text message *Message* + the sender's *PK* and produces as output a pair of integers *R*, *s*. The signing procedure of EdDSA works as following Eq. (14).

$$EdDSA_{signature} = (Message, PK) = (R, s) \tag{14}$$

The steps followed for creating the signature in EdDSA.

(a) Calculate the *PU* as in the given Eq. (14).

(b) Generate a secret integer *r* as in following Eq. (15).

$$r = hash(hash(PK) + Message) modq \tag{15}$$

(c) Calculate the *PU* point behind *r* as in following Eq. (16).

$$R = r * G \tag{16}$$

(d) Calculate the hash value *h* as in following Eq. (17).

$$h = hash(R + PU + Message) modq \tag{17}$$

(e) Calculate the *s* as in following Eq. (18).

$$s = (r + h * PK) modq \tag{18}$$

(f) Return the signature in the form (R,s).

The EdDSA signature holds a compressed point *R* + the integer *s* that confirms the sender is authenticated and he knows the message and the private key.

3. *Verifying the signature*: The EdDSA signature verification algorithm takes as input a text message *Message* + (*PU*) + (*R*, *s*). It produces the output as Boolean value, i.e., valid or invalid signature is as shown in Eq. (19). The verification steps of the EdDSA signature is as follows.

$$EdDSA_{signverify}(Message, PU, (R, s)) = valid/invalid \tag{19}$$

(a) Calculate the hash value h as described in the following Eq. (20).

$$h = hash(R + PU + Message) mod q \tag{20}$$

(b) Calculate *P1* as defined in the following Eq. (21).

$$P1 = s * G \tag{21}$$

(c) Calculate *P2* as defined in the following Eq. (22).

$$P2 = R + h * PU \tag{22}$$

(d) Return *P1* == *P2*. As defined in the above give Eq. (21), the point *P*1 is calculated. Put *P*1 value in Eq. (18) by multiplying the *G* on both sides. Further calculations are as defined in the following Eq. (23) where Eqs. (16) and 13 are used.

$$\begin{aligned} s * G &= (r + h * PK) mod q * G \\ P1 &= r * G + h * PK * G \\ P1 &= R + h * PU \\ P1 &== P2(valid) \end{aligned} \tag{23}$$

If these points *P*1 and *P*2 are the same on EC point, this proves that the point *P*1, calculated by the *PK* matches the point *P*2, calculated by its corresponding *PU*.

5. Difference between EdDSA and ECDSA

Signature creation in EdDSA is deterministic in nature whereas ECDSA requires high-quality randomness for each and every signature to be safe. If it uses low-quality randomness, then an attacker can compute the private key. The process by which the actual parameters of EdDSA have been chosen is completely open. So, there is a secret backdoor only known to the organization that selected the parameters. A short comparison between EdDSA signatures and ECDSA signatures is as shown in Table 3.

Table 3 Comparison between EdDSA and ECDSA signature algorithms.

Parameters	EdDSA	ECDSA
Fullform	Edward curve digital signature algorithm	Elliptic curve digital signature algorithm
Performance	8% faster	8% slower
Private key length	32 bytes (256 bits = 251 variable bits + 5 predefined)	32 bytes (256 bits)
Public key length	32 bytes (256 bits = 255-bit y-coordinate + 1-bit x-coordinate)	33 bytes (257 bits = 256-bit x-coordinate + 1-bit y-coordinate)
Signature size	64 bytes	64 bytes or 65 bytes (513 bits)
Public key recovery	Not possible (signature verification involves hashing of the public key)	Possible (with 1 recovery bit added in the signature)
Security level	128 bit (more precisely 125.8)	128 bit (more precisely 127.8)
Safe-curves security	11 of 11 tests passed	7 of 11 tests passed

References

[1] R.C. Merkle, A certified digital signature, in: Conference on the Theory and Application of Cryptology, Springer, 1989, pp. 218–238.

[2] A. Salomaa, Public-key Cryptography, Springer Science & Business Media, 2013.

[3] D.W. Kravitz, Digital signature algorithm, 1993. US Patent 5,231,668.

[4] D. Johnson, A. Menezes, S. Vanstone, The elliptic curve digital signature algorithm (ECDSA), Int. J. Inform. Security 1 (1) (2001) 36–63.

[5] R. Amin, S.K.H. Islam, G.P. Biswas, M.K. Khan, N. Kumar, An efficient and practical smart card based anonymity preserving user authentication scheme for TMIS using elliptic curve cryptography, J. Med. Syst. 39 (11) (2015) 180.

[6] Z. Zhang, Q. Qi, N. Kumar, N. Chilamkurti, H.-Y. Jeong, A secure authentication scheme with anonymity for session initiation protocol using elliptic curve cryptography, Multimedia Tools Appl. 74 (10) (2015) 3477–3488.

[7] S. Josefsson, I. Liusvaara, Edwards-curve digital signature algorithm (EdDSA), in: Internet Research Task Force, Crypto Forum Research Group, RFC, vol. 8032, 2017.

About the authors

Shubhani Aggarwal is pursuing PhD from Thapar Institute of Engineering & Technology (Deemed to be University), Patiala, Punjab, India. She received the BTech degree in Computer Science and Engineering from Punjabi University, Patiala, Punjab, India, in 2015, and the ME degree in Computer Science from Panjab University, Chandigarh, India, in 2017. She has many research interests in the area of Blockchain, cryptography, Internet of Drones, and information security.

Neeraj Kumar received his PhD in CSE from Shri Mata Vaishno Devi University, Katra (Jammu and Kashmir), India in 2009, and was a postdoctoral research fellow in Coventry University, Coventry, UK. He is working as a Professor in the Department of Computer Science and Engineering, Thapar Institute of Engineering & Technology (Deemed to be University), Patiala (Punjab), India. He has published more than 400 technical research papers in top-cited journals such as IEEE TKDE, IEEE TIE, IEEE TDSC, IEEE TITS, IEEE TCE, IEEE TII, IEEE TVT, IEEE ITS, IEEE SG, IEEE Netw., IEEE Comm., IEEE WC, IEEE IoTJ, IEEE SJ, Computer Networks, Information sciences, FGCS, JNCA, JPDC, and ComCom. He has guided many research scholars leading to PhD and ME/MTech. His research is supported by funding from UGC, DST, CSIR, and TCS. His research areas are Network Management, IoT, Big Data Analytics, Deep Learning, and Cybersecurity. He is serving as editor of the following journals of repute: ACM Computing Survey, ACM·IEEE Transactions on Sustainable Computing, IEEE·IEEE Systems Journal, IEEE·IEEE Network Magazine, IEE·IEEE Communication Magazine, IEE·Journal of Networks and Computer Applications, Elsevier Computer Communication, Elsevier International Journal of Communication Systems, Wiley. Also, he has been a guest editor of various international journals of repute such as IEEE Access, IEEE ITS, Elsevier CEE, IEEE Communication Magazine, IEEE Network Magazine, Computer Networks, Elsevier, Future Generation Computer Systems, Elsevier, Journal of Medical Systems, Springer, Computer and Electrical Engineering, Elsevier, Mobile Information Systems, International Journal of Ad Hoc and Ubiquitous Computing, Telecommunication Systems, Springer, and Journal of Supercomputing, Springer. He has also edited/authored 10 books with international/national publishers like IET, Springer, Elsevier, CRC: Security and Privacy of Electronic Healthcare Records: Concepts, Paradigms and Solutions (ISBN-13: 978-1-78561-898-7), Machine Learning for Cognitive IoT, CRC Press, Blockchain, Big Data and IoT, Blockchain Technologies Across Industrial Vertical, Elsevier, Multimedia Big Data Computing for IoT Applications: Concepts, Paradigms and Solutions (ISBN: 978-981-13-8759-3), Proceedings of First International Conference on Computing, Communications, and Cyber-Security (IC4S 2019) (ISBN 978-981-15-3369-3). One of the edited text-book entitled, "Multimedia Big Data Computing for IoT Applications: Concepts, Paradigms, and Solutions" published in Springer in 2019 is having 3.5 million downloads till June 6, 2020. It attracts attention of the researchers across the globe (https://www.springer.com/in/book/9789811387586). He has been a workshop chair at IEEE Globecom 2018 and IEEE ICC 2019 and TPC Chair and member for various international conferences such as IEEE MASS 2020 and IEEE MSN 2020. He is a senior member of the IEEE. He has more than 12,321 citations to his credit with current h-index of 60 (September 2020). He has won the best papers award from IEEE Systems Journal and ICC 2018, Kansas City in 2018. He has been listed in the highly cited researcher of 2019 list of Web of Science (WoS). In India, he is listed in top 10 position among highly cited researchers list. He is an adjunct professor at Asia University, Taiwan, King Abdul Aziz University, Jeddah, Saudi Arabia, and Charles Darwin University, Australia.

CHAPTER FIVE

Signature primitives☆

Shubhani Aggarwal and Neeraj Kumar
Thapar Institute of Engineering & Technology, Patiala, Punjab, India

Contents

Abstract

In cryptography, a ring signature is a type of digital signature, which can be used by any member of a user's group that each have keys. Therefore, a message signed with a ring signature is endorsed by someone in a particular group of people. To enhance the privacy and anonymity of transactions, some advanced signature primitives such as ring signature and multisignatures are widely used. In this chapter, we discuss the ring signatures, one-time signatures, multisignatures, Borromean ring signatures in detail. The detailed working and functionality of these signature primitives are described.

Chapter points

- In this chapter, we discuss the signature primitives that can be used in blockchain for security and anonymity.
- Here, we discuss the signing and verification process of the signature primitives in detail.

☆ Cryptographic primitives used in blockchain.

Advances in Computers, Volume 121
ISSN 0065-2458
https://doi.org/10.1016/bs.adcom.2020.08.005

1. Ring signatures

Ring signatures [1] are digital signatures that allow the number of group members to sign the message in a secure and anonymous way. No one except the original signer knows who the member of the group actually signed the message. This method was developed by Ron Rivest, Adi Shamir, and Yael Tauman in 2001. The creation and verification of the ring signature are defined as follows.

1. *Creating the signature*: In ring signature, firstly define the group of N members and each member have their own key-pairs, i.e., $(P_1, S_1), (P_2, S_2), (P_3, S_3)...(P_N, S_N)$. If any member ($i$) of the group wants to sign the message (*Msg*) then, he/she uses its own private key (S_i), but the public keys of the other members in the group should be (Msg, S_i, $P_1, P_2, ...P_N$). There should be a possibility to check the validity of the group by knowing the public keys of the group members. But, without any knowledge of the private keys of the group members, there should not possible to check the validation of the signature.

   Let us assume one example in which four group members are present, *i.e., Bob, Alice, Raman, Karan*. From which, *Raman* wants to sign the message. Then, he follows the following step equations to create the

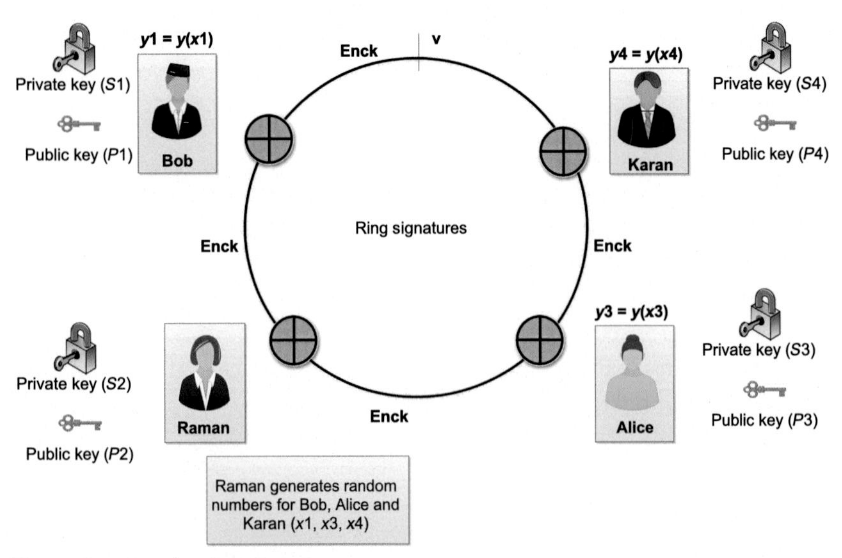

Fig. 1 Creating the ring signature.

signature using ring signatures. The pictorial representation of creating the signature using this method is as shown in Fig. 1.

(a) Generate the encryption of the message *Msg* as shown in the following Eq. (1).

$$Enc_k = Hash(Msg) \tag{1}$$

(b) Generate a random value u.

(c) Encrypt the value u to find v in the following Eq. (2).

$$v = Enc_k(u) \tag{2}$$

(d) Except for the signer, every member of the group computes the following values shown as under.

- Calculate the value f shown in the following Eq. (3).

$$f = Si^{P_i}(modN_i) \tag{3}$$

where S_i is the random number generated for the private key of the i_{th} party and P_i is the corresponding public key.

- Calculate the value v as shown in the following Eq. (4).

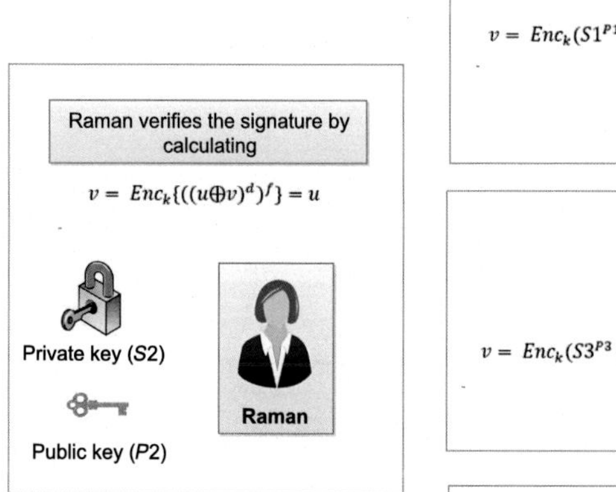

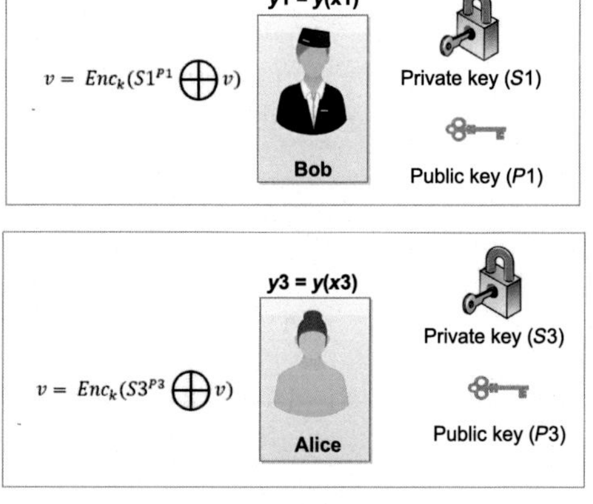

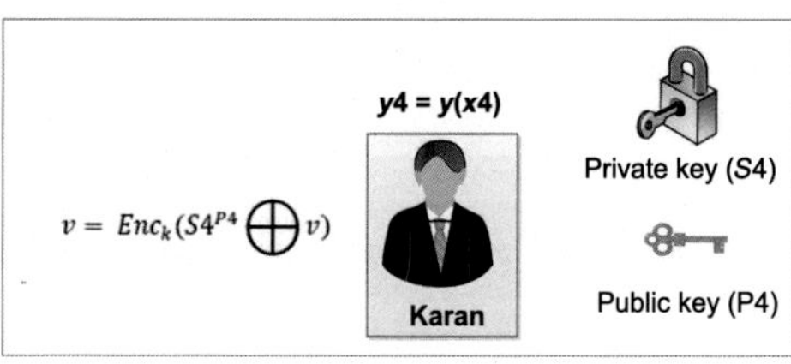

Fig. 2 Verifying the ring signature.

$$v' = v \oplus f \tag{4}$$

(e) For the signed member (i), calculate s_i as shown in the following Eq. (5).

$$s_i = (v \oplus u)^d (mod N_i) \tag{5}$$

where d is the secret key of the signing party.

2. *Verifying the signature*: The verification of the ring is as following and as shown in the Fig. 2.

(a) Generate the encryption of the message Msg as calculated in the above given Eq. (1).

(b) Generate a random value u.

(c) Encrypt the value u to find v in the above given Eq. (2).

(d) Calculate the secret key for Bob as shown in the following Eq. (6).

$$v = Enc_k(S_1^{P_1} \oplus v) \tag{6}$$

(e) Calculate the secret key for Alice as shown in the following Eq. (7).

$$v = Enc_k(S_3^{P_3} \oplus v) \tag{7}$$

(f) Calculate the secret key for Karan as shown in the following Eq. (8).

$$v = Enc_k(S_4^{P_4} \oplus v) \tag{8}$$

(g) Raman verifying the signature by calculating the equation as shown following Eq. (9).

$$v = Enc_k((u \oplus v)^d)^f \tag{9}$$

(h) If the result of the Eq. (9) is same as u value then, the signature is true otherwise, not.

2. One-time signatures

One-time signature [2] is used to secure the keys of the group by enabling the usage of "one by once" by any of the group members. This is conceptually similar to the one-use session key.

2.1 Lamport one-time signature

For security proofs, most of the hash functions or signatures rely on one-time signatures or one-way functions. In the Lamport scheme, the signature was relying on a one-time signature to ensure the security. As shown in Eq. (10), x and y both are integers and used to sign only a single bit.

$$\begin{aligned} h(x)|h(y) &= publickey \\ (x, y) &= privatekey \end{aligned} \tag{10}$$

if it is 0, publish the value x.
if it is 1, publish the value y.
For example, with SHA-256, there is a need for 256 private key-pairs as shown in the following Eq. (11).

$$\begin{aligned} h(x_0)|h(y_0)|h(x_1)|h(y_1)|h(x_2)|h(y_2)\ldots|h(x_{256})|h(y_{256}) &= publickey \\ (x_0, y_0, x_1, y_1, x_2, y_2\ldots x_{256}, y_{256}) &= privatekey \end{aligned} \tag{11}$$

If we want to sign (*101101...*).

Then, the result is, $(y_0, x_1, y_2, y_3, x_4, y_5\ldots)$.

2.2 Winternitz one-time signature

Robert Winternitz of the Stanford Mathematics Department has developed the one-time signatures unlike Lamport. He has published $h^w(x)$ instead of publishing $(h(x)|h(y))$ as shown in the following Eq. (12).

$$x \rightarrow h(x) \rightarrow h(h(x)) \rightarrow h(h(h(x)))\ldots \rightarrow h^w(x) \tag{12}$$

where x is the private and $h^w(x)$ is the public key.

3. Multisignatures [3]

When multiple users create one signature, it is known as a multisignature. It is used to obtain a compact signature for a group, as the individual signatures are comparatively large enough. There are two schemes N/N and $(N-1)/N$ to create multisignatures. The notations used in the following schemes are: private key is represented as PK, a public key is represented as PU. PU is derived from the PK as shown in the following Eq. (13).

$$PU = PK * G \tag{13}$$

where G is a point on the elliptic curve. The multiplication of a PK by G yields a PU, which is also a point on the same curve.

3.1 *N/N* scheme

Firstly, the participants (*Alice and Bob*) share their own private view and public spend keys with each other. Then, calculate their respective sums. The sum of the private view keys becomes the private view key for the new wallet. Similarly, the wallet's public spend key is calculated by summing the participant's public spend keys. The process of creating an N/N in terms of two participants, i.e., *2/2* wallet is shown in Fig. 3.

3.2 $(N-1)/N$ scheme

If the participants select the $N-1/N$ scheme instead of N/N. Then, also they would still have to share their private view and public spend keys with each other. They must multiply all public spend keys received with their own private spend key and hashed them. Thus, a new set of private spend keys is created, which is called the multisignature key for each participant, as shown in Fig. 4. These multisignature keys have one important property as defined as the following Eq. (14).

$$PK_1' * PU_1 = PK_1' * PK_2' * G = PK_2' * PU_1 \tag{14}$$

From Eq. (14), multiply a private key with a public key, the indices can be moved as one would like without affecting the result. This means that every multisignature key is shared between exactly two participants. To calculate a common public spend key for the participants is by summing all the unique public multisignature keys and shares the result with the participants. Similarly, the private view is by summing all the private view keys of the participants. The step-wise description of $(N-1)/N$ scheme in terms of 2/3 is as shown in Fig. 4.

4. Borromean

The Borromean ring signatures [4] can be described as a straightforward generalization, where the ordinary ring signatures take a set of verification keys $(v_i)_{i=1}^{n}$ and then, describe a signature signed with $s_1 OR s_2 OR \ldots OR s_n$, where $s_i$is the secret key corresponding to v_i. Borromean ring signatures can describe signatures signed with arbitrary functions of the signing keys. The pictorial representation of a Borromean ring signature for $(P_0||P_1||P_2) and (P_0'||P_3||P_4)$ as shown in Fig. 5.

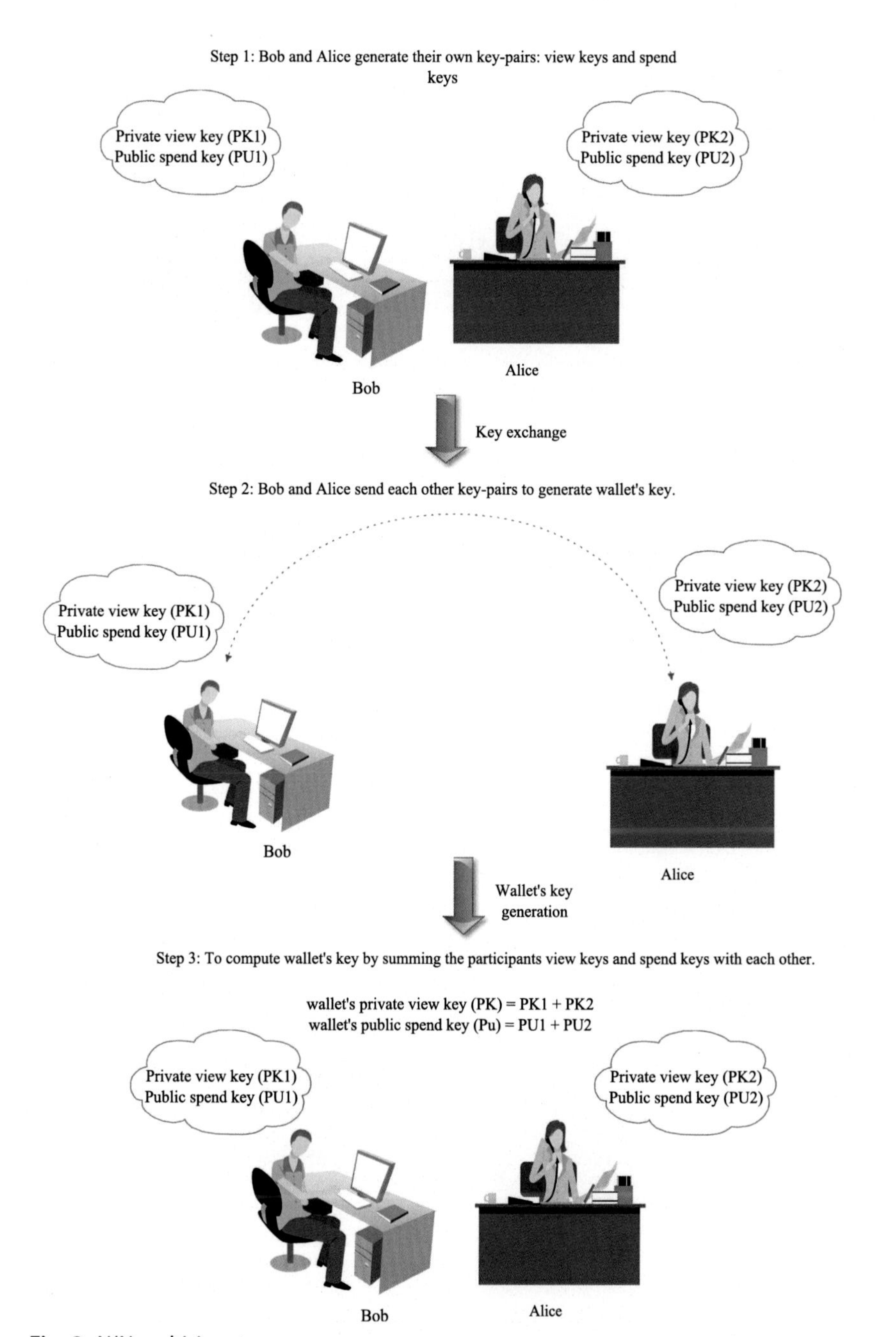

Fig. 3 *N/N* multisignatures.

Step 1: Bob, Coral, and Alice generate their own key-pairs: view keys and spend keys

Step 2: Everyone send his private view key and public spend key to each other.

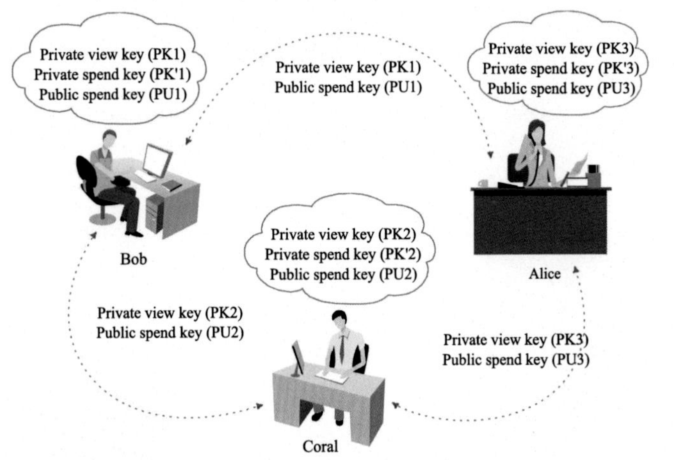

Step 3: Each participant multiplies its private spend key with each other public spend keys and hashes the product.

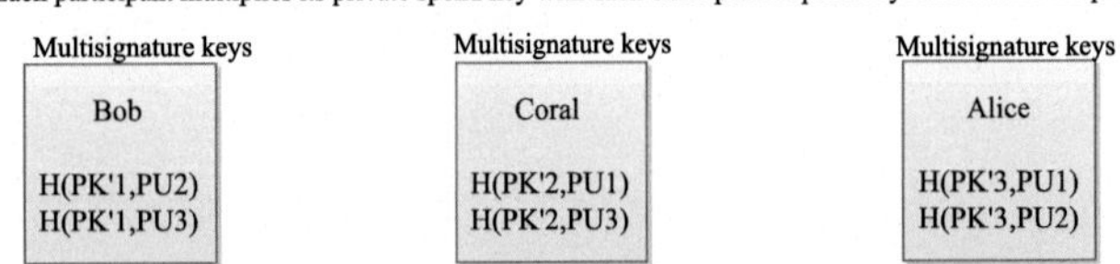

Step 4: Each participant multiplies its multisignatures with G and share it with each others.

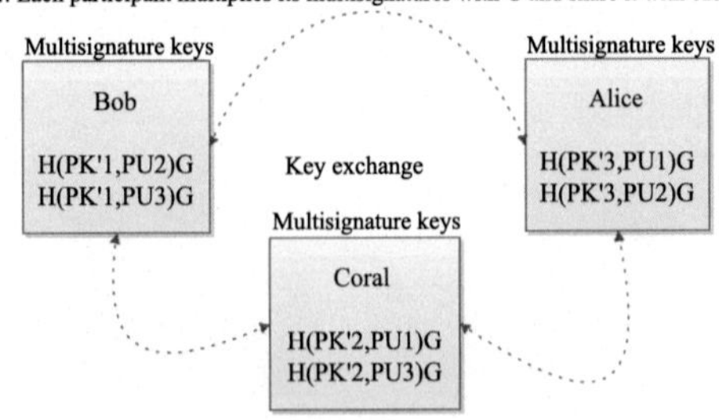

Step 5: Generates common private view key and public spend key.

Common private view key (PK) = PK1 + PK2 + PK3
Common public spend key (Pu) = H(PK'1,PU3)G + H (PK'3,PU2)G + H(PK'2,PU3)G
Any two participants is enough.

Fig. 4 $(N-1)/N$ multisignatures.

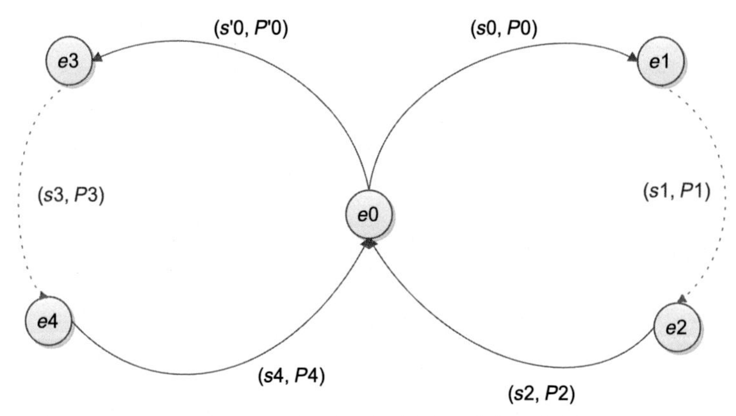

Fig. 5 A Borromean ring signature for $(P_0||P_1||P_2) and (P'_0||P_3||P_4)$.

4.1 Signing of the Borromean

A signer has a number of verification keys $P_{I,j}$ for $0 \leq i \leq n-1, 1 \leq j \leq m_i - 1$ and wants to create a signature of knowledge of the n keys $(P_{i,j_i^*})_{i=0}^{n-1}$ where, j_i^* are fixed and unknown indices, P_{i,j_i^*} denotes the private key by x_i. The step-wise description of signing the Borromean ring signatures are as follows.

1. Compute the hash of the message *Message* to be signed.
2. Compute the set of verification keys as described in the text.
3. For each $0 \leq i \leq n-1$.
   (a) Choose a uniform random scalar k_i.
   (b) Set $e_{i,j_i^*+1} = Hash(Message||k_iG||i||j_i^*)$ where G is the generalized point.
   (c) For j such that $j_i^* \leq j < m_i - 1$, choose randomly $s_{i,j}$ and calculate the $e_{i,j+1}$ as the following Eq. (15).

$$e_{i,j+1} = Hash(Message||s_{i,j}G - e_{i,j}P_{i,j}||i||j) \tag{15}$$

4. Choose s_{i,n_j} for each random value of i and set the following Eq. (16).

$$e_0 = Hash(s_{i,m_j}G||e_{i,m_j}P_{i,j}||\ldots||s_{n,m_j}G - e_{n,m_j}P_{i,j}) \tag{16}$$

   where e_0 commits to various s-values, each one from the different ring.
5. For each $0 \leq i \leq n-1$.
   (a) For j such that $0 \leq j < j_i^* - 1$, choose the random value of $s_{i,j}$ and calculate the following Eq. (17).

$$e_{i,j+1} = Hash(Message||s_{i,j}G - e_{i,j}P_{i,j}||i||j) \tag{17}$$

   where means e_0, which is identical to the Eq. (15).

(b) Set the value of $s_{i,\, j_i^*}$ as the following Eq. (18).

$$s_{i,\, j_i^*} = k_i + x_i e_{i,\, j_{i-1}^*} \tag{18}$$

The resulting signature consists of m, i.e.,
$\sigma = e_0, s_{i,\, j} : 0 \leq i \leq n,\ 0 \leq j \leq m_i$

4.2 Verification of the Borromean

Since verification does not depend on which specific keys are known, it avoids the "two-phase" structure of signing and is therefore much simpler. We assume, we have a message *Message*, a collection $(P_{i,\, j})$ of verification keys whose indices range as before, and a signature σ whose notation is the same as before. The step-wise description of verification of the Borromean ring signatures act as follows.

1. Compute the hash of the message *Message* to be signed.
2. Compute the set of verification keys as described in the text.
3. For each $0 \leq i \leq n - 1$ and for each $0 \leq j \leq m_j - 1$, calculate the following Eqs. (19) and 20.

$$R_{i,j+1} = s_{i,j}G + e_{i,j}P_{i,j} \tag{19}$$

$$e_{i,j+1} = Hash(Message||R_{i,j+1}||i||j + 1) \tag{20}$$

4. Calculate the following Eq. (21).

$$e_0' = Hash(R_{0,m_0}||\ldots||R_{n,m_n}) \tag{21}$$

5. Return *1* if $e_0' = e_0$, otherwise, return *0*.

The use of signature primitives used in blockchain platforms is as shown in Table 1.

Table 1 Signature primitives used in blockchain platforms.

Blockchain platforms	Ring signature	One-time signature	Borromean	Multisignature
Bitcoin	No	No	No	Yes
Ethereum	No	No	No	No
Dash	No	No	No	Yes

Table 1 Signature primitives used in blockchain platforms.—cont'd

Blockchain platforms	Ring signature	One-time signature	Borromean	Multisignature
Litecoin	No	No	No	Yes
Zcash	No	Yes	No	No
Zcoin	No	No	No	Yes
Ripple	No	No	No	Yes
Blackcoin	No	No	No	No
Qtum	No	No	No	Yes
BitConnect	No	No	No	Yes
Komodo	No	Yes	No	No
Dogecoin	No	No	No	Yes
Ark	No	No	No	Yes
Byteball	No	No	No	Yes
Naivecoin	No	No	No	No
RScoin	No	No	No	No

References

[1] E. Fujisaki, K. Suzuki, Traceable ring signature, in: International Workshop on Public Key Cryptography, Springer, 2007, pp. 181–200.
[2] M. Abe, B. David, M. Kohlweiss, R. Nishimaki, M. Ohkubo, Tagged one-time signatures: tight security and optimal tag size, in: International Workshop on Public Key Cryptography, Springer, 2013, pp. 312–331.
[3] J. Pieprzyk, H. Wang, C. Xing, Multiple-time signature schemes against adaptive chosen message attacks, in: International Workshop on Selected Areas in Cryptography, Springer, 2003, pp. 88–100.
[4] K.S. Chichak, S.J. Cantrill, A.R. Pease, S.-H. Chiu, G.W.V. Cave, J.L. Atwood, J.F. Stoddart, Molecular Borromean rings, Science 304 (5675) (2004) 1308–1312.

About the authors

Shubhani Aggarwal is pursuing PhD from Thapar Institute of Engineering & Technology (Deemed to be University), Patiala, Punjab, India. She received the BTech degree in Computer Science and Engineering from Punjabi University, Patiala, Punjab, India, in 2015, and the ME degree in Computer Science from Panjab University, Chandigarh, India, in 2017. She has many research interests in the area of Blockchain, cryptography, Internet of Drones, and information security.

Neeraj Kumar received his PhD in CSE from Shri Mata Vaishno Devi University, Katra (Jammu and Kashmir), India in 2009, and was a postdoctoral research fellow in Coventry University, Coventry, UK. He is working as a Professor in the Department of Computer Science and Engineering, Thapar Institute of Engineering & Technology (Deemed to be University), Patiala (Punjab), India. He has published more than 400 technical research papers in top-cited journals such as IEEE TKDE, IEEE TIE, IEEE TDSC, IEEE TITS, IEEE TCE, IEEE TII, IEEE TVT, IEEE ITS, IEEE SG, IEEE Netw., IEEE Comm., IEEE WC, IEEE IoTJ, IEEE SJ, Computer Networks, Information sciences, FGCS, JNCA, JPDC, and ComCom. He has guided many research scholars leading to PhD and ME/MTech. His research is supported by funding from UGC, DST, CSIR, and TCS. His research areas are Network Management, IoT, Big Data Analytics, Deep Learning, and Cybersecurity. He is serving as editor of the following journals of repute: ACM Computing Survey, ACM·IEEE Transactions on Sustainable Computing, IEEE·IEEE Systems Journal, IEEE·IEEE Network Magazine, IEE·IEEE Communication Magazine, IEE·Journal of Networks and Computer Applications, Elsevier Computer Communication, Elsevier

International Journal of Communication Systems, Wiley. Also, he has been a guest editor of various international journals of repute such as IEEE Access, IEEE ITS, Elsevier CEE, IEEE Communication Magazine, IEEE Network Magazine, Computer Networks, Elsevier, Future Generation Computer Systems, Elsevier, Journal of Medical Systems, Springer, Computer and Electrical Engineering, Elsevier, Mobile Information Systems, International Journal of Ad Hoc and Ubiquitous Computing, Telecommunication Systems, Springer, and Journal of Supercomputing, Springer. He has also edited/authored 10 books with international/national publishers like IET, Springer, Elsevier, CRC: Security and Privacy of Electronic Healthcare Records: Concepts, Paradigms and Solutions (ISBN-13: 978-1-78561-898-7), Machine Learning for Cognitive IoT, CRC Press, Blockchain, Big Data and IoT, Blockchain Technologies Across Industrial Vertical, Elsevier, Multimedia Big Data Computing for IoT Applications: Concepts, Paradigms and Solutions (ISBN: 978-981-13-8759-3), Proceedings of First International Conference on Computing, Communications, and Cyber-Security (IC4S 2019) (ISBN 978-981-15-3369-3). One of the edited text-book entitled, "Multimedia Big Data Computing for IoT Applications: Concepts, Paradigms, and Solutions" published in Springer in 2019 is having 3.5 million downloads till June 6, 2020. It attracts attention of the researchers across the globe (https://www.springer.com/in/book/9789811387586). He has been a workshop chair at IEEE Globecom 2018 and IEEE ICC 2019 and TPC Chair and member for various international conferences such as IEEE MASS 2020 and IEEE MSN 2020. He is a senior member of the IEEE. He has more than 12,321 citations to his credit with current h-index of 60 (September 2020). He has won the best papers award from IEEE Systems Journal and ICC 2018, Kansas City in 2018. He has been listed in the highly cited researcher of 2019 list of Web of Science (WoS). In India, he is listed in top 10 position among highly cited researchers list. He is an adjunct professor at Asia University, Taiwan, King Abdul Aziz University, Jeddah, Saudi Arabia, and Charles Darwin University, Australia.

CHAPTER SIX

Accumulators☆

Shubhani Aggarwal and Neeraj Kumar
Thapar Institute of Engineering & Technology, Patiala, Punjab, India

Contents

Abstract

The accumulator is a one-way function that can provide membership proofs without revealing any individual member in the underlying set. It can be used in many other cryptographic primitives such as commitments and ring signatures for building blocks. The main functionalities for the accumulator, i.e., membership witness, nonmembership witness, and dynamics. In this chapter, we summarize the existing accumulator schemes according to the properties and requirements. The existing accumulator schemes can be divided into three categories, i.e., RSA-based, pairing-based, and hash-based. The brief description and functionality of these three categories are described in this chapter.

Chapter points

- In this chapter, we discuss the accumulators which can be used in blockchain for security and privacy of data.
- Here, we discuss the three main accumulators that can be used in a blockchain.

The accumulator is a one-way function that provides participation security without announcing the individual participant in a defined set [1]. It is a building block for many cryptographic primitives used in blockchain such as ring signatures. Furthermore, it can be used for implementing range proofs in the blockchain. For instance, a typical range proof of Zerocoin consensus mechanism used in blockchain as shown in the following Eq. (1).

☆ Cryptographic primitives used in blockchain

Advances in Computers, Volume 121
ISSN 0065-2458
https://doi.org/10.1016/bs.adcom.2020.08.006

$$(c = c_1) \vee (c = c_2) \vee \ldots. \vee (c = c_n) \tag{1}$$

which is implemented by a member witness of the accumulator for the member set as $(c_1, c_2, \ldots, c_n)$. There are three functionalities that have been used for the accumulator, i.e., membership witness, nonmembership witness, and dynamics. First, the *membership witness* is the process that allows the prover to work and generate the membership witness efficiently for any element in the set. Second, the *nonmembership witness* process allows the prover to compute and manage the nonmembership witness efficiently for any element that is not present in the set. The accumulator having the nonmembership witness functionality is known as universal accumulator. The last *dynamics* functionality of the accumulator empowers the prover that dynamically update, restore, and control the elements in the set and the corresponding witnesses. After the functionalities of the accumulator, the security requirements of the accumulator are defined as *one-wayness, indistinguishability, collision-resistance, and undeniability*. The first two requirements for the security of the accumulator are associated with the information leakage from the witness and the other two requirements for the security of the accumulator are related to the generation and updating of the witness. The security requirements of the accumulator are defined as follows.

1. The *one-wayness* security requirement that guarantees the witness cannot announce any information about the members or participants present in the set.
2. The *indistinguishability* security requirement assures that anyone cannot tell a member x from set Q_0 or Q_1 with the witness $x \in Q_0 \cup Q_1$, where Q_0 and Q_1 are two disjoint sets.
3. The *collision-resistance* security requirement guarantees that no one generates and updating the membership witness for a member that is not present in the set.
4. The *undeniability* security requirement ensures that no one generates the membership witness and the nonmembership witness for the same member.

   The other two properties for the accumulator, i.e., *trust-setup-free* and *trapdoor-free* can be easily understood.

The existing accumulator schemes can be classified into three categories, i.e., RSA-based, pairing-based, and hash-based. As shown in Fig. 1 that depicts a summary of techniques for building accumulators and their description is defined in Sections 1–3.

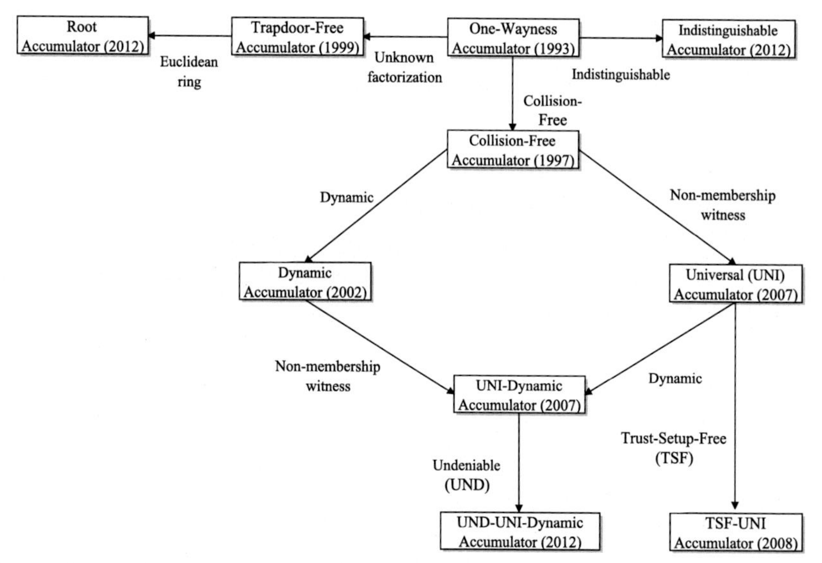

Fig. 1 The evolution of techniques in accumulators.

1. RSA-based accumulators

Due to the popularity of Rivest, Shamir, Adleman (RSA) cryptosystem, RSA-based accumulators are used for many cryptographic primitives. Based on the trapdoor in the RSA problem, the first RSA-based accumulator scheme only satisfies one-wayness security requirements which are not efficient and sufficient for a hostile environment [2]. To solve the problem of a hostile environment, Barić and Pfitzmann [3] and Sander [4] developed the collision-freeness as another security requirement. This accumulator scheme is static in nature. So, the concept of the dynamic accumulator and a concrete dynamic accumulator scheme is developed where the update cost is independent of the size of the member set. With the development of the accumulator, the non-membership witness is considered as the desired requirement [5].

2. Pairing-based accumulators

The first pairing-based accumulator scheme was developed by Nguyen in 2005 that can be used for building the constant size of the signature by using the identity-based ring signature. This scheme requires knowledge of the master private key to update and control the accumulator and it does not provide nonmembership witnesses. These problems are solved by Li [6] in 2007 and Au [7] in 2009. Further, these problems are solved by Acar and Nguyen [8] in 2011.

3. Hash-based accumulators

The first hash-based accumulator scheme was developed by Nyberg [9] in 1996. This accumulator scheme is more efficient than the RSA-based accumulator as it has trapdoor-free and undeniability security requirements. By using hash functions and hash trees in accumulators that provide undeniable universal dynamic accumulator schemes with a trusted party and updates the accumulator. The indistinguishability security requirement of a hash-based accumulator scheme has been more popularized instead of other security requirements accumulators because it provides a hostile environment to the accumulators. Recently, Libert [10] in 2016 described a new construction based on the hash trees that supports the zero-knowledge argument of knowledge (idea to protect the privacy and anonymity of the transaction is to make transactions unlinkable).

References

[1] D. Derler, C. Hanser, D. Slamanig, Revisiting cryptographic accumulators, additional properties and relations to other primitives, in: Cryptographers Track at the RSA Conference, Springer, 2015, pp. 127–144.
[2] J. Benaloh, M. De Mare, One-way accumulators: a decentralized alternative to digital signatures, in: Workshop on the Theory and Application of of Cryptographic Techniques, Springer, 1993, pp. 274–285.
[3] N. Barić, B. Pfitzmann, Collision-free accumulators and fail-stop signature schemes without trees, in: International Conference on the Theory and Applications of Cryptographic Techniques, Springer, 1997, pp. 480–494.
[4] T. Sander, Efficient accumulators without trapdoor extended abstract, in: International Conference on Information and Communications Security, Springer, 1999, pp. 252–262.
[5] J. Camenisch, A. Lysyanskaya, Dynamic accumulators and application to efficient revocation of anonymous credentials, in: Annual International Cryptology Conference, Springer, 2002, pp. 61–76.
[6] J. Li, N. Li, R. Xue, Universal accumulators with efficient nonmembership proofs, in: International Conference on Applied Cryptography and Network Security, Springer, 2007, pp. 253–269.
[7] M.H. Au, P.P. Tsang, W. Susilo, Y. Mu, Dynamic universal accumulators for DDH groups and their application to attribute-based anonymous credential systems, in: Cryptographers Track at the RSA Conference, Springer, 2009, pp. 295–308.
[8] T. Acar, L. Nguyen, Revocation for delegatable anonymous credentials, in: International Workshop on Public Key Cryptography, Springer, 2011, pp. 423–440.
[9] D. Gollmann, Fast Software Encryption Third International Workshop Cambridge, UK, February 21–23 1996 Proceedings, in: Conference proceedings FSE, Springer, 1996, p. 261.
[10] B. Libert, S. Ramanna, M. Yung, Functional Commitment Schemes: From Polynomial Commitments to Pairing-Based Accumulators From Simple Assumptions, 2016.

About the authors

Shubhani Aggarwal is pursuing PhD from Thapar Institute of Engineering & Technology (Deemed to be University), Patiala, Punjab, India. She received the BTech degree in Computer Science and Engineering from Punjabi University, Patiala, Punjab, India, in 2015, and the ME degree in Computer Science from Panjab University, Chandigarh, India, in 2017. She has many research interests in the area of Blockchain, cryptography, Internet of Drones, and information security.

Neeraj Kumar received his PhD in CSE from Shri Mata Vaishno Devi University, Katra (Jammu and Kashmir), India in 2009, and was a postdoctoral research fellow in Coventry University, Coventry, UK. He is working as a Professor in the Department of Computer Science and Engineering, Thapar Institute of Engineering & Technology (Deemed to be University), Patiala (Punjab), India. He has published more than 400 technical research papers in top-cited journals such as IEEE TKDE, IEEE TIE, IEEE TDSC, IEEE TITS, IEEE TCE, IEEE TII, IEEE TVT, IEEE ITS, IEEE SG, IEEE Netw., IEEE Comm., IEEE WC, IEEE IoTJ, IEEE SJ, Computer Networks, Information sciences, FGCS, JNCA, JPDC, and ComCom. He has guided many research scholars leading to PhD and ME/MTech. His research is supported by funding from UGC, DST, CSIR, and TCS. His research areas are Network Management, IoT, Big Data Analytics, Deep Learning, and Cybersecurity. He is serving as editor of the following journals of repute: ACM Computing Survey, ACM·IEEE Transactions on Sustainable Computing, IEEE·IEEE Systems Journal, IEEE·IEEE Network Magazine, IEE·IEEE Communication Magazine, IEE·Journal of Networks and Computer Applications, Elsevier Computer Communication, Elsevier

International Journal of Communication Systems, Wiley. Also, he has been a guest editor of various international journals of repute such as IEEE Access, IEEE ITS, Elsevier CEE, IEEE Communication Magazine, IEEE Network Magazine, Computer Networks, Elsevier, Future Generation Computer Systems, Elsevier, Journal of Medical Systems, Springer, Computer and Electrical Engineering, Elsevier, Mobile Information Systems, International Journal of Ad Hoc and Ubiquitous Computing, Telecommunication Systems, Springer, and Journal of Supercomputing, Springer. He has also edited/authored 10 books with international/national publishers like IET, Springer, Elsevier, CRC: Security and Privacy of Electronic Healthcare Records: Concepts, Paradigms and Solutions (ISBN-13: 978-1-78561-898-7), Machine Learning for Cognitive IoT, CRC Press, Blockchain, Big Data and IoT, Blockchain Technologies Across Industrial Vertical, Elsevier, Multimedia Big Data Computing for IoT Applications: Concepts, Paradigms and Solutions (ISBN: 978-981-13-8759-3), Proceedings of First International Conference on Computing, Communications, and Cyber-Security (IC4S 2019) (ISBN 978-981-15-3369-3). One of the edited text-book entitled, "Multimedia Big Data Computing for IoT Applications: Concepts, Paradigms, and Solutions" published in Springer in 2019 is having 3.5 million downloads till June 6, 2020. It attracts attention of the researchers across the globe (https://www.springer.com/in/book/9789811387586). He has been a workshop chair at IEEE Globecom 2018 and IEEE ICC 2019 and TPC Chair and member for various international conferences such as IEEE MASS 2020 and IEEE MSN 2020. He is a senior member of the IEEE. He has more than 12,321 citations to his credit with current h-index of 60 (September 2020). He has won the best papers award from IEEE Systems Journal and ICC 2018, Kansas City in 2018. He has been listed in the highly cited researcher of 2019 list of Web of Science (WoS). In India, he is listed in top 10 position among highly cited researchers list. He is an adjunct professor at Asia University, Taiwan, King Abdul Aziz University, Jeddah, Saudi Arabia, and Charles Darwin University, Australia.

Basics of blockchain☆

Shubhani Aggarwal and Neeraj Kumar
Thapar Institute of Engineering & Technology, Patiala, Punjab, India

Contents

Abstract

A blockchain is a distributed network, which can be used as a digital ledger as well as a mechanism that enables the secure transfer of assets without a central authority. Like the Internet is a technology that facilitates the digital flow of information, blockchain is a technology that facilitates the digital exchange of units of value. Anything from currencies to land titles to votes can be tokenized, stored, and exchanged on a blockchain network. In addition to the secure transfer of value, blockchain technology provides a permanent record of transactions and a single version of the truth called a network state, which is fully transparent in real-time for the benefit of all participants. In this chapter, we briefly describe the overview of blockchain and their characteristics that defines a blockchain is a prominent technology.

Chapter points

- In this chapter, we briefly discuss the centralized and distributed systems that is used in real-time applications.
- Here, we discuss the overview of blockchain and their characteristics. With this, we have also discussed the scope and motivation of blockchain.

☆ Introduction to blockchain.

Advances in Computers, Volume 121
ISSN 0065-2458
https://doi.org/10.1016/bs.adcom.2020.08.007

From the last few decades, the rapid advancements in information and communication technology (ICT) gives an exponential increase in the smart devices usage results in the demand of one of the most popular technologies called as the *Internet of things* (IoT). It refers to the ever-growing network of physical connected objects using the Internet. It consists of billions of interconnected objects to provide many services like e-healthcare, smart transportation, smart sensing, to name a few to the end-users. In such an environment, the bulk amount of data is being generated from these heterogeneous and graphically distributed devices. Michael Kanellos and Vernon Turner described the statistics in their report [1] that approximately 80 billion devices will be connected to the Internet by 2025. They also described that the total amount of digital data generated worldwide would be 4.4 Zettabytes in 2013 to 180 Zettabytes by 2025 as shown in Fig. 1. They also proved that currently connected devices per person is greater than the world population and is increasing exponentially as shown in Fig. 2.

From the above-mentioned facts, we found that the transmission of the huge amount of data and information to the core platform requires security and privacy as this whole information is shared using an open channel, i.e., Internet. In addition, it is also efficient for processing and storage of data at the cloud repository. However, the cloud systems are not found to be trustworthy as the global ecosystem has witnessed an exponential in the number of attacks in various applications like the financial system, supply chain management, healthcare, to name a few. Fig. 3 shows the growth trend of some of the recent attacks in various domains. To handle this situation, the dependability on cyber security mechanisms has increased manifold. There exist many

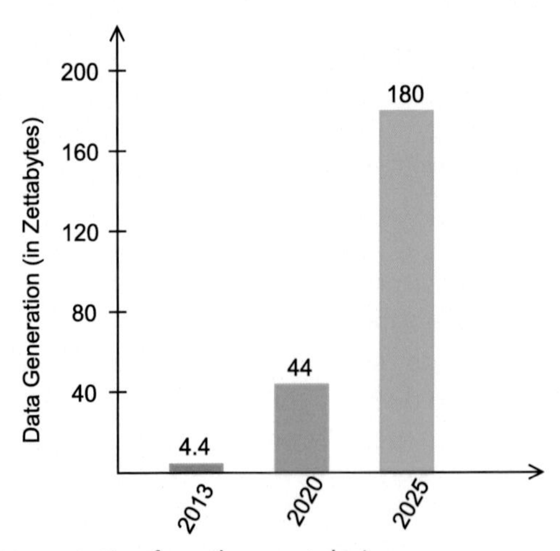

Fig. 1 Digital data generation from the smart devices.

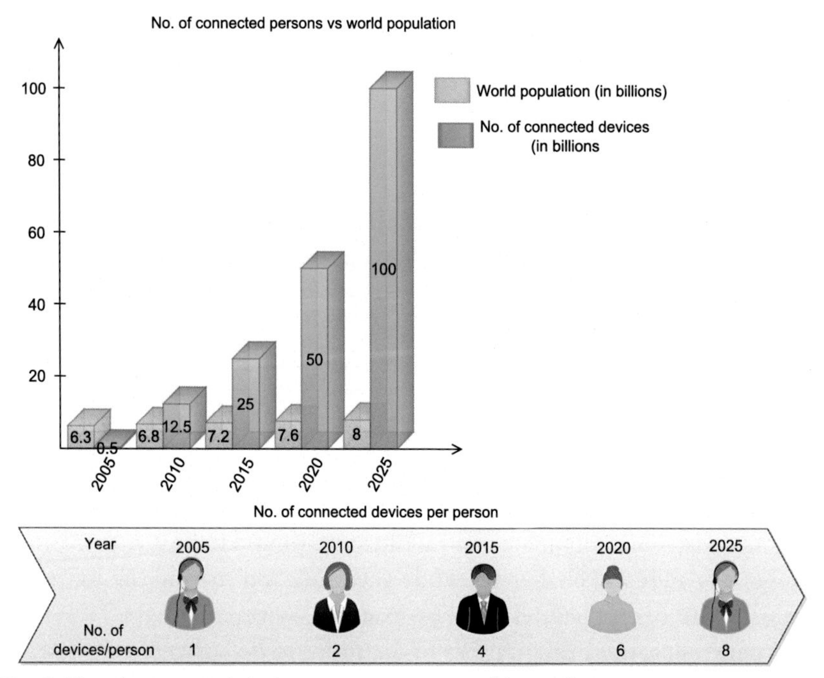

Fig. 2 No. of connected devices per person vs world population.

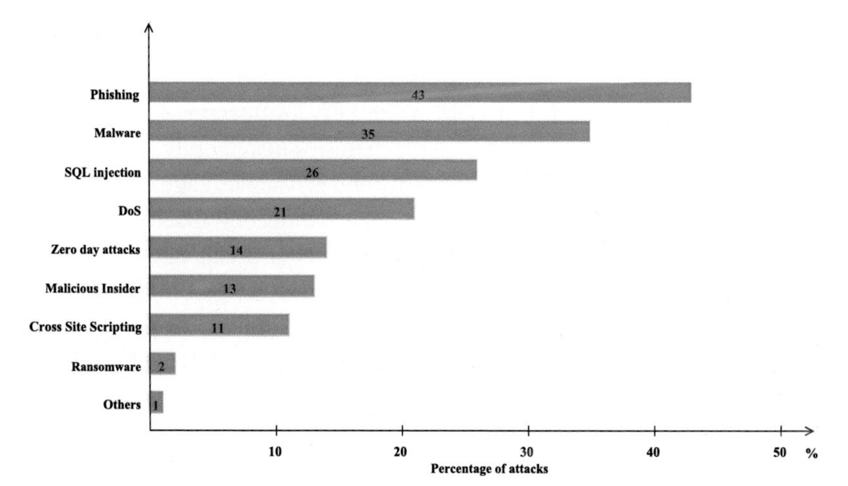

Fig. 3 Cyber attacks.

solutions to handle different type of cyber attacks on these applications [2]. However, the need of ubiquitous services in various applications on the move creates hurdles in front of the existing security solutions for many attacks such as replay attack, distributed denial-of-service (DDoS), data modification, eavesdropping. Therefore, to tackle these attacks in a distributed environment, blockchain technology can be an effective solution due to its properties of easy transactions tracing, immutability, and decentralization.

1. Overview of blockchain

From the past few years, there has been an exponential increase in the usage of ICT for quick, efficient, and secure data exchange among different devices across the globe. With the emergence of the Internet, digital transformations between different entities emerged which results in data and information interchange among them. This interchanging is done through online transactions such as financial transactions for making payment and receive funds from different users. This entire communication and transactional system is validated using a trusted third party (TTP) system as shown in Fig. 4. This TTP system not only guarantees the safe and secure data delivery but also ensures the accurate updates in multiple accounts at the same time. However, there are a number of issues and challenges while accessing and sharing the information over the open channel [3]. For example, with the single TTP network controller, there are number of questions to be answered such as (1) What if the TTP becomes a fraud and cannot be trusted? (2) What if this party is hacked by a hacker or attacker gets hold of all the data which may act as a single point of failure? (3) Each time a TTP is used which may occur additional delay in a communication network? (4) The authenticity and validation of each transaction are very important.

Moreover, the existing traditional mechanisms heavily rely on TTP, i.e., the centralized network, where anonymity and privacy leakage are the major concerns. For this reason, it is important to have decentralized/distributed and secure system which can execute contracts and can handle transactions during the communication among devices. To tackle these issues, a secure communication technique is required, which can provide data security, integrity, authentication, non-repudiation, and confidentiality while transmitting the data generated by the various smart devices [4].

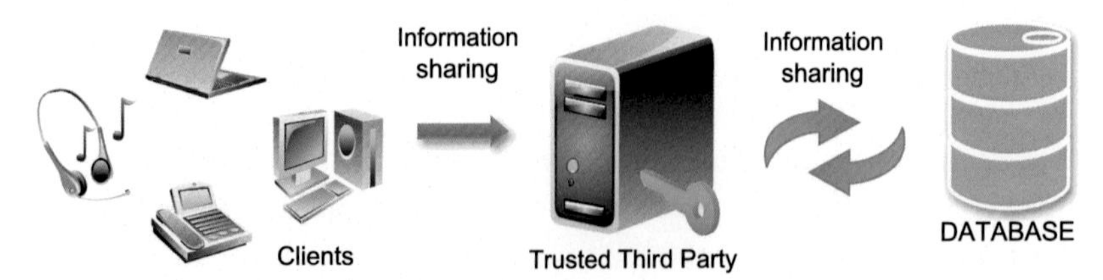

Fig. 4 Scenario of trusted third party system.

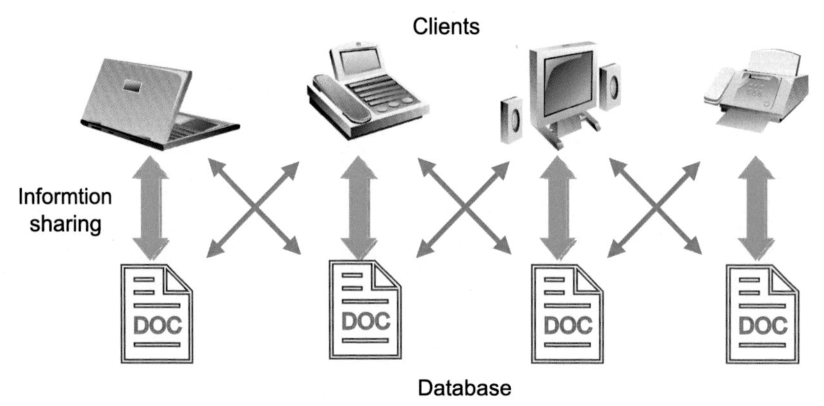

Fig. 5 Scenario of distributed system.

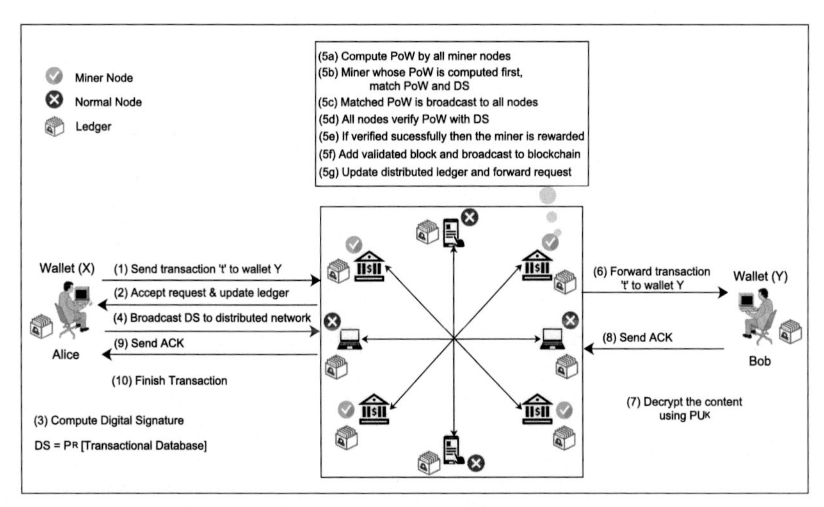

Fig. 6 Bitcoin transaction volume.

The solution to all these problems can be provided by distributed ledger technology called *BLOCKCHAIN* as shown in Fig 5. It was invented by Satoshi Nakamoto by introducing the first-ever decentralized cryptocurrency called "Bitcoin" [5, 6]. The popularity of bitcoin is evident in the bitcoin transaction volume, which is presented in Fig. 6. It is a distributed network having the potential for creating and using smart contract in IoT-based smart applications to ensure security and to remove a single point of failure. In these smart applications, there can be many network attacks as depicted in Fig. 7. To defend against the attacks such as DDoS, malware, server message block, brute force, etc., blockchain technology is one of

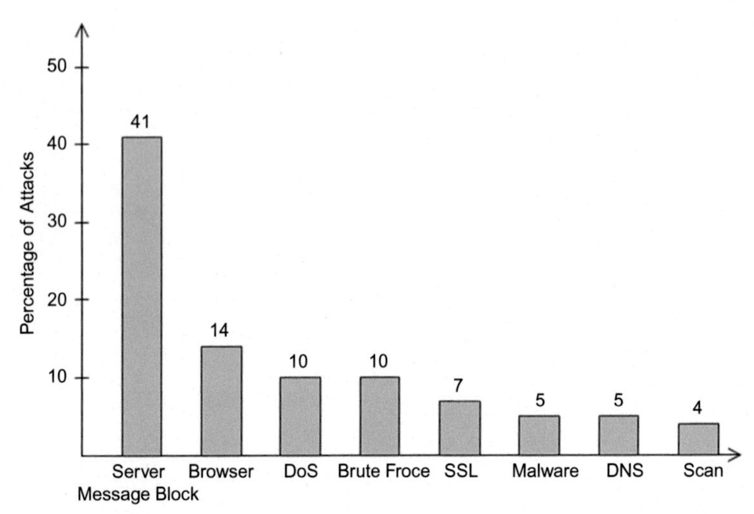

Fig. 7 Network attacks.

the possible solutions. It ensures security and privacy to the system and provides the properties such as scalability, edibility, reliability, authorization, identity management, data integrity, confidentiality, nonrepudiation, authentication, and accountability.

In the recent past, there came a few pioneering technologies. Also, previously discarded technologies get a fresh life because of a dazzling array of distinct improvisations in the information technology (IT) infrastructures domain and also due to the steady emergence of integrated and insightful platforms. Artificial intelligence (AI) with a bombardment of a heavily improved machine learning and deep learning (ML/DL) algorithms is garnering a lot of attraction and attention these days. With a host of digitization and edge technologies gaining uninhibited prominence and dominance, the aspects of extreme connectivity and deeper integration among digitized entities and devices are seeing the reality. The IoT and cyber-physical systems (CPS) concepts are traversing through surging popularity these days. Another highly visible trend is to club together multiple proven technologies to present new and clustered technology. The blockchain technology is a neat combination of several pioneering technologies and this blended paradigm is being pronounced as one of the most significant technologies capable of making solid and spectacular impacts.

Blockchain is a part of a distributed software system which is one of the most popular technologies of security of the modern era. To handle the security, blockchain technology can be used, which is based on the concept of the cryptocurrency system. It provides secrecy and privacy to the users using cryptographic primitives to authenticate the communication among the nodes. It is a chain of*digital signatures* comprising of nodes which are connected to each other in a mesh topology.

The most important feature of blockchain is that it is a decentralized computation and information sharing platform that enables multiple authoritative domains to combine and come to a common platform for providing security and privacy for different transactions. These multiple domains can coordinate, cooperate, and collaborate to develop applications at business intelligence process. In this way, this technology can be useful for rational decision-making in management and engineering applications [7].

2. Traditional centralized systems

In a traditional system, the way of sharing the documents is by using the document file. For example *Alice* wants to share the document with Bob. Firstly, *Alice* writes down all the content inside the document then she will share the document to *Bob*. After that, *Bob* updates the document according to the needs and share the document again to her. So, this was the traditional way of coordination and cooperation between *Alice* and Bob when they want to read or write the document. The scenario of sharing between *Alice and Bob* is as shown in Fig. 8.

The particular scenario is as shown in Fig. 8, there are many problems. The first fundamental problem of this architecture is that *Alice and Bob* both are not able to edit the document simultaneously. The replaceable architecture of Fig. 8 is as shown in Fig. 9. In this architecture, a shared document platform has been used. Here, *Alice and Bob* both can write or read the document simultaneously. But, this architecture is still centralized and have some major drawbacks as follows. The first drawback of this system is that it may have single point of failure. The second drawback is that if something happens or server crash down, then the entire information or data get lost.

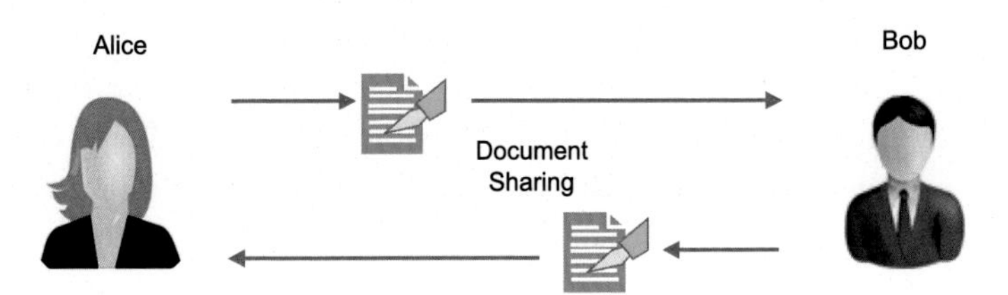

Fig. 8 Traditional way of sharing the document 1.

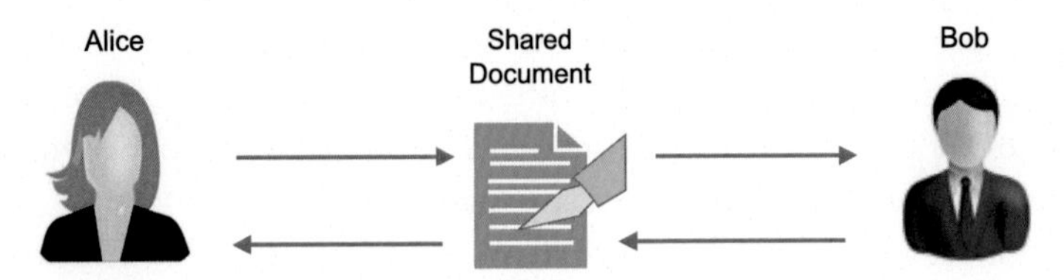

Fig. 9 Traditional way of sharing the document 2.

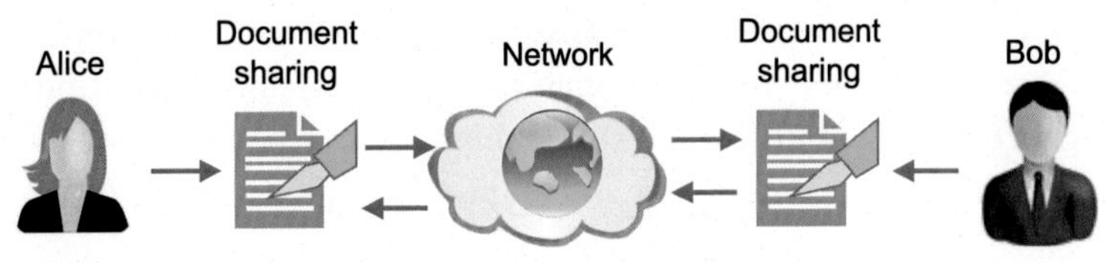

Fig. 10 Sharing the document using blockchain network.

From the aforementioned reasons, researchers move from the centralized platform to a distributed platform that provides scalability and reliability as shown in Fig. 10.

3. Types of systems

There are three types of systems, i.e., centralized, decentralized, and distributed systems that are used in real-time applications. Their description is described in the following sections

3.1 Centralized system

In the case of centralized architecture (see Fig. 11), there is a central coordination system in which each node is connected to that system and whatever they want to share the information will be shared by that system. In this system, there is a problem if central coordination system fails (single point of failure) than all of these individual nodes will get disconnected and the network becomes shut down.

3.2 Decentralized system

In the case of the decentralization system (see Fig. 12), there is a number of coordinators instead of single and all the coordinators and their individual nodes are cooperative with each other. If any particular or multiple nodes fail then they can connect to the other coordinators or to the other individual nodes. They can share their information or perform the tasks using available coordinators. In this architecture, if a simultaneous failure of nodes

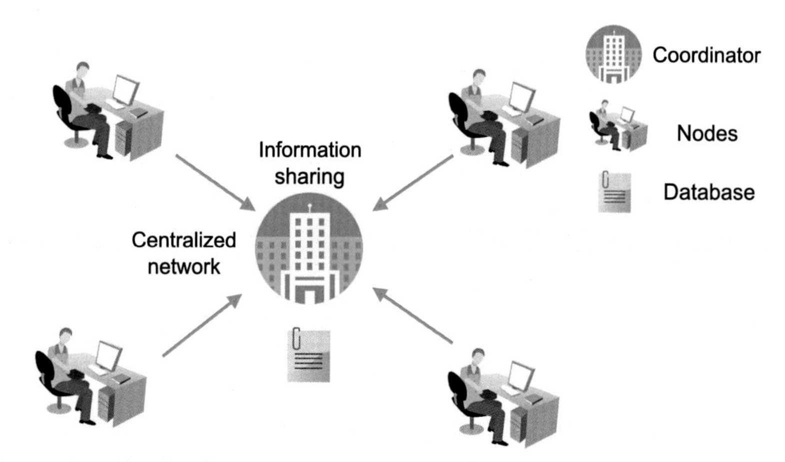

Fig. 11 Centralized architecture.

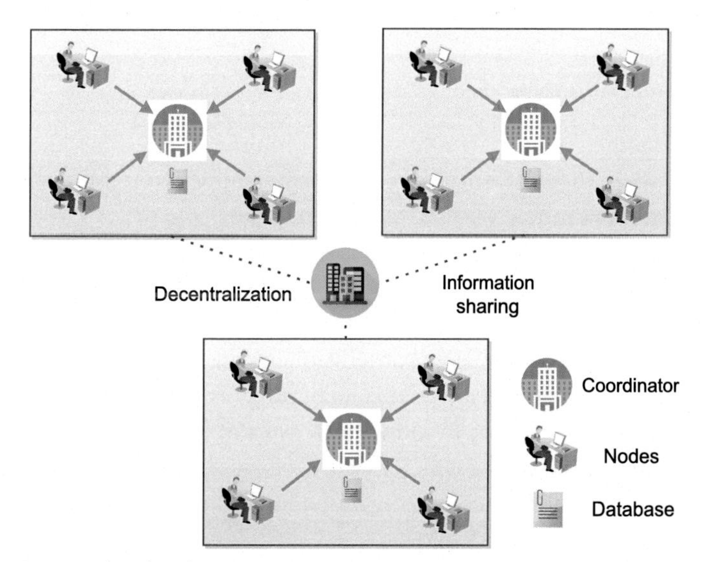

Fig. 12 Decentralized architecture.

occur then individual nodes may not be able to cooperate with each other and the network gets disconnected. So, to solve the problem of a decentralized system, distributed architecture comes into the picture.

3.3 Distributed system

In a distributed architecture (see Fig. 13), there is no need for a central authority. In this system architecture, each node is connected and coordinated with every node. They can collectively share the information and

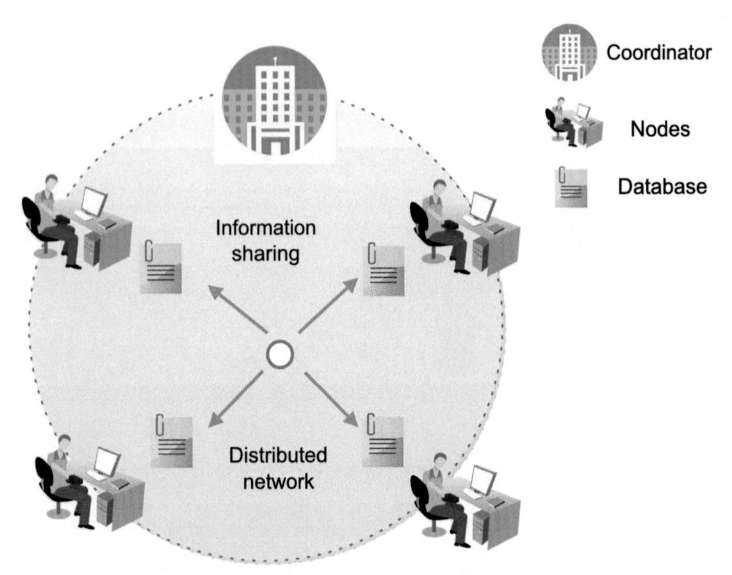

Fig. 13 Distributed architecture.

worked together. In this, if any single or multiple nodes fails then other nodes can share their information or perform the operation using coordination and cooperation.

So, from the aforementioned facts it is clear that, blockchain is a platform which helps us to provide a decentralized and distributed platform where we can share the information among others in a trusty manner. Using this technology, both *Bob and Alice* can edit or update the document simultaneously. The main advantage of this technology is that it does not rely on the centralized system. So, by definition, we can say that blockchain is a platform that provides consistency in the database. It also supports whatever information *Alice* and *Bob* are writing individually on the document, they get synchronized over the network.

4. Scope of the blockchain

Blockchain, a relatively recent technological trend, is a decentralized and chain of cryptographic blocks linked together to form a P2P network which is distributed in nature [8, 9]. This technology can be leveraged to achieve authorization, accountability, authentication (AAA), integrity, security, privacy, confidentiality and nonrepudiation for real-time applications which may not be provided by the centralized systems effectively and efficiently [10–14].

It is a technology which is the combination of three technologies: public-key cryptography, P2P network, and the program. It has shown its revolution in terms of digital cryptocurrency that removes the requirement of an intermediatory expert in the field of registration and distribution. It has also provided the most popular product, i.e., Bitcoin which is a type of cryptocurrency and work as a public ledger for all the transactions done on the network. It has resolved the problems like double-spending, unauthorized accessing, etc. and thus improves the security and privacy of the network. An incredible scope of this technology has been observed in various application like smart grid, voting system, financial system, supply chain management, etc. [15]. Incorporating blockchain with digital transactions gives many benefits such as time and money can be saved which is used for validating and processing the transactions. Its function on a distributed database makes the operation smoothly, ensuring tight security, and made it safe from the cyber attacks. It is one of the most consistent technologies when it requires to keep track of the digital properties. It also has the abilities to add distinct features like security and privacy in the company's structure.

The future scope of the blockchain technology in different sectors is described as under.

1. *Blockchain in digital advertising*: Presently, digital advertising faces a lot of problems such as fraud domain, lack of transparency, data tampering, etc. Due to the issues, incentives are not affiliated but blockchain has provided a solution to ensure transparent, tamper-proof data to the network as it executes trust in a trustless environment.
2. *Cyber security*: The blockchain data is stored in the public ledgers which is verified and encrypted using cryptographic primitives. So, the data and information cannot to be tampered or attacked without any involvement of central authority.
3. *No single point of failure*: It is a decentralized technology in which each node is connected to every node without any involvement of the third party which may act as a single point of failure. It provides a public ledger that consists of verifiable and validated transaction data which lowers the risk of data modification and trust issues on the network.
4. *Supply chain management*: It record all the information or data into the public distributed ledgers and supervise them more transparently. It also helps to minimize human errors and time delays. It is also used to monitor costs, employment, and releases at each point of the supply chain.

5. *Beyond the world of computing*: Currently, most of the countries are developing their blockchain strategies for usage in the future. But still, there are many issues such as security and privacy in various sectors like finance where blockchain can be used to address various problems like data modification. It can also be used to generate a database for medical purpose, to manage insurance policies, etc.
6. *Internet of things and networking*: The different companies such as IBM, Samsung, etc. are utilizing blockchain technology to create a new distributed network of IoT devices. It will improve the requirement of central authority to manage the central database among all communicating parties.
7. *Cloud storage*: Data stored on the central cloud storage can be exposed to hacking, loss of data or human error. With this technology, it is possible to make cloud storage more protected and robust against various type of attacks.

5. Characteristics of blockchain

In this section, we discuss the main features of blockchain technology. These features provide major benefits in most of the applications due to the properties discussed below (Fig. 14).

1. *Immutability*: Immutability means something that cannot be altered or changed. To create the immutability, ledgers is one of the main values of blockchain technology. All the centralized systems and database can be corrupted and requires a TTP to ensure integrity. But, the blockchain-based decentralized database provides transparency and tamper-proof data.
2. *Decentralization*: In a traditional centralized system, each transaction needs to be validated through TTP which may act as a single point of failure. In contrast to this, the decentralized-based blockchain system provides a trustless environment between the different nodes who do not trust each other. These nodes can coordinate and cooperate with each other in a rational decision-making process.
3. *Digital*: All the information on the blockchain is stored as digital that eliminates the need for manual documentation. This digital information is stored in the blocks of the blockchain.
4. *Distributed ledger*: The indistinguishable copies of all the information are shared on the blockchain. A public and distributed ledgers store every information about a transaction and the participant. So, every node can act as verifiers of the ledger. Each node can participate individually

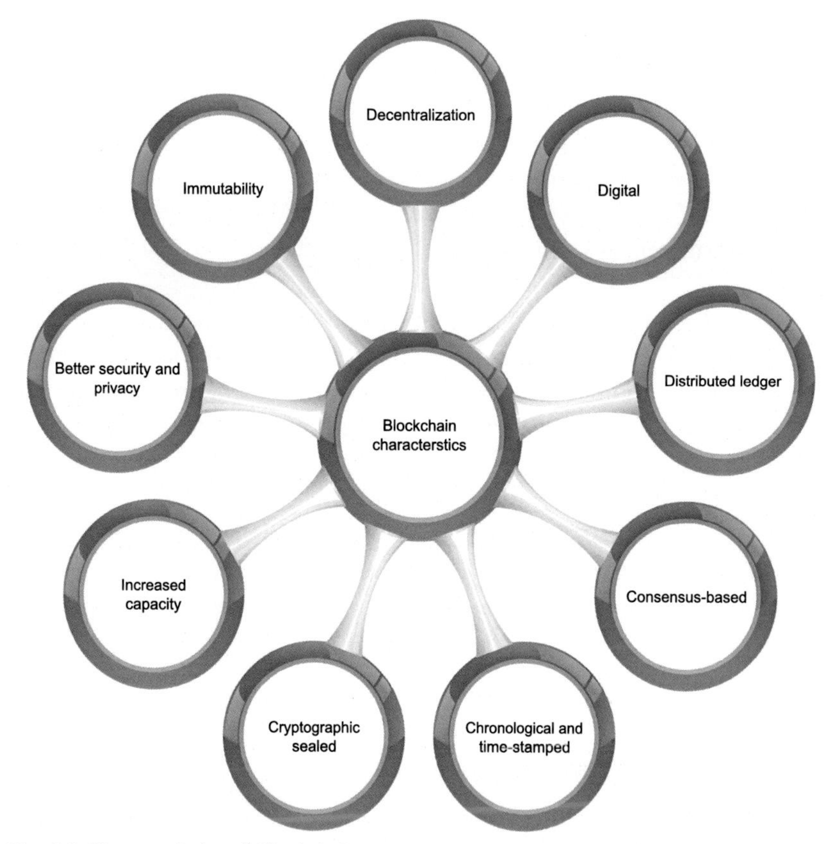

Fig. 14 Characteristics of Blockchain.

and validate the transaction independently without any involvement of central authority. If any node fails, the remaining nodes can continue to operate and ensure no disruption.

5. *Consensus-based*: In simple words, the consensus is a decision-making process for a group of nodes that are active on the network. Here, the nodes can come to an agreement as fast as possible because when million of nodes are validating a transaction, a consensus-based agreement is necessary for the system. The transaction on the blockchain can be executed only if all the nodes on the network anonymously approve it.
6. *Chronological and time-stamped*: Blockchain is a chain of blocks each being a repository that stores information pertaining to a transaction and also links to the previous block in the same transaction. These connected blocks form a chronological chain by providing a stream of the underlying transactions.

7. *Cryptographically sealed*: The blocks in the blockchain are created and combined together using cryptographic primitives in the chain. So, it is very difficult to edit, delete or update the already created blocks and it on the blockchain network. Thereby, creating digital assets ensures a high level of robustness and trust. Also, it is known as *failure-resistant*, because, if the failure of a large number of nodes in the network, then blockchain remains available and eliminates the problem of a single point of failure.
8. *Increased capacity*: This feature of the blockchain technology increases the capacity of the entire network. Because there are thousands of computer which are working together as a whole in the blockchain network and have great power than a few computers worked in the centralized system.
9. *Better security and privacy*: Blockchain provides better security and privacy to the network than a centralized system because there is no single point of failure to shut down the whole network. It also provides privacy because every information on the blockchain is hashed cryptographically that acts as a firewall for attacks.

6. Key motivations for blockchain

Many technical, business, and use cases are emerging for experimenting with and embracing the blockchain technology, which is seen largely as a disruptive one by many. Today banks play a very vital role in facilitating financial transactions. Banks provide much-needed trust when we exchange money with our friends and colleagues across the world. But this model is very expensive and slow. These intermediary-based financial transactions stifle the innovation culture. The centralized computing model being adopted by banks comes in the way of exploiting the distinct benefits of distributed and decentralized computing. If the centralized database gets broken, then all kinds of customer, confidential, and corporate data get stolen. This model is vulnerable to DDoS attacks. Hence, the brewing trend is that we need trust but at the same time, we want to discard banks and other intermediaries forthwith. Employing banks makes the whole process a bit complicated. For legal transactions, lawyers and judges play that intermediary role. Contract documents are being made and used for eliminating any kind of confusion and cheatings. Precisely speaking, trust and formal contracts have been the predominant ways and means for all kinds of transactions. Unfortunately, both trust and contract are not optimal solutions. The blockchain technology offers a third option, which is affordable, inventive, fast, and secure.

Second, we are heading toward the connected era. We are being bombarded with a dazzling array of slim and sleek, trendy and handy devices. These embedded and networked devices come handy in producing next-generation services and applications not only for business houses but also for common people. That is, we are on the verge of entering into the era of the IoT and cyber-physical systems (CPS). The well-known challenge and concern here is the security, safety, and privacy of IoT devices and data. The traditional security solution approaches are found incompetent in guaranteeing the utmost security due to the participation of resource-constrained devices. Any everyday environment stuffed and equipped with intelligent sensors and actuators is typically dynamic and liable for instability. Herein, the arrival of the blockchain paradigm is being applauded. That is, futuristic environments (personal, industrial, social, and professional) are going to be not only smart but also secure and safe.

Finally, suppose you are a retailer souring from over 100 different fast. Moving consumer goods (FMCG) companies, including over 10,000 stock keeping units, in one store, and your B2B transactions amount to over 1000 high transactions a day. Now, these 1000 transactions are just from one store. You may have ten or more stores. Each of these transactions gets recorded on a combination of paper and digital methods. Now the problem here is that your purchase department can under-report transactions and bill you for the ordered quantity. So, while your stock shows the right amount, your store shelves are never stacked up after a sale. This presents a problem because it hurts the top line and bottom-line revenues. Further on, there can also be price changes very often being made by the FMCG company. That details cannot be recorded by the retailer.

Now enter blockchain, assuming that the retailer has moved all transactions digitally, which creates blocks of each transaction like Vendor A supplied X, Vendor B supplied Y and Vendor C supplied YY and in the process the system ropes in the finance and the purchasing team together with each receipt generated. Now each transaction becomes part of a block and gets recorded in the ledger of all parties involved. Any changes made to a price or purchase agreement, another block gets added, which has to be verified by all parties. The algorithm secures each block of transactions and it cannot be tampered with because all parties have the keys to make the transactions legitimate. The technology brings transparency and traceability.

Newer technologies, architectural patterns, and tools are forthcoming in order to assist deeper and decisive automation in business operations. Also, disruptive applications can be realized through fresh technology-inspired approaches and algorithms. Decentralized applications (DApps) and smart contracts are being demanded these days. Blockchain has shown the initial exuberance and the innate strength to set up and sustain DApps across multiple industry verticals. There are a number of distinct advantages of distributed databases and decentralized applications. IT, business, and device services can do certain actions automatically through smart contracts. That is, contract logic gets embedded inside the application and service code in order to automate the contractual obligations. Thus, eliminating third-party involvement through automation is the core strength of this new technological paradigm. Several other business verticals are keenly experimenting and exploring to gain something new out of this concept.

References

[1] Forbes, 2016, Available: https://www.forbes.com/sites/michaelkanellos/2016/03/03/152000-smart-devices-every-minute-in-2025-idc-outlines-the-future-of-smart-things/ (accessed 26 July 2019).
[2] R. Iqbal, F. Doctor, B. More, S. Mahmud, U. Yousuf, Big data analytics and computat. ional intelligence for cyber-physical systems: recent trends and state of the art applications, Futur. Gener. Comput. Syst. 105 (2017) 766–778.
[3] D. Puthal, S.P. Mohanty, P. Nanda, U. Choppali, Building security perimeters to protect network systems against cyber threats [future directions], IEEE Consumer Electron. Mag. 6 (4) (2017) 24–27.
[4] N. Komninos, E. Philippou, A. Pitsillides, Survey in smart grid and smart home security: issues, challenges and countermeasures, IEEE Commun. Surv. Tutorials 16 (4) (2014) 1933–1954.
[5] R. Grinberg, Bitcoin: an innovative alternative digital currency, Hast. Sci. Technol. Law J. 4 (2012) 159.
[6] S. Nakamoto, Bitcoin: A Peer-to-Peer Electronic Cash System, Working Paper, 2008.
[7] Q. Feng, D. He, S. Zeadally, M.K. Khan, N. Kumar, A survey on privacy protection in blockchain system, J. Netw. Comput. Appl. 126 (2019) 45–58.
[8] Y.-L. Gao, X.-B. Chen, Y.-L. Chen, Y. Sun, X.-X. Niu, Y.-X. Yang, A secure cryptocurrency scheme based on post-quantum blockchain, IEEE Access 6 (2018) 27205–27213.
[9] D. Puthal, N. Malik, S.P. Mohanty, E. Kougianos, C. Yang, The blockchain as a decentralized security framework [future directions], IEEE Consumer Electron. Mag. 7 (2) (2018) 18–21.
[10] N. Kshetri, J. Voas, Blockchain in developing countries, IT Professional 20 (2) (2018) 11–14.
[11] V. Gatteschi, F. Lamberti, C. Demartini, C. Pranteda, V. Santamaría, To blockchain or not to blockchain: that is the question, IT Professional 20 (2) (2018) 62–74.
[12] A. Anjum, M. Sporny, A. Sill, Blockchain standards for compliance and trust, IEEE Cloud Comput. 4 (4) (2017) 84–90.

[13] A. Jindal, G.S. Aujla, N. Kumar, SURVIVOR: a blockchain based edge-as-a-service framework for secure energy trading in SDN-enabled vehicle-to-grid environment, Comput. Netw. 153 (2019) 36–48.
[14] A.N. Mencias, D. Dillenberger, P. Novotny, F. Toth, T.E. Morris, V. Paprotski, J. Dayka, T. Visegrady, B. O'Farrell, J. Lang, An optimized blockchain solution for the IBM z14, IBM J. Res. Dev. 62 (2/3) (2018) 4-1.
[15] S.V. Akram, P.K. Malik, R. Singh, G. Anita, S. Tanwar, Adoption of blockchain technology in various realms: opportunities and challenges. Security Privacy (2020). https://doi.org/10.1002/spy2.109.

About the authors

Shubhani Aggarwal is pursuing PhD from Thapar Institute of Engineering & Technology (Deemed to be University), Patiala, Punjab, India. She received the BTech degree in Computer Science and Engineering from Punjabi University, Patiala, Punjab, India, in 2015, and the ME degree in Computer Science from Panjab University, Chandigarh, India, in 2017. She has many research interests in the area of Blockchain, cryptography, Internet of Drones, and information security.

Neeraj Kumar received his PhD in CSE from Shri Mata Vaishno Devi University, Katra (Jammu and Kashmir), India in 2009, and was a postdoctoral research fellow in Coventry University, Coventry, UK. He is working as a Professor in the Department of Computer Science and Engineering, Thapar Institute of Engineering & Technology (Deemed to be University), Patiala (Punjab), India. He has published more than 400 technical research papers in top-cited journals such as IEEE TKDE, IEEE TIE, IEEE TDSC, IEEE TITS, IEEE TCE, IEEE TII, IEEE TVT, IEEE ITS, IEEE SG, IEEE Netw., IEEE Comm., IEEE WC, IEEE IoTJ, IEEE SJ, Computer Networks, Information sciences, FGCS, JNCA, JPDC, and ComCom. He has guided many research scholars leading to PhD and ME/MTech.

His research is supported by funding from UGC, DST, CSIR, and TCS. His research areas are Network Management, IoT, Big Data Analytics, Deep Learning, and Cybersecurity. He is serving as editor of the following journals of repute: ACM Computing Survey, ACM·IEEE Transactions on Sustainable Computing, IEEE·IEEE Systems Journal, IEEE·IEEE Network Magazine, IEE·IEEE Communication Magazine, IEE·Journal of Networks and Computer Applications, Elsevier Computer Communication, Elsevier International Journal of Communication Systems, Wiley. Also, he has been a guest editor of various international journals of repute such as IEEE Access, IEEE ITS, Elsevier CEE, IEEE Communication Magazine, IEEE Network Magazine, Computer Networks, Elsevier, Future Generation Computer Systems, Elsevier, Journal of Medical Systems, Springer, Computer and Electrical Engineering, Elsevier, Mobile Information Systems, International Journal of Ad Hoc and Ubiquitous Computing, Telecommunication Systems, Springer, and Journal of Supercomputing, Springer. He has also edited/authored 10 books with international/national publishers like IET, Springer, Elsevier, CRC: Security and Privacy of Electronic Healthcare Records: Concepts, Paradigms and Solutions (ISBN-13: 978-1-78561-898-7), Machine Learning for Cognitive IoT, CRC Press, Blockchain, Big Data and IoT, Blockchain Technologies Across Industrial Vertical, Elsevier, Multimedia Big Data Computing for IoT Applications: Concepts, Paradigms and Solutions (ISBN: 978-981-13-8759-3), Proceedings of First International Conference on Computing, Communications, and Cyber-Security (IC4S 2019) (ISBN 978-981-15-3369-3). One of the edited text-book entitled, "Multimedia Big Data Computing for IoT Applications: Concepts, Paradigms, and Solutions" published in Springer in 2019 is having 3.5 million downloads till June 6, 2020. It attracts attention of the researchers across the globe (https://www.springer.com/in/book/9789811387586). He has been a workshop chair at IEEE Globecom 2018 and IEEE ICC 2019 and TPC Chair and member for various international conferences such as IEEE MASS 2020 and IEEE MSN 2020. He is a senior member of the IEEE. He has more than 12,321 citations to his credit with current h-index of 60 (September 2020). He has won the best papers award from IEEE Systems Journal and ICC 2018, Kansas City in 2018. He has been listed in the highly cited researcher of 2019 list of Web of Science (WoS). In India, he is listed in top 10 position among highly cited researchers list. He is an adjunct professor at Asia University, Taiwan, King Abdul Aziz University, Jeddah, Saudi Arabia, and Charles Darwin University, Australia.

CHAPTER EIGHT

History of blockchain-Blockchain 1.0: Currency☆

Shubhani Aggarwal and Neeraj Kumar
Thapar Institute of Engineering & Technology, Patiala, Punjab, India

Contents

Abstract

Bitcoin is a cryptocurrency, which is not backed by any country's central bank or government. It can be traded for goods or services with vendors who accept bitcoins as payment. These bitcoins are the blocks of secure data. This data is transferred from one person to another and verifying the transaction, i.e., spending the money that requires high computing power to safely verify the individual transactions. The P2P network monitors and verifies the transfer of bitcoins between users. It can be used to book hotels, shopping, financial transactions, buy video games, etc. In this chapter, we describe the evolution of bitcoin cryptocurrency to evolution of blockchain and their usage in real-world entities.

☆ Introduction to blockchain.

Advances in Computers, Volume 121
ISSN 0065-2458
https://doi.org/10.1016/bs.adcom.2020.08.008

Chapter points

- In this chapter, we firstly discuss the bitcoin cryptocurrency and describes the working ofbitcoin cryptocurrency. Then, we discuss the problems of double-spending and Byzantine General's Computing in bitcoin. After that, we discuss the evolution ofblockchain from Bitcoin cryptocurrency.
- Here, we also discuss the fundamentals and potential impact ofBlockchain in real-time world.

1. Bitcoin cryptocurrency

Bitcoin is a decentralized digital currency that enables instant payments to anyone, anywhere in the world. The primary concern of the bitcoin is a cross country payment transaction and other concern is that there is no need for a central authority or no government organization will have to control over it. It uses P2P technology and supports different level of securities so that the entire system becomes tamper-proof.

The two main operations of bitcoin cryptocurrency is as follows.

1. *Transaction management*: In this, everyone can transfer of bitcoins from one user to another. For example, anyone can buy something by utilizing the bitcoins make transfer of bitcoins from India to any other country.
2. *Money issuance*: It regulates the monetary base ofbitcoins like economical aspects of a coin base of digital cryptocurrency whereas, in our banking system, there is a central authority that regulates the money inside the country.

Bitcoin cryptocurrency must be in controlled and in limited supply to have value. The value of bitcoin increases due to the following reasons.

- It is decentralized technology.
- It is limited in number (21 billion).
- It is P2P network.
- It can be anonymous.
- It is transparent in nature (open source).
- It is easy to buy and sell.
- It is irreversible (no-charge backs).
- It is difficult and expensive to hack.
- It has increased in value over 10,000 percent in a short period of time.

Traditionally, people deposit their money into a bank account. But now, they can store their currency into bitcoin wallets. It can store every transaction has to be done in the network. These transactions are verified by the bitcoin miners and get rewarded by bitcoin currency. Any malicious

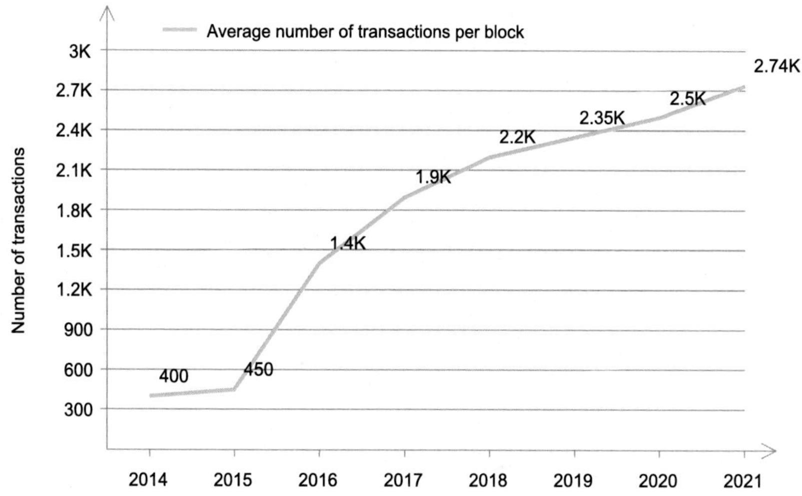

Fig. 1 Increase in average number of bitcoin transactions per block as per year advances.

currency needs to be rejected and only to accept the actual currency that is flowing through the network. According to the study as shown in Fig. 1, it has been observed that an average number of bitcoin transactions per block is increased as per years advance.

1.1 Creation of bitcoins

In bitcoin standard architecture, the creation of the block rate is adjusted after every 2016 blocks. This block creation takes 2 weeks of the period. After every 2016 blocks, there is readjustment for bitcoins that have been created from mining the blocks. So, this process still continues and to again manage the number of bitcoins generated from the mining blocks and is decreased geometrically.

For example, the number of bitcoins that are rewarded during the mining process gets decreased with a rate of 50 % after every 2,10,000 blocks creation. It approximately takes 4 years to complete this amount of blocks creation. From 2008 to 2012, when bitcoin cryptocurrency was developed by Satoshi Nakamoto, the block reward fees for one block is 50 bitcoins. After then, from 2012 to 2016 it gets reduced to 12.50 bitcoins per block which is again gradually decreased to 6.25 bitcoins per block for the next 4 years. In this way, the reward for mining the block will get close to 0. After some mathematical calculations, it has been found that when the number of bitcoins will reach up to 21 billion in the network, then further no reward will get to the miners from the system. So, by the time passes the

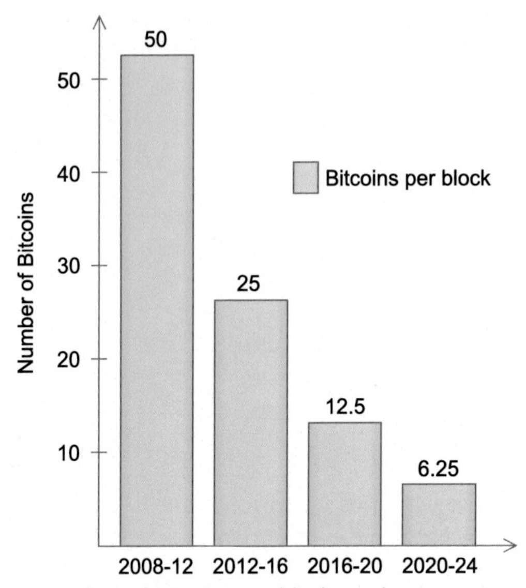

Fig. 2 Reduction in number of bitcoins per block can be seen as years advance.

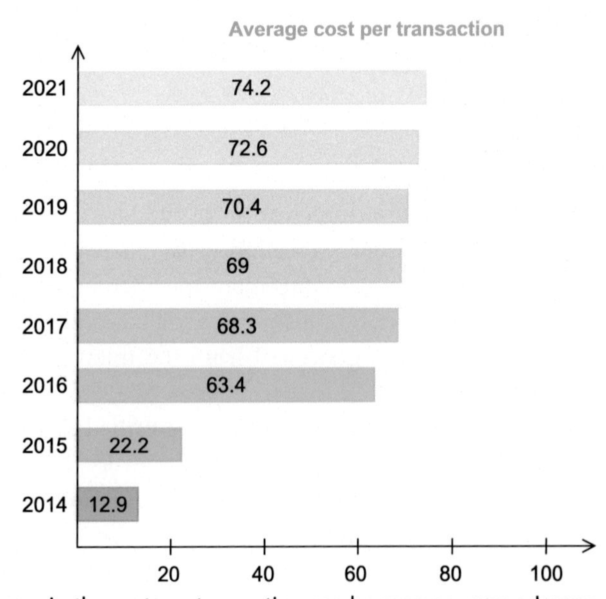

Fig. 3 Increase in the cost per transaction can be seen as years advance.

number of bitcoins per block will decrease and transaction fees will be increased. Over the past few years, a per year reduction in number of bitcoins per block and an increase in the cost per bitcoin transaction have been observed are as shown in Figs. 2 and 3.

1.2 Control of bitcoin network

Nobody controls the bitcoin network. It is controlled by their own bitcoin users across the globe. All the users are free to choose what software and version they want to use in the bitcoin network. It can work only when strong consensus agreement is present among all the users. By this, all users and developers have incentives irrespective of the security to the consensus agreement.

1.3 Bitcoin transaction system

From the user perspective view, bitcoin is nothing more than a personal wallet of the user and allows a user to send and receive bitcoins with each other. This is how it works for all users. Let us explain with example of *Alice* and *Bob*. All nodes in the network have their own digital wallets to store the bitcoin cryptocurrency on their computers. Basically, bitcoin wallets are public ledgers that provide access to all multiple bitcoin addresses. An address is a string of letters and numbers like "*KULP2589gcdg5UKUD*" and each address has its own balance of bitcoins. Now, *Bob* creates a new bitcoin address for accepting the bitcoin cryptocurrency from *Alice*. The scenario of bitcoin address is as shown in Fig. 4.

To verify the legitimate account user in the bitcoin network, the digital signatures with Public-key cryptography has been used. The bitcoin address represents a unique private or secret key and their corresponding public key

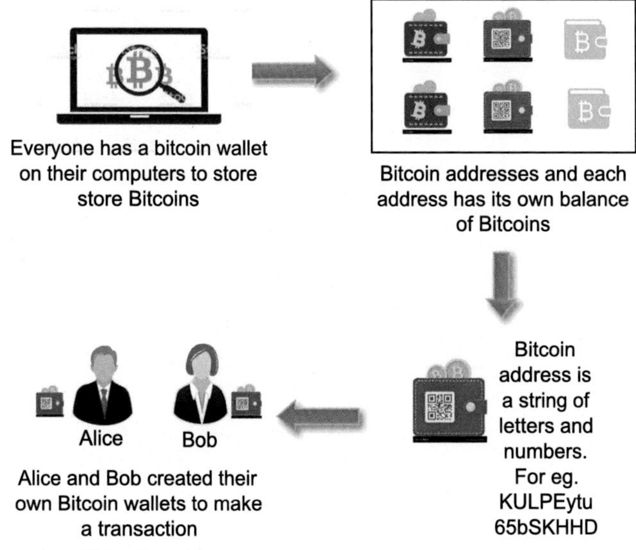

Fig. 4 Scenario of bitcoin address.

are stored in the bitcoin address. A *secret key* is used to sign a message for security and matching *public key* is used to verify the signed message it is valid or not. Now, to make a secure bitcoin transaction between *Alice* and *Bob*, *Alice* wants to transfer some amount of bitcoins to *Bob*. He holds his bitcoin address with a private key and broadcast this transaction to the blockchain network. Anyone on the network can use the public key of *Alice* to verify the transaction request, which is actually coming from the legitimate account user. By the use of digital signatures, we can provide authentication and integrity to the network. The scenario of the digital signatures is as shown in Fig. 5 and how it can be used in bitcoin system is as shown in Fig. 6.

The next step after the authenticity of the user is to verify the transactional data using cryptographic hash functions such as SHA-1 and SHA-2. These cryptographic primitives are used to secure data transmission between *Alice* and *Bob*. These hash functions transform the collection of data into an alphanumeric string with a fixed length called hash value and it is impossible for an adversary to predict the initial data that will create specific value. The transactional data, previous hash value, and the nonce (random number) create a new hash value that is completely different from

Concept of Digital signatures

Different messages have completely different signatures
- Sig (Message, secret key) = Signature
- Verify (Message, signature, public key) = True/False

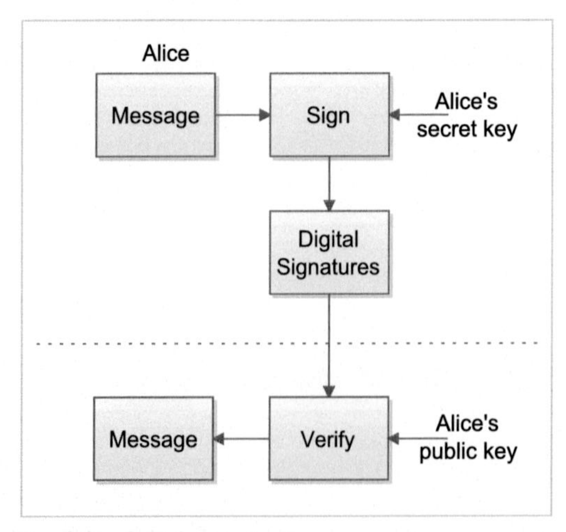

Fig. 5 The scenario of the digital signatures.

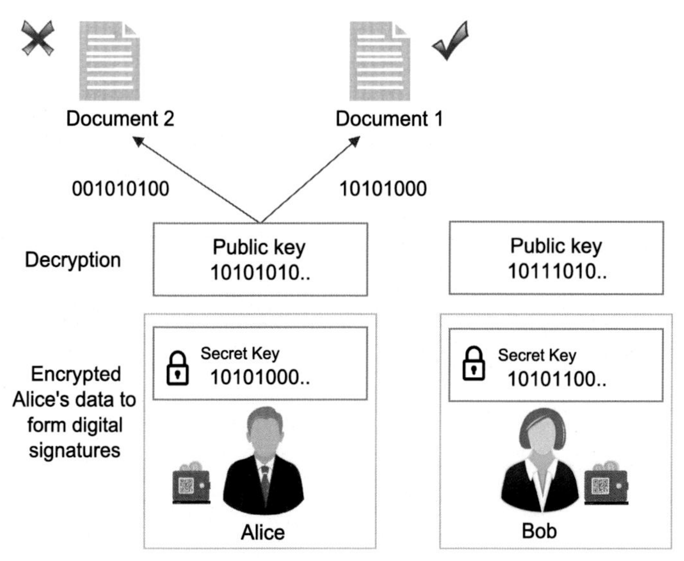

Fig. 6 Digital signatures used in bitcoin transaction system.

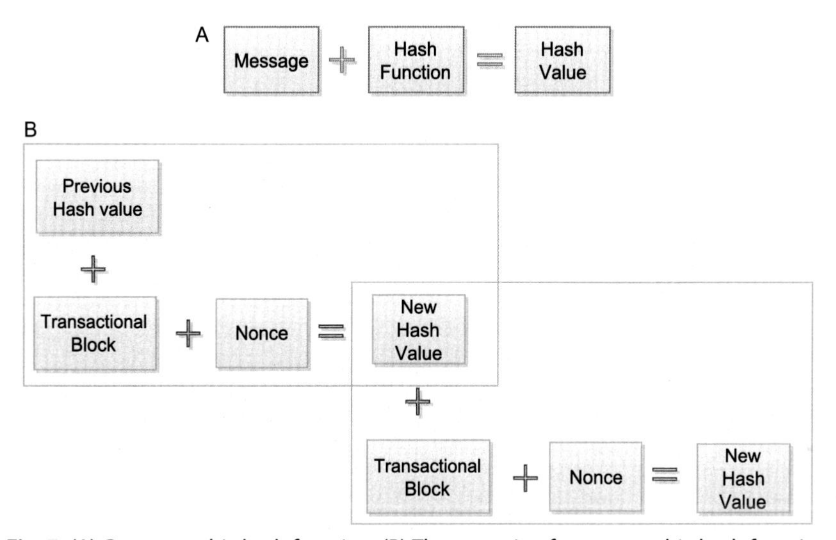

Fig. 7 (A) Cryptographic hash function. (B) The scenario of cryptographic hash function in bitcoin system.

other hash values. A little change in the input value will completely change the output hash value because each new value of hash contains information about previous hash value. The scenario of cryptographic hash function in bitcoin system is as shown in Fig. 7.

However, creating hashes is computationally important but in the bitcoin system, it must start with a certain number of zeroes like '*0000000000000... 5876335859358*'. After creating the hash value, the transactional data is stored into the blocks where bitcoin miners calculate a new hash value for verification on the basis of present information (previous hash value, block, and nonce). The miners have no idea to predict which nonce will produce a correct hash value with a required number of zeroes. So, they will produce different hash value with different nonces until they reach on one that works. They use their maximum computational resources and power to find out the exact hash value during this process. After the mining process has been done, the first miner who finds the computational hash correctly get rewarded with bitcoin cryptocurrency in its wallet. The block is added into the blockchain network and each node of the network updates their local copies. As time passes, *Alice's* bitcoin transfer to *Bob* is done successfully and *Bob* receives the bitcoins. The bitcoin transaction between *Alice* and *Bob* is as shown in Fig. 8.

In this bitcoin transaction process, if anyone wants to modify the details then he or she would have to redo all the tasks and process as bitcoin miners did. Any little changes require or completely different nonce value. So, it is impossible for a hacker to hack transactional data that has to be done in the blockchain network.

1.4 Advantages of bitcoin cryptocurrency

The main advantages of bitcoin cryptocurrency over traditional systems are described as follows.

- *Payment freedom*: Bitcoin cryptocurrency is used for online purchases, payments, and investments. It is easily possible to send and receive bitcoins anywhere anytime in the world. There is no borders, bureaucracy, bank holidays, etc. to make payment transactions. It allows its users to be in full control of their currency. They can use their currency in their own way.
- *Choose our own fees*: In the bitcoin cryptocurrency system, there is no particular fees to send or receive bitcoins from anywhere at any time. In this, the transaction fee is not related to the number of bitcoins that are being sent or received by the users. However, it depends on how secure the transaction is? The more bitcoin transaction fee represents the high security and fast confirmation of the transaction.
- *Fewer risks of wholesaler*: Bitcoin transactions are irreversible, secure and do not contain personal information. It protects the wholesaler from frauds or fraudulent charge-backs. The net results of this currency are lower fees, larger markets, and fewer administrative costs than fiat currency.

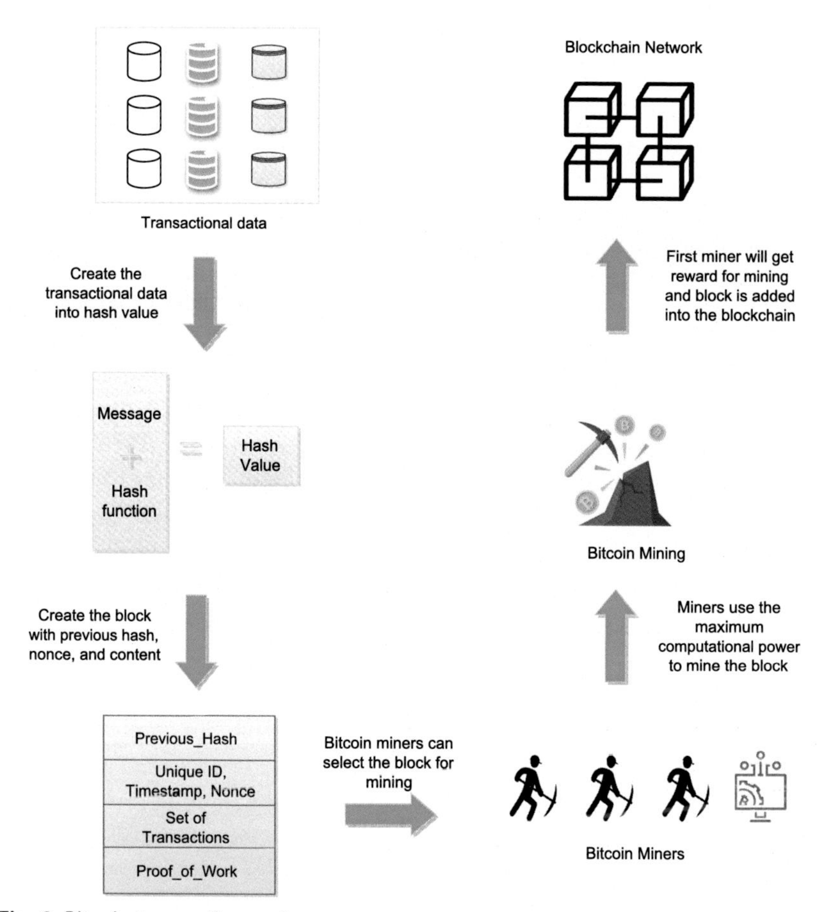

Fig. 8 Bitcoin transaction system.

- *Security and control*: Bitcoin transactions are full control under the bitcoin users. So, it is impossible for a hacker to hack or to force unwanted changes in this transaction. It can be made without any personal information of the user which provides strong protection against identity theft.
- *Transparent and neutral*: All the bitcoin transactions are passed through the blockchain network where every node can verify the transaction and use in a real-time. There is no individual control and manipulation on the bitcoin protocol because it is secured by cryptography. This allows the core of bitcoin to be trusted for being neutral, tamper-proof, and transparent.

After the study of the advantages of the bitcoin cryptocurrency, there is a number of disadvantages that needs to be resolved for ongoing development. Many people are still aware of this cryptocurrency and their software and tools are still in active development with some incomplete features. These features and services are being developed to make the bitcoin more secure and accessible to masses. Most of the bitcoin-based companies are new and small in number in the process of maturing.

2. Double-spending problem

Double-spending is a problem in which the same digital currency can be spent more than once. At the same time, it is an instance in which the transaction uses the same input as another transaction that has already been broadcast on the network. This is the major flaw present in the digital cryptocurrency. The bitcoin system is designed to prevent double-spending in a decentralized environment where there is no central authority to interfere with the disputes. However, as bitcoin cryptocurrency is neither physical money nor transmitted via centralized database, it raises challenges on preventing this issue. In this system, a ledger of each node is the most important concept. It is used to record the transactions and contain information of how much amount of currency each account still has.

To understand the concept, let us take an example as shown in Fig. 9. Suppose *Alice* wants to pay some currency to *Bob* in bitcoin cryptocurrency. This transaction will be sent to the P2P network and be recorded in the ledgers of the network as a type of verification process that takes some time to complete. That means, during verification time *Alice* might be able to send another payment to *Charlie* using same bitcoin currency token that was spent

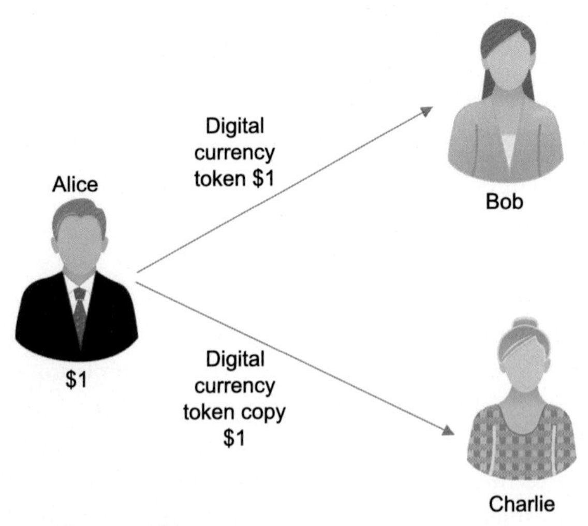

Fig. 9 Double-spending problem.

in the *Alice-Bob* transaction. It means, one bitcoin token being paid twice. This problem is called double-spending problem.

2.1 Bitcoin prevents the double-spending problem

Fig. 10 shows the trading of cryptocurrency between a *buyer* and a *seller*. The *buyer* advised the miners to make a bitcoin payment to *seller* while the *seller* delivers the service to the *buyer* simultaneously. The*buyer* can make secret

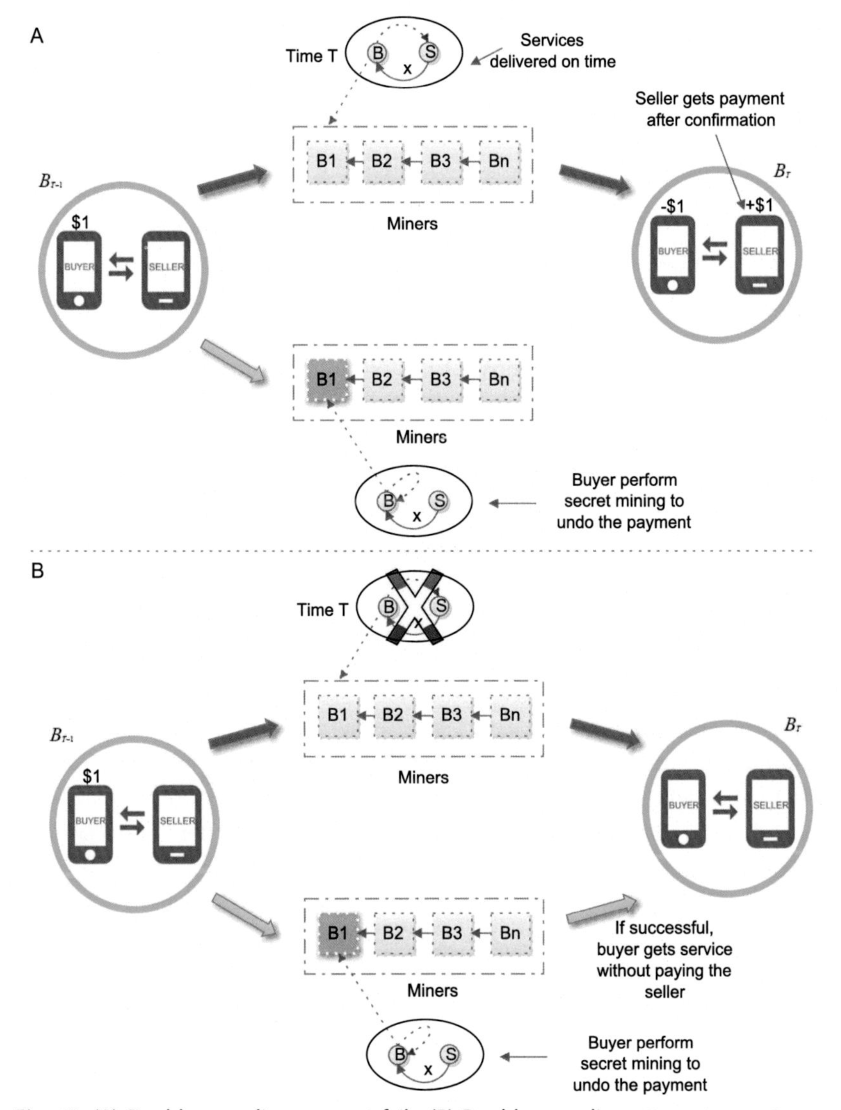

Fig. 10 (A) Double-spending attempt fails. (B) Double-spending attempt accepts.

mining for an alternative history where the funds are not transferred. The outcome of the transaction depends on which the payment transaction is incorporated into the network first. If the payment transaction is incorporated, then the double-spending attempt fails and the *seller* receives the payment. If the transaction is accepted, then double-spending attempt accepts and the *seller* selling the service to the *buyer* without paying.

3. Byzantine Generals' problem

Byzantine Generals' is a problem that symbolizes the difficulty of having a coordinated conversation when nontrusted parties are involved. They must agree on the single strategy to avoid complete failure but somewhere involved parties are corrupt and sharing false information.

Let us explain with the help of an example. On each side of the enemy city, there are two armies present to attack at the same time. But the city is strong enough to safe by itself against one of the armies attacks but not so strong to defend against both the armies at the same time. If they do not attack at the same time, they lose the enemy city's attack. So, the Generals' of both the armies must agree on the same time of when to attack. They can communicate with each other by sending a messenger through the enemy city. There is no other way to communicate with each other. By this, *General A* send the message "*Hey General B, we are going to attack this city on Monday. Can we count on you with us on this attack?*" The messenger then walks through the enemy city and delivers the message to *General B. General B* responds back to the *General A* with a message "*No, we can not do attack on Monday. We will not free on that day. So, what about Friday? If we attack on Friday, will you join us on Friday to attack?*" Then, again the messenger walks through the city and delivers the message to General again and so on. However, during message delivery, here is the attacker, where the messenger could potentially get caught and replaced by the fake-news messenger in the city. The fake-news messenger tries to change with the other General to attack the city at the wrong time. So, In this scenario, there is no way to check that the delivered message is authentic or not in a trust-less environment, which is called the Byzantine Generals problem. The scenario of this problem is as shown in Fig. 11.

3.1 Solution of Byzantine Generals with proof-of-work

PoW consensus mechanism has been used to solve the Byzantine Generals problem as it achieves a majority agreement without any involvement of

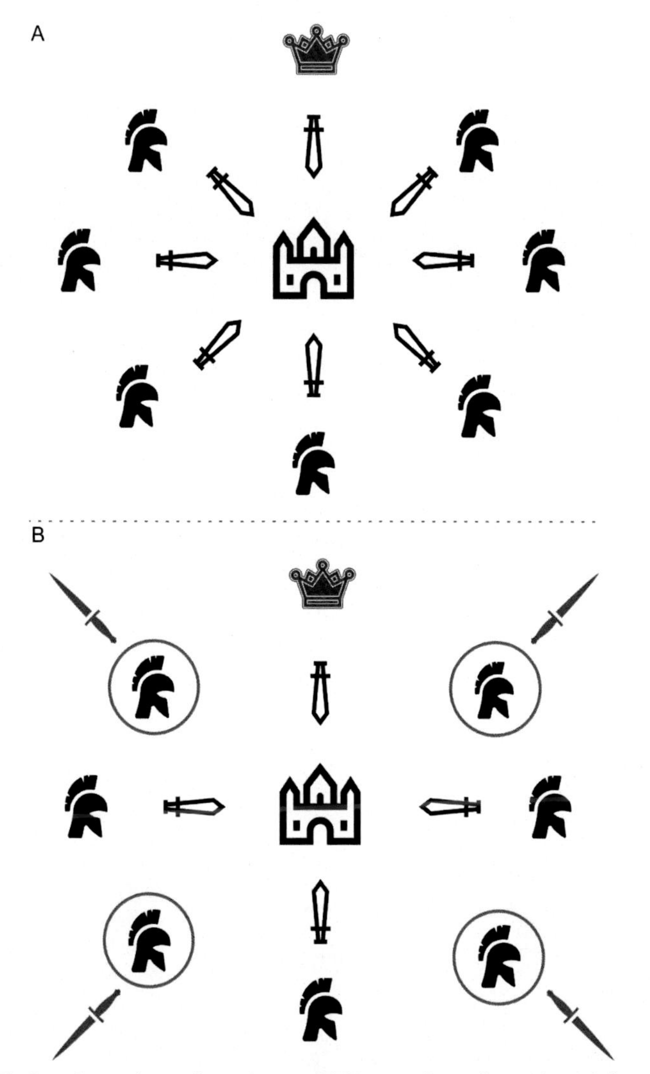

Fig. 11 (A) Coordinated attack = victory. (B) Uncoordinated attack = defeat.

central authority and in-spite the use of untrusted or unknown parties. It is the original blockchain mechanism that enables the user on the network to reach a trusty consensus. It is essentially an answer to a complex mathematical problem. It takes a lot of time and work to create but make the validate process easy for others. It allows the distributed and uncoordinated Generals to come to a consensus agreement as under.

1. The Generals agree on the first plan that has been accepted as a plan by all Generals.
2. A number of Generals solve the PoW problem, creating a block, and broadcast it to the network so that all Generals have received it.
3. Each General verifies the block and works on solving the next PoW problem so that their plan add it to the previous information.
4. Each time General solves the PoW problem, a block is generated, and the chain begins to grow.

Fig. 12 shows the consensus between armies using PoW. By this consensus mechanism in Byzantine Generals problem, the Generals can arrive at a state where they know when to attack the city and can estimate their chances of successfully doing so. In this way, they can prevent different messages and signals coming from many Generals to attack being sent simultaneously. This mechanism also prevents the system from malicious actors, i.e., traitors that destroy the network by changing the message with historic messages. In this mechanism, every information is stored in the ledgers as the hash values. This value is stored as a previous hash in every new block. So, a little change to an earlier block will fully change the hash value of all the successive blocks. This would take a huge amount of computing power that ensures the ledger is secure and tamper-proof.

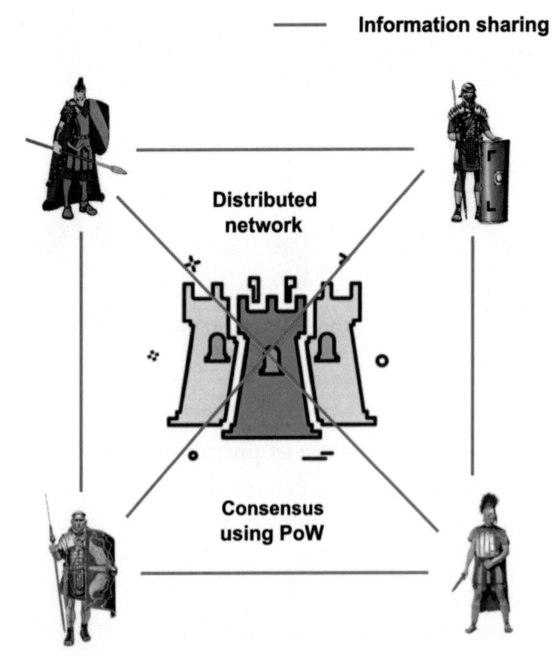

Fig. 12 Solving the Byzantine Generals problem using PoW.

4. Evolution of blockchain

From January 2009, blockchain technology has been developing and emerging. In the beginning, the blockchain technology supporting bitcoin that produced the following properties.

- Decentralization of the bitcoin cryptocurrency and financial transactions.
- Decentralization of the data storage using a distributed database.
- Eliminates the central authority that verify the transactions in a centralized system.
- Support for tamper-proof data and transparency in a P2P network.
- Introduces the PoW concept which makes the blockchain technology more unique because it combines with high computational power by the use of nodes connected to the network. These nodes verify the transactions and secure the public ledger.

4.1 Understanding blockchain technology

The blockchain is a time-stamped series of an immutable record of data that is distributed and managed by a cluster of computers. Each of these blocks is combined together using cryptographic primitives to make a secure chain. As it is a decentralized system, there is no third-party validator to manage all the records and data information at blockchain network. So, it is a shared immutable public ledger that is open and accessible for everyone to see. Hence, the data present on the blockchain is transparent in nature and everyone can be involved for their accountable reasons. The understanding of blockchain is as shown in Fig. 13.

The blockchain technology is a simple yet creative way of transmitting the information among the nodes of the network in a fully automated and secure manner. The nodes of the network create a block that contains transaction information. Then, this block is verified by thousands maybe millions of distributed computer nodes around the network. The verified block is then added to the chain across the network by creating a unique record with a unique history. The main advantage of the blockchain is that if falsifying a single record then, the entire chain of blockchain is false in millions of instances which is impossible in practice. It carries no transaction cost but has high infrastructure cost. It can not use only for making transactions but also replace bitcoin models and processes that rely on charging a payment for a transaction. For example, The gig economy hub Fivver charges $0.5 on five transactions between the buyer and the seller services. Bitcoin itself use

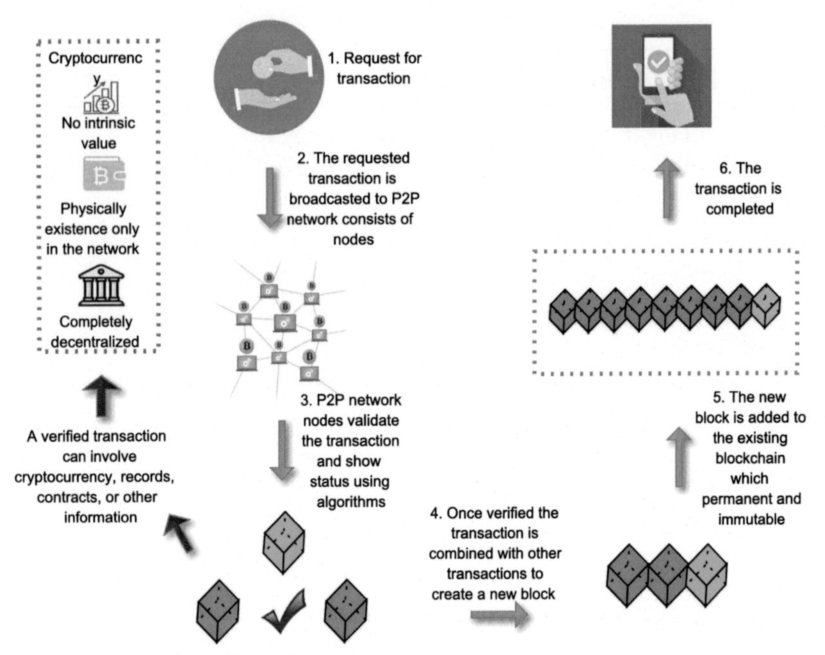

Fig. 13 Understanding blockchain.

this model to make the transaction secure but it can also be used in other ways. The main reasons why blockchain has gained popularity are as follows.

- It is not kept by only one single entity means there is no involvement of the third-party validator that may leads to a single point of failure.
- The data stored on the blocks of the blockchain is cryptographically secured.
- It is immutable in nature so no one can tamper the data that is inside the blockchain.
- It is transparent in nature so anyone can track the data if they want to.

5. Fundamentals of blockchain

The three main properties of blockchain technology that helped it to gain recognition all over the world are as follows.

1. *Decentralization*: Before bitcoin cryptocurrency, all the users were used to centralized services that depend on the third-party validator. The centralized system stores all type of data and information. The user would interact with this system whatever information and data they want to use it or required at any-time. For example, Banks, that stores all the

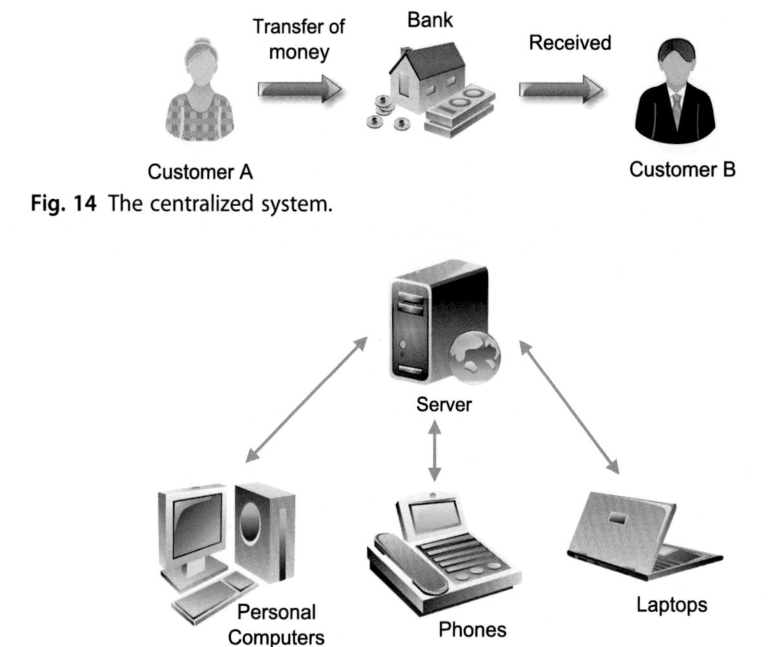

Fig. 14 The centralized system.

Fig. 15 The client–server model.

currency and money in a centralized way. The only way to pay someone is by going through banks as shown in Fig. 14.

The other example is the client–server model as shown in Fig. 15 in which when someone is searching on the Google then, he or she sends a query to server who gets back at you after few seconds with relevant information. Hence, there are some vulnerabilities in a centralized system that are mentioned below.

- In a centralized system, all the data and information is stored in one place where anyone can attack or change the data.
- If the centralized system goes through some software up-gradations then, it would halt the entire system.
- If in between the centralized system shut-down then, no one will be able to do a task or to access information.
- If the centralized system is malicious or corrupted then, all the data inside it will be compromised.

So, from the above-mentioned reasons are the causes that change the entire system from centralized to decentralized. In the decentralized system, everyone owns the information by itself in the network. If someone wants to interact with a particular node then, he or she can directly

Fig. 16 Transaction without central authority.

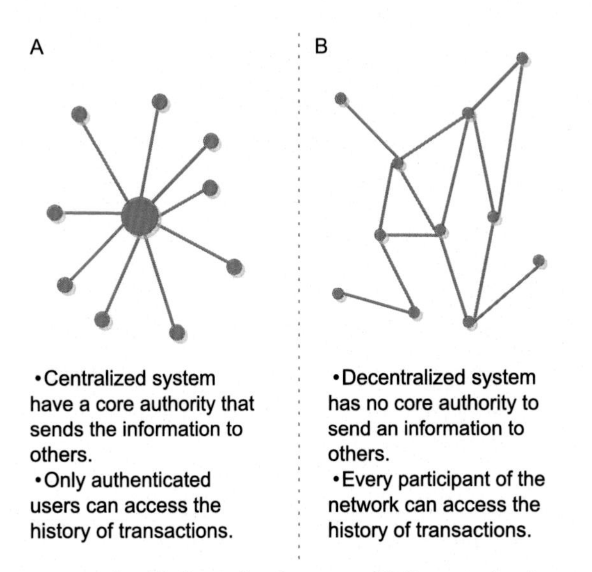

Fig. 17 The new networks. (A) Centralized system; (B) Decentralized system.

interact with that node without going through a third-party validator. In this, we are the only one for the charge of our cryptocurrency. Anyone can send or receive currency from anybody without the involvement of third-party as shown in Fig. 16. The new networks instead of using centralized system is as shown in Fig. 17.

2. *Transparency*: The most important and interesting concept in blockchain technology is transparency. In the system, a person's personal information is hidden using complex cryptographic primitives and is only represented by their public address. For example, Fig. 18 shows the person's transaction history and Fig. 19 shows the detail information of person in which their personal information (address) is secured by hash cryptography.
3. *Immutability*: It means that once something has been written or done into the blockchain, it can not be tampered or changed because it is secured by a cryptographic hash function. In this context, the transaction is taken as an input and run through a hashing algorithm (SHA-256) which gives an output of a fixed length as shown in Fig. 20. Even, a small change in

Transaction View information about a bitcoin transaction

1f619bd71cc2246bb7335f30801feed6e97f94f89ca24590f5d37fa23d9c425c

14QNGkpvyg7CmX9FX1Cuxip7Es9DZeDpKP (211.46264042 BTC - Output) → 1C5APLYTuCkkMXDyp8pMzCbgrDGCgvcfq4 - (Unspent) 0.0737729 BTC
14QNGkpvyg7CmX9FX1Cuxip7Es9DZeDpKP - (Spent) 211.38876356 BTC

SPONSORED Crypto Credit | 3 Confirmations | 211.46253646 BTC

Summary	
Size	225 (bytes)
Weight	900
Received Time	2019-09-09 10:12:22
Included In Blocks	593992 (2019-09-09 10:12:22 + 0 minutes)
Confirmations	3
Visualize	View Tree Chart

Inputs and Outputs	
Total Input	211.46264042 BTC
Total Output	211.46253646 BTC
Fees	0.00010396 BTC
Fee per byte	46.204 sat/B
Fee per weight unit	11.551 sat/WU
Estimated BTC Transacted	0.0737729 BTC
Scripts	Hide scripts & coinbase

Activate W
Go to Setting

Fig. 18 Transaction history.

Fig. 19 The detail information of person's transaction.

SHA256 Hash Copyright Notice

Data: Hi, i am Alice. I am sending 10 bitcoins to Bob.

Hash: d0458db03b961ff7f94564db6aa8e61fe76871ebdbc0db21cec2a200fcea3322

Fig. 20 Cryptographic hash function 1.

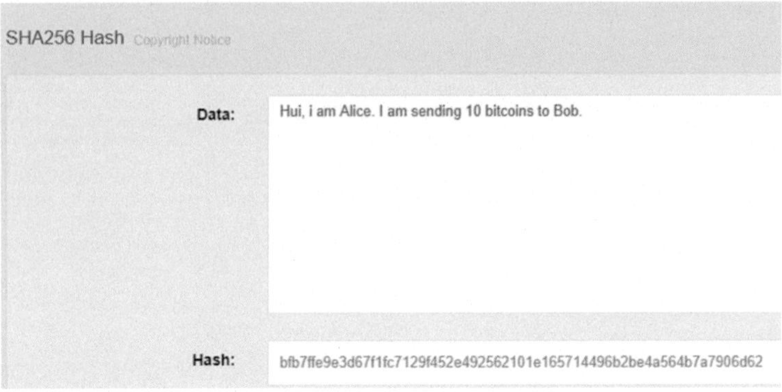

Fig. 21 Cryptographic hash function 2.

the input data that directly reflects the output value. For example, the Fig. 21 shows a small change in the letter "*Hi*" to "*Hui*" that reflects the full hash value of the output. Because of this property, it has been used in blockchain technology. In this technology, a link list is formed that contains data and a hash pointer which directs to a previous hash and creating a chain called blockchain.

6. Potential impact of blockchain

The various number of identified use cases show that the potential applications of the blockchain-based framework have been recommended for virtually all stages of the chain value [1]. Blockchain, the distributed ledger technology (DLT), continues to expand in various field of applications such as banking sector, insurance services, IoT, governance, etc. It improves privacy and confidentiality, enhances user safety, and provides a high level of care services to customers [2]. It gives Internet users the ability to authenticate digital information. It provides revolutionary change in new business applications that are described as under.

1. *Cryptocurrencies*: At present, cryptocurrency is the most important use-case in blockchain technology. Using this technology, hundreds of cryptocurrencies like bitcoin, litecoin, ethereum, etc. exist. The secured nature of smart contracts helps to make cryptocurrency transactions without the involvement of any central authority. All these transactions are authorized by blockchain network nodes. So, the increasing demand for famous cryptocurrencies will help to attract the crowd toward blockchain technology.

2. *Smart contracts*: DLT provides simple codes for smart contracts that will execute when sufficient conditions are met. At the current development level of blockchain technology, smart contracts can be programmed to perform or to develop various type of applications, functions, and programs. For this, Ethereum has been used which is an open-source blockchain project. This platform has the potential to take advantage of the usefulness of blockchain technology in a trust-less environment.
3. *Authentication/notary services*: The secured and protected nature of smart contracts in blockchain technology gives online notary services to the customers. These smart contracts can be used by people to authorize, authenticate, and accountable the other people's actions as a part of the evidence. With this technology, it is also possible to check the authentication and authorization of a document or a file.
4. *Crowdfunding*: This is the important concern in blockchain technology because the use of digital cryptocurrencies start-up companies can crowdfund or collect money from the public for their upcoming journey in an early stage [3]. For example, Kickstarter and Gofundme are companies that doing advance work for P2P economy. Regardless, decentralized autonomous organization (DAO) suggests that blockchain has the potential to guide in "*a new paradigm of economic cooperation*."
5. *Governance*: The transparent and public-accessible nature of DLT would bring full transparency in several applications like day-to-day governance operations, online voting system, hiring of personnel, providing identities, hospitality management, etc. [4]. The secured nature of smart contracts in Ethereum-based platform gives high-security by automating the process in blockchain technology. This means governance provides transparency and ensure smoothness to digital assets, equity or information.
6. *Internet-of-things*: With the generation and development of IoT, blockchain technology can be used to keep track of individual history by maintaining the record of data transmission between devices, web-services, and end-users. This technology would be useful in IoT applications as a security and privacy purpose because the transactions recorded in blockchain are secured by cryptographic primitives. So, it is very difficult for hackers to manipulate the information that exchanges between the devices and end-users.
7. *Transport sharing*: There is a classic and standard case to utilize blockchain technology for transport sharing or ride-sharing. During transport

sharing, people make payments to each other using cryptocurrencies that do not need any central authority to deal between them [5].

8. *Energy sharing*: Blockchain technology empowers the buying and selling of the renewable energy generated by microgrids. The automatic nature of Ethereum-based smart contracts redistributes the energy when solar panel makes excess amount of energy. Using these smart contracts, the microgrids can consume or contribute their energy to and from the grids easily [6]. They can make their payments in a P2P network using cryptocurrencies. For example, A "*transactive grid*," Brooklyn, working with the distribution of energy outfit called "*intelligent grid*" in an IoT functionality environment.
9. *Data storage*: The decentralized way to store and share the files using blockchain technology provides security and privacy from getting hacked or lost. Using this technology, it is possible to generate a network infrastructure that stores unalterable data at the nodes and remove duplicated files from the network. It is also used to obtain public address of network nodes for accessing storage facilities to search a file in the blockchain network.
10. *Financial services*: Blockchain technology can be applied for banking and insurance operations between the participating nodes without requiring the intervention of third-party validator. The transactions are based on the level of trust between the network nodes in a particular blockchain. The participating nodes may pay premiums in the form of cryptocurrencies and insurance policy can be concerned in the form of smart contracts. They may act as a validator or authenticator for claim. So, blockchain services can be very helpful in financial services.

References

[1] A. Kumari, R. Gupta, S. Tanwar, N. Kumar, Blockchain and AI amalgamation for energy cloud management: challenges, solutions, and future directions. J. Parallel Distrib. Comput. (2020). https://doi.org/10.1016/j.jpdc.2020.05.004.

[2] D. He, K.-K.R. Choo, N. Kumar, A. Castiglione, IEEE access special section editorial: research challenges and opportunities in security and privacy of Blockchain technologies, IEEE Access 6 (2018) 72033–72036.

[3] C. Lin, D. He, S. Zeadally, N. Kumar, K.-K.R. Choo, SecBCS: a secure and privacy-preserving blockchain-based crowdsourcing system, Sci. China Inf. Sci. 63 (3) (2020) 1–14.

[4] U. Bodkhe, P. Bhattacharya, S. Tanwar, S. Tyagi, N. Kumar, M.S. Obaidat, Blohost: Blockchain enabled smart tourism and hospitality management, in: 2019 International Conference on Computer, Information and Telecommunication Systems (CITS) IEEE, 2019, pp. 1–5.

[5] X. Li, Y. Wang, P. Vijayakumar, D. He, N. Kumar, J. Ma, Blockchain-based mutual-healing group key distribution scheme in unmanned aerial vehicles Ad-Hoc network, IEEE Trans. Vehicular Technol. 68 (11) (2019) 11309–11322.
[6] A. Miglani, N. Kumar, V. Chamola, S. Zeadally, Blockchain for Internet of energy management: review, solutions, and challenges. Comput. Commun. 151 (2020) 395–418, https://doi.org/10.1016/j.comcom.2020.01.014.

About the authors

Shubhani Aggarwal is pursuing Ph.D. from Thapar Institute of Engineering and & Technology (Deemed to be University), Patiala, Punjab, India. She received the B.Tech degree in Computer Science and Engineering from Punjabi University, Patiala, Punjab, India, in 2015, and the M.E. degree in Computer Science from Panjab University, Chandigarh, India, in 2017. She has many research interests in the area of Blockchain, cryptography, Internet of Drones, and information security.

Neeraj Kumar received his Ph.D. in CSE from SMVD University, Katra (Jammu and KashmirJ & K), India, and was a postdoctoral research fellow in Coventry University, Coventry, UK. He is working as an Associate Professor in the Department of Computer Science and Engineering, Thapar Institute of Engineering & Technology, Patiala (Punjab.), India since 2014. Dr. Neeraj is an internationally renowned researcher in the areas of VANET & CPS Smart Grid & IoT Mobile Cloud computing & Big Data and Cryptography. He has published more than 150 technical research papers in leading journals and conferences from IEEE, Elsevier, Springer, John Wiley, and Taylor and & Francis.

CHAPTER NINE

Architecture of blockchain[☆]

Shubhani Aggarwal and Neeraj Kumar
Thapar Institute of Engineering & Technology, Patiala, Punjab, India

Contents

Abstract

A blockchain technology allows untrusting parties with common interests to cocreate a permanent, unchangeable, and transparent record of exchange and processing without relying on a central authority. It serves as an immutable ledger that allows transactions to take place in a decentralized manner and secured by cryptographic primitives. Its exceptional characteristics include irreversibility, decentralization, persistence, and anonymity. With these advantages, it has found applications in almost all fields requiring data sharing among multiple parties but with secure authentication, anonymity, and permanence. Some of the applications are finance, real estate, smart grid, transportation system, and IoT. In this chapter, we have described the working and structure of blockchain in the real-world. We described the mining process during validation and verification on the blockchain network.

Chapter points

- In this chapter, we discuss the architecture of blockchain, technology behind blockchain, and need of blockchain in the real-world.
- Here, we discuss the detailed working and functionality of blockchain in real-time applications.

☆ Introduction to blockchain.

Advances in Computers, Volume 121
ISSN 0065-2458
https://doi.org/10.1016/bs.adcom.2020.08.009

1. Technology behind blockchain

Governments and corporates all over world are slowly and steadily realising the value in blockchain, which is a new kid on the block. It originated from the financial circles and is permeating into the business world. Its ardent supporters sincerely applaud it for its unbreakable and impenetrable security features. The shortest definition for blockchain is it is a distributed ledger. That is, it stores any list of transactions in a P2P network. Data in a blockchain is stored in fixed structures called "*blocks*."

- *Cryptocurrency* is a digital currency in which encryption techniques are used to regulate the generation of units of currency, and to verify the transfer of funds operating independently of a central bank.
- *Distributed ledger* is a database that is consensually shared and synchronised across the network spread across multiple sites, institutions or geographies. It allows transactions to have public "witnesses," thereby making a cyberattack more difficult.
- *Smart contracts* are agreements that are encoded in a computer program and automatically executed upon certain criteria being met. Advantages of smart contracts include improved quality, reduced contract execution costs, and increased speed. Smart contracts can be stored on the blockchain.
- *Miners* are the people keeping the blockchain running by providing a huge amount of computing resources competing to solve a cryptographic puzzle and upon solving the puzzle, they generate a block and they also get rewarded. Miners compete with each other to generate a valid block of transactions. Miners collect all pending transactions from the decentralized network then they guess a random number (nonce) to solve cryptographic puzzle, on successfully solving the puzzle they generate a block then they push that block into the network for verification from other nodes so that other nodes after verification can add that block in their copy of blockchain.
- *Nonce*: The cryptographic puzzle that miners solve is to identify the value of nonce. A nonce is a random number which can be used only one time. Mostly it is a random number with combination of some data. Blockchain adds a value called nonce in each block. This nonce is like a salt added to the contents of a block. By adding nonce, the hash output of the contents of the block will change.
- *Hash*: Hashing is a cryptographic technique which maps input data to data of a fixed size output. Bitcoin uses the *SHA-256* algorithm for it.

SHA-256 output a fixed length number. A slight change in input would change the complete output but the output would always of the same length.

Consensus algorithm is a process in computer science used to achieve agreement on a single data value among distributed systems. Consensus algorithms are designed to achieve reliability in a network involving multiple nodes. Consensus algorithms are capable of doing two things: ensuring that the next block in a blockchain is the one and only version of the truth, and keeping powerful adversaries from derailing the system and successfully forking the chain.

A blockchain can be defined as an anonymous online ledger that uses the data structure to simplify the way we transact. Blockchain allows the users to manipulate the ledger in a secure way and without the help of any third party. A blockchain is anonymous, thus protecting the identities of the users. This makes blockchain a more secure means to carry out financial transactions in the extremely and deeply connected world. The algorithm used in blockchain reduces the dependence on the people to verify the transactions. A blockchain is a kind of transparent, independent, and permanent database coexisting in multiple locations and shared by a common community.

Bitcoin is the first and foremost application of the blockchain technology. We would like to explain the blockchain nitty-gritty through the famous bitcoin application. As we all know, bitcoin is the well known and increasingly used application of the blockchain technology. We often hear and read about blockchain-compliant cryptocurrencies. Bitcoin is a digital currency that can be used to exchange products and services over, just like we use our paper currencies such as Indian Rupees (INR), United States Dollar (USD), Euro (EUR), etc.

1.1 Briefing of the blockchain system elements

We know that any blockchain system is simply a distributed and decentralised ledger. The ledger is a chain of blocks. Let us have a look at what a block comprises of. Every block in a blockchain comprises of three core elements.

1. *Data inside the block*: The data inside each block changes as per the technology used. For example, in a bitcoin blockchain, the blocks contain the sender's ID, the receiver's ID as well as the amount of bitcoins being transferred. The blocks on the bitcoin blockchain are 1 MB of data each. At this point of time, there are around *525,000* blocks. The total size of

Fig. 1 Blocks on the blockchain are 1 MB of data each.

the bitcoin blockchain is around *525,000 MB*. The primary data on the bitcoin blockchain is transaction data of all the bitcoin transactions as shown in Fig. 1.

Block 1 chronologically describes the first transactions that have occurred up to 1 MB. The subsequent transactions weighing *1 MB* would be stored in block 2 and so on. These blocks are now being chained together. To chain them, every block gets a unique (digital) signature that corresponds to exactly the string of data in that block. If there is any change inside a block, the block will get a new digital signature. This is done through the activity called hashing, which is explained below in detail.

Let us say block 1 registers two transactions: transaction 1 and transaction 2. Assume that these transactions make up a total of *1 MB*. This block of data now gets a signature for this specific string of data. Let us say the signature is "*X32*." This is pictorially displayed below.

A single digit change to the data in block 1 will produce a new signature. The data in block 1 is now linked to block 2 by adding the signature of block 1 to the data of block 2. With this arrangement, the signature of block 2 depends on the signature of block 1. The linkage between block 1 and 2 is vividly illustrated in Fig. 2.

The signatures link the blocks together and help in realizing a chain of blocks, which are a collection of multiple transactions. The blockchain is displayed in Fig. 3.

As indicated above, if there is a slight modification of data in block 1, then the block 1 gets a new digital signature and hence block 1 and 2 cannot be joined to each other as illustrated in Fig. 4.

This gives an indication to other users of this blockchain that some data gets changed in block 1. They reject this unauthorized change by shifting back to a previous record of the blockchain where all the blocks

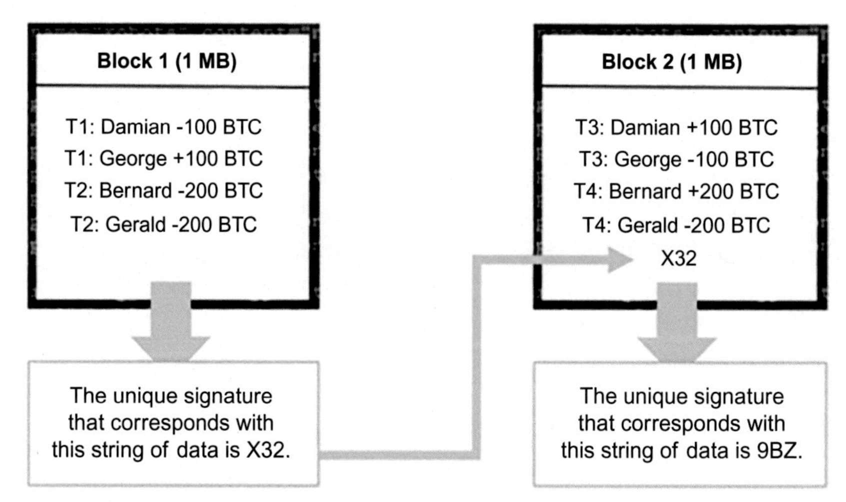

Fig. 2 Linkage between the blocks.

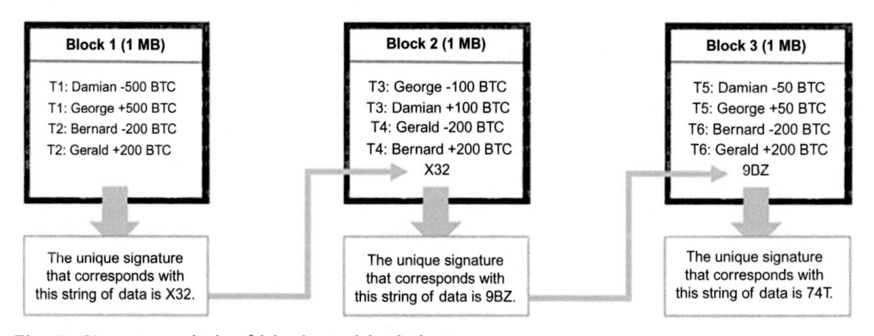

Fig. 3 Signatures link of blocks in blockchain.

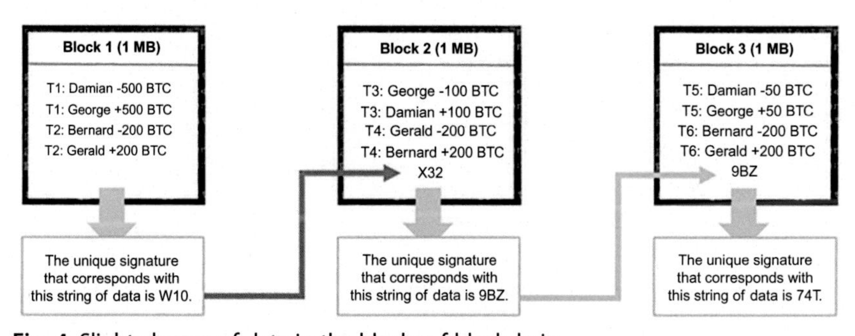

Fig. 4 Slight change of data in the blocks of blockchain.

are still chained together. The only way for an alteration to be accepted or undetected is if all the blocks stay chained together.

2. *The hash of a block:* As mentioned above, the hash of a block is its digital signature and the hash technique plays out a pivotal role in shaping up the concept of blockchain. Each block in a blockchain contains a unique hash and can be generally compared to a fingerprint. If a hash of a block is changed, then the block loses its original form. The hash of the previous block contributes for the formation of the chain between the blocks. It acts as a valid proof of concept for the further blocks to act upon and is the foremost security measure of a blockchain against tampering is as shown in Fig. 5.

   We now understand what forms a block in a blockchain and how blocks get limed up together to form a chain. This link is established and sustained through the hash concept. Now if in case, the hash of any block gets changed in a blockchain, the consecutive blocks no longer recognize it as a valid block. Thus, any kind of tampering makes the block invalid. This is a prominent security mechanism provided natively by blockchain.

3. *How the signature (hash) is created:* Block 1 is a record of only one transaction *A* sends *100* bitcoin to *B*. This specific string of data now requires a signature. In blockchain, this signature is created by a cryptographic hash function, which is a very complicated formula that takes any string of input and turns it into a unique *64-digit* string of output.

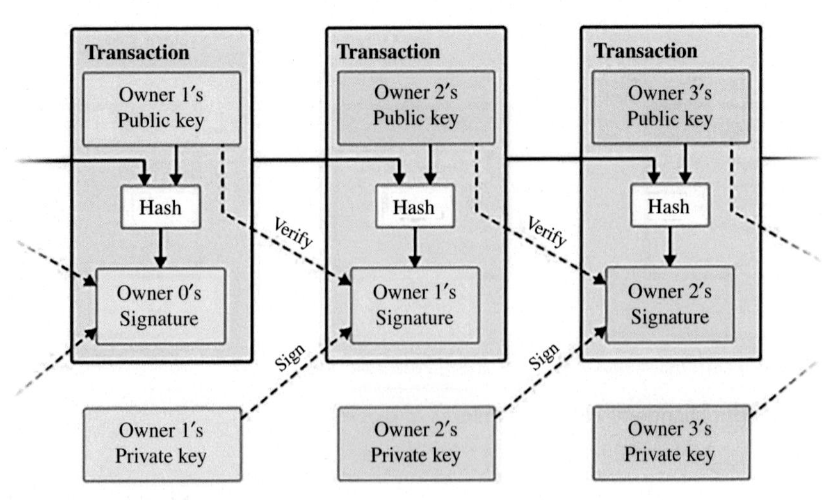

Fig. 5 Hash of a block.

For example, if we insert the word "*Jinglebells*" into this hash function, the output for this specific string of data is.

761A7DD9CAFE34C7CDE6C1270E17F773025A61E511A56F 700D415F0D3E199868

If a single digit of the input changes, including a space, changing a capital letter or adding a period, for example, the output will be totally different. A cryptographic hash function always gives the same output for the same input, but always a different output for different input. This cryptographic hash function is used by the bitcoin blockchain to give the blocks their signatures. The input of the cryptographic hash function in this case is the data in the block, and the output is the signature that relates to that. Let us have a look at block 1 again. *Thomas* sends *100* Bitcoin to *David* is as shown in Fig. 6.

Block 1 Thomas −100 David +100

If this string of data is inserted in the hashing algorithm, the output (signature) will be this:

BAB5924FC47BBA57F4615230DDBC5675A81AB29E2E0FF 85D0C0AD1C1ACA05BFF

This signature is now added to the data of block 2. Let us say that *David* now transfers *100* bitcoin to *Jimi*. The blockchain is displayed as shown in Fig. 7.

Block 2 David −100 Jimi +100

BAB5924FC47BBA57F4615230DDBC5675A81AB29E2E0FF85D 0C0AD1C1ACA05BFF

If this string of data is inserted in the hashing algorithm, the output (signature) will be this:

25D8BE2650D7BC095D3712B14136608E096F060E32CEC73 22D22E82EA526A3E5

And so, this is the signature of block 2.

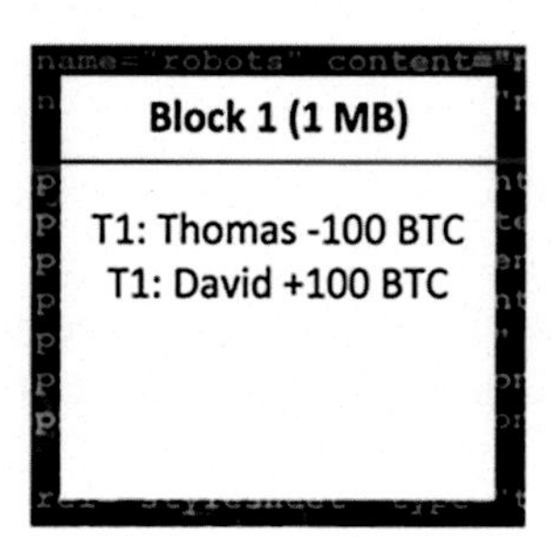

Fig. 6 Bitcoin transaction.

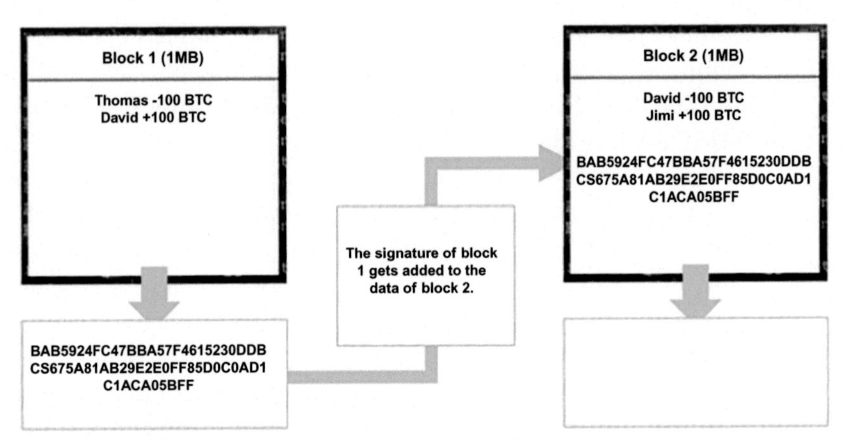

Fig. 7 Bitcoin transfers from David to Jimi.

There are several hash algorithms. For bitcoin application, *SHA-256* hashing algorithm is preferred to generate digital signature for each of the blocks in any blockchain.

The question here is how the signatures stop someone from simply inserting a new signature for each block after altering one. As mentioned above, any change goes unnoticed if all the participating blocks are properly linked. The answer is that only hashes (signatures) that meet certain requirements are accepted on the blockchain and is defined in the next section.

2. Structure of blockchain

Blockchain is a peer-to-peer technology that uses a series of digital signatures to provide security using lightweight cryptographic methods. Each node in the system acts as a client and a server in the blockchain. In nutshell, blockchain is a decentralized system that secures communication among various smart devices. The transactions are grouped into blocks in a sequential way and each block is linked to the previous one by hash function as illustrated in Fig. 8.

A signature does not always qualify. A block will only be accepted on the blockchain if its digital signature starts with a consecutive number of zeroes. For example; only blocks with a signature starting with at least 10 consecutive zeroes qualify to be added to the blockchain. If the signature (hash) of a block does not start with 10 zeroes, the string of data of a block needs to be changed repeatedly until a specific string of data is found that leads to a signature starting with ten zeroes. Because the transaction data and metadata

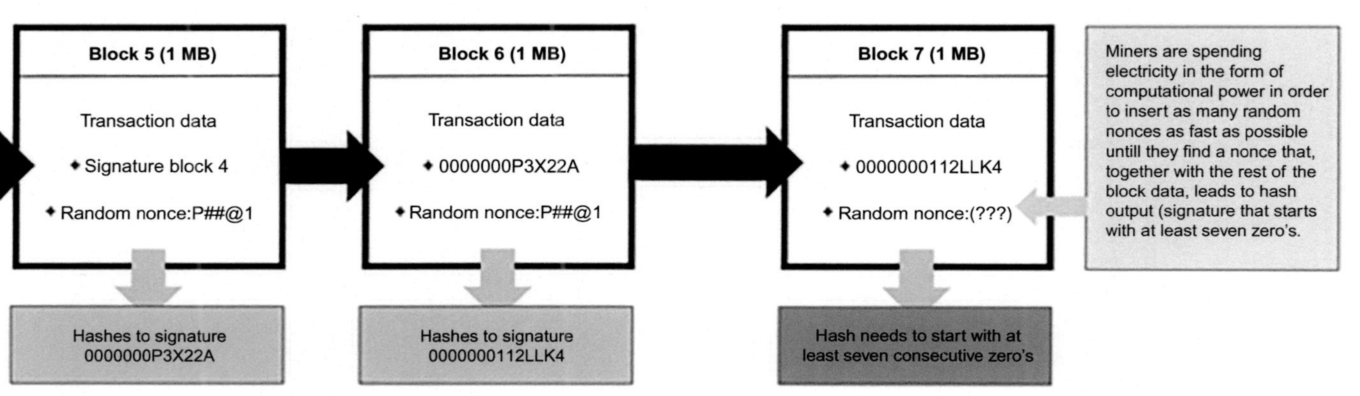

Fig. 8 Structure of blockchain.

(block number, timestamp) need to stay the way they are, a small specific piece of data is added to every block that has no purpose except for being changed repeatedly in order to find an eligible signature. This piece of data is called the nonce of a block. The nonce is completely random and could literally form any set of digits, ranging from spaces to question marks to numbers, periods, capital letters, and other digits.

To summarize, a block now contains (1) transaction data, (2) the signature of the previous block, and (3) a nonce. The process of repeatedly changing the nonce to find an eligible signature is called mining and is what miners do. Miners spend electricity in the form of computational power in order to constantly try different nonces. The more computational power they have, the faster they can insert random nonces and the more likely they are to find an eligible signature faster. Thus, the blockchain technology is a combination of multiple proven technologies in order to bring in utmost security for all kinds of digital data and assets in the extremely online world.

3. Benefits of blockchain technology

Having seen the benefits of a shared ledger, many banks are teaming up together to create and sustain private blockchains. Enterprise blockchain solutions are gaining higher acceptance and importance. The financial services industry is the first and foremost to embrace the blockchain paradigm. Other industries are showing exemplary interest in leveraging the blockchain technology. The various proof of concepts (PoCs) and pilots are being initiated across industry verticals in order to deeply and decisively understand its distinct competencies. On the other hand, there are some limitations and risks with it. Researchers and enthusiasts are working in unison to bring forth technological solutions to surmount those identified issues. Futuristically speaking, there are industry case studies in support of this disruptive paradigm. The widely exclaimed advantages is described as follows.

- Users have complete control of the value they own. There is no intermediary and third-party authenticator to hold their value. There is no one to limit their access to it.
- The cost to perform a value transaction from and anywhere on the planet earth is very low and hence this blockchain solution allows micropayments.
- The value can be transferred in a few minutes and the transactions can be considered secure in a few hours, not days or weeks.

- Since anyone at any point in time can verify every transaction made on the blockchain, the much-needed transparency is granted.
- The blockchain technology contributes immensely to build decentralized applications that simplify information management and value transfer in a secure and faster manner.
- The concept of smart contracts is fast emerging and with the maturity and stability of right technologies, a number of activities can be easily automated and accelerated programmatically.

The bitcoin cryptocurrency can reduce or eliminate the need for certain intermediaries and automate manual tasks. Distributed ledger stores the entire ownership history of an asset. The other use cases are described as follows.

1. Know your customer (KYC),
2. Antimoney laundering (AML),
3. Trade surveillance,
4. Collateral management,
5. Settlement and clearing,
6. The ability to capture historical and current ownership of high-value items.

Blockchain already has a significant impact on industries on how they react, adapt and change. It has the inherent potential to revolutionize how businesses operate.

- Blockchain is a true P2P network that will reduce reliance on some types of third-party intermediaries like banks, lawyers, and brokers.
- Blockchain can speed up process execution in multiparty scenarios and allow for faster transactions that are not limited by office hours.
- Information in blockchain is viewable by all participants and cannot be altered. This will reduce risk and fraud and create trust.
- Distributed ledgers will provide quick return on investment (ROI) by helping businesses create leaner, more efficient, and more profitable processes.
- The distributed and encrypted nature of blockchain means it will be difficult to hack. This shows promise for business and IoT security.
- Blockchain is programmable, which will make it possible to automatically trigger actions, events, and payments once conditions are met.

The blockchain technology is still evolving and there are many product and platform vendors contributing immensely for making it penetrative, persuasive, and pervasive. Business houses are experimenting with this new entrant and exploring the possibility of leveraging this unique idea to optimize business processes to realize bigger and better applications. With cloud centres are being recognized as the highly optimized and organized IT environments to deploy and deliver blockchain platforms and services, this nascent technology is finding a solid, strong and stimulating base to grow faster and glow in the days to unfurl.

4. Working of blockchain

Blockchain technology is the world's leading technology for digital advantage. It is basically a list of records called blocks that are growing continuously which are linked and secured by the cryptography system. When two or more nodes communicate with each other in a network system than number of security issues and attacks can come during the communication. The main attacks that can come during communication are replay attack, DoS, data modification, misrepresentation of data, eavesdropping, etc. So, for providing security and privacy to the public networks or channel communication from these type of attacks between the nodes, the blockchain technology is the best solution. It provides security, AAA, integrity, and confidentiality to the network without the central authority. The detail description of the working of blockchain is described in the following subsections.

4.1 Bitcoin cryptocurrency

The flow of transactions in bitcoin cryptocurrency is shown in Fig. 9 and is discussed below. Let us say that there are two parties *Alice and Bob* who want to communicate with each other for funds transfer over an insecure channel, Internet. Then following sequence of activities are performed between two parties.

- If *Alice* wants to send some coins from her wallet*X* to *Bob's* wallet *Y*, then a request of transactional data "*t*" is sent to *Bob*. This request is broadcasted in the entire network.
- The distributed nodes accept the request and update their ledgers with the transactional information of *Alice–Bob*.
- After updating ledgers, Alice computes digital signature (*DS*) and broadcasts it in the network.
- A miner node is selected to verify and validate the transaction. It computes proof-of-work (PoW) to match the *DS* received. If PoW is successfully matched with *DS*, then the result is broadcast to all the nodes for verification and validation.

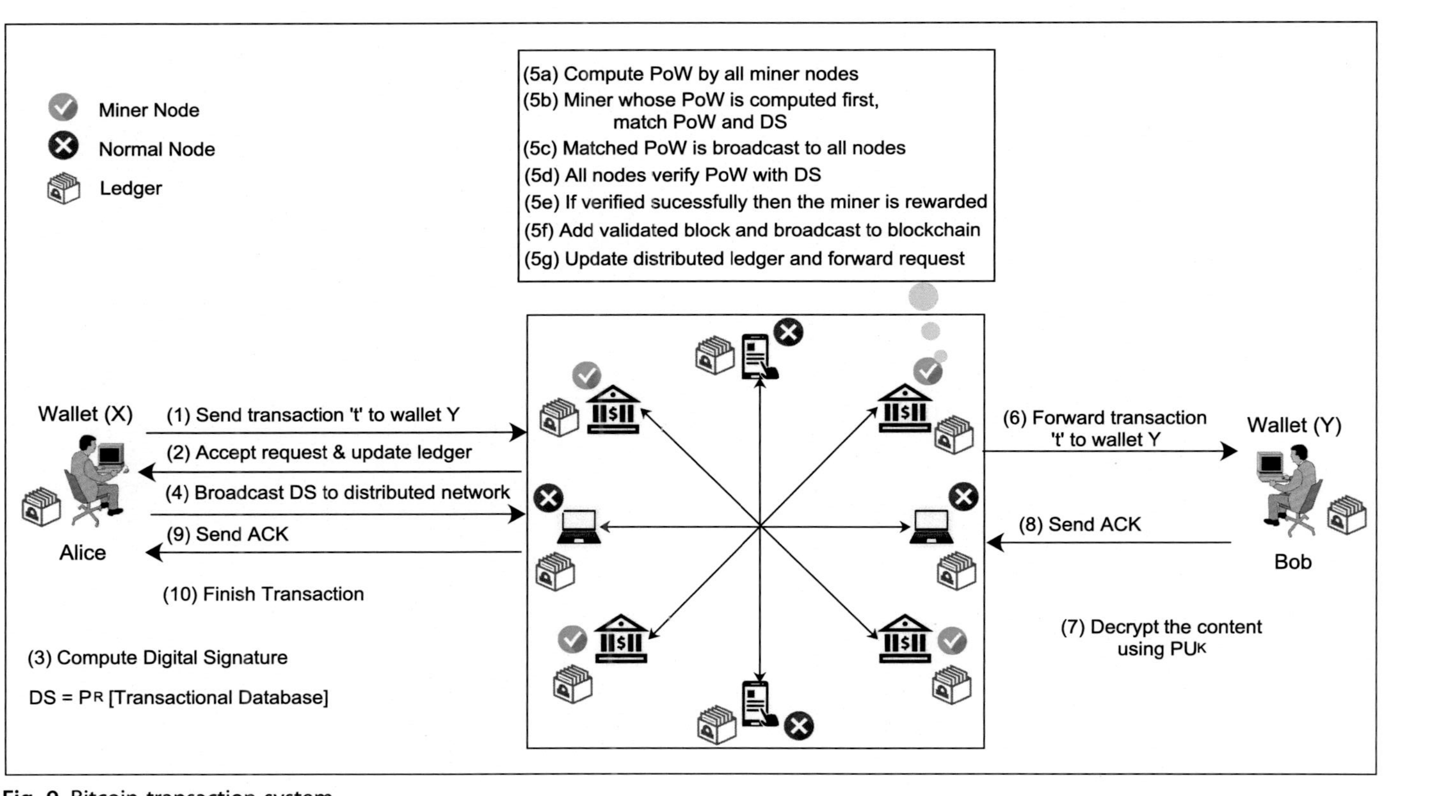

Fig. 9 Bitcoin transaction system.

- The other miner nodes also verify the PoW with *DS*. If the verification is successful, then the miner node is (financially) rewarded for computing the PoW.
- The validated block is added in the validated chain and the transaction is broadcasted to the entire blockchain.
- Using the validated transaction "*t*," the bitcoins are added to wallet*Y* of *Bob*.
- *Bob* decrypts the content using the paired public key (*PUK*) of Alice and sends the acknowledgment (*ACK*) to *Alice*.
- The transaction is finished once Alice receives the transaction acknowledgment.

4.2 Block creation and validation

Blockchain is a P2P network in which all the nodes are anonymous to each other and each node has its own ledger to store the history of the transactions. During the communication process in the blockchain technology, a consensus agreement is established for the transmission of information between the nodes. This agreement is appended with the transaction into the ledger, known as the block. In this technique, a cryptographic public key is used by the nodes to transfer the requests during the communication. Once the transactions are authenticated by the miner nodes, these are broadcasted into the network [1]. Then, these transactions are appended into a block that consists of a *block header, policy header, and content*. The block header contains the hash of the previous block which is immutable. The policy header is used to authenticate and authorize the transactions. It also contains the information of the requester and requisite, who are responsible for giving the response, identity information and the action that is to be performed. Content is used to store the transactions which are responsible for communication between the nodes as shown in Fig. 10.

It is to be noted that the transactional data is added to the content of the block only if it is verified as secure and confidential by the miner nodes. Once the transactions are added into the block, there is no mechanism to delete, update or modify the data from the block, i.e., it is immutable. Once a block is full, it is pushed into the blockchain and a mining process is performed on it. Each node in the system acts as a client and a server in the blockchain network as shown in Fig. 11.

In summary, a blockchain is a decentralized system that secures the communication among various smart devices. The transactions are grouped into blocks in a sequential way and each block is linked to the previous one by hash function as shown in Fig. 12.

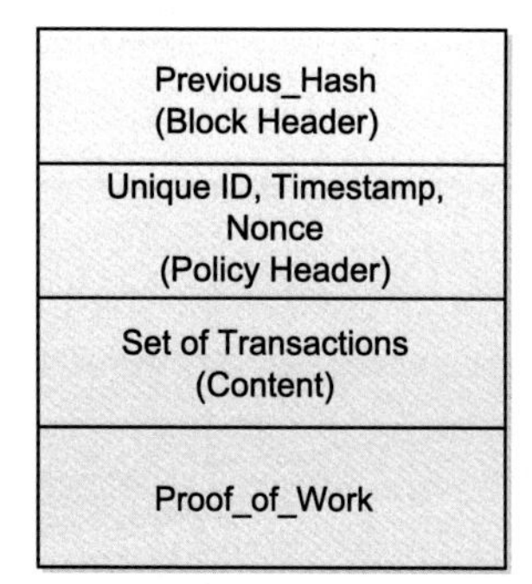

Fig. 10 Structure of block.

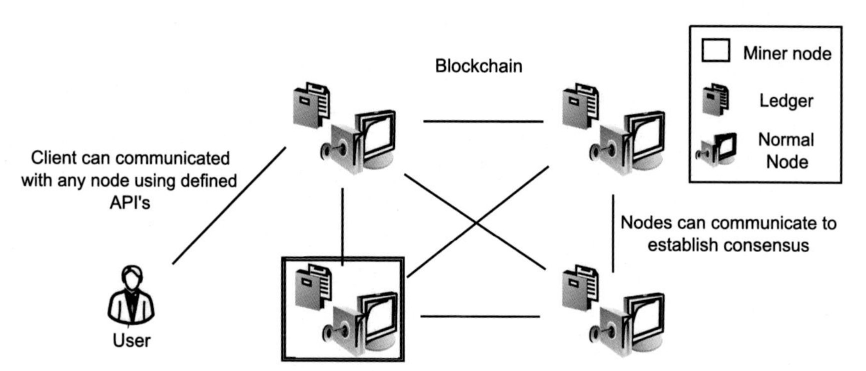

Fig. 11 Components of blockchain.

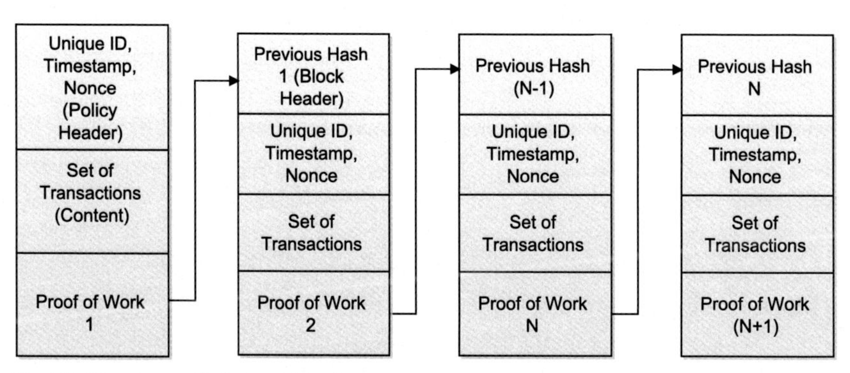

Fig. 12 Structure of chained blocks.

4.2.1 A. Transaction verification using elliptic curve cryptography

There are two verifications done in the blockchain network, one is of block and other is the transactions with in the block. For both the verification, the existing cryptographic primitives are used. For example, elliptic curves of the elliptic curve cryptography (ECC) are used in public-key encryption for digital signature generation which can be used for block verifications.. The use of elliptic curves in cryptography was given independently by Koblitz and Miller in 1985. An elliptic curve is a plane curve over a finite fieldL which consists of the points satisfying Eq. (1)

$$y^2 = x^3 + ax + b \tag{1}$$

along with a distinguished point at infinity, denoted as infinity. This set together with the group operation of elliptic curves is an abelian group, with the point at infinity as an identity element. The structure of the group is inherited from the divisor group of the underlying algebraic variety. A Baireto Naehring (BN) curve defines an elliptic curve which can be used for pairings for high security and efficiency in various online transactions.

Let us illustrate it with an example. Here, we use pairings over a *256*-bit BN-curve and derives a signature for a message where *Alice and Bob* can generate a shared key. Firstly, *Alice*takes the transactional data and hash it to the point on the elliptic curve (P). Then, take the private key (PK) by choosing a random 256-bit value and generates the signature by multiply$P$ with the PK to give ($PK * P$). Then, she generates his public key (PU) by multiplying the generator (G) with her PK to give ($PK * G$). She also takes the signature and multiplies it with G to give $value1 = (G * P * PK)$.

After that, *Bob* takes the message and hashes it to the point on the elliptic curve (P'). Then, he multiplies *Alice's PU* with the P' to give $value2 = (P' * PU)$. If the signature is correct then the two generated values, *value1* and *value2* are matched. Like *Bob*, other normal and miner nodes also create the value2' and match with the value1 to verify the confirmation of the transaction (Fig. 13).

4.2.2 B. Block verification using proof-of-work

In block verification process, Alice computes the digital signatures using private key (P_K), which contains transactional database (TD) having the source IP address, destination IP address, unique identity, action performed, and the content having a set of transactions. Digital signatures is created by cryptographic hash algorithms such as SHA-512, SHA-256. The DS is computed by *Alice* using Eqs. (2) and 3 as follows.

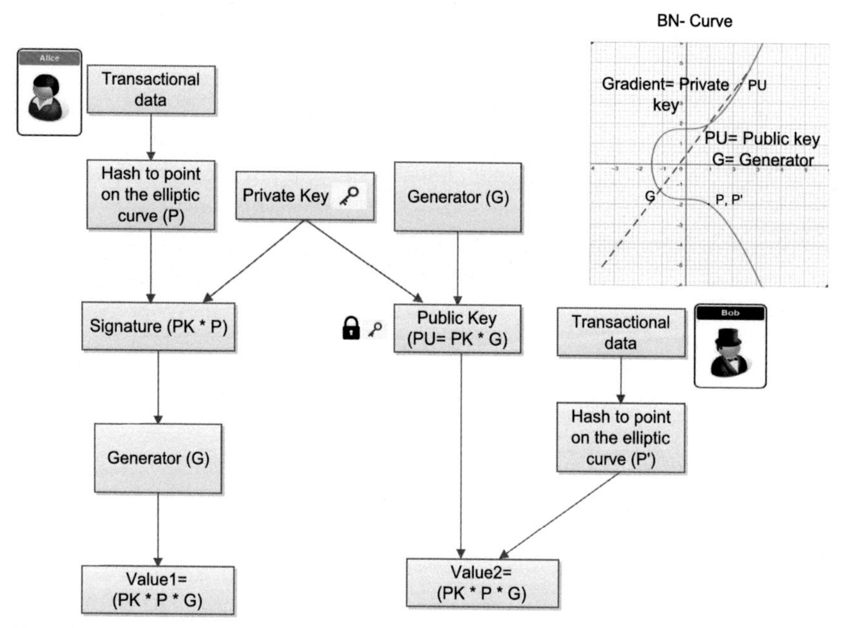

Fig. 13 Transaction verification using ECC.

$$TD = SHA - 512(H||C) \tag{2}$$

$$DS = P_K(TD) \tag{3}$$

where H represents header, DS represents the digital signatures, and C denotes content.

Now, the DS is sent to all the distributed nodes for authentication. Miner node computes the PoW for the authentication of DS using SHA-512. Here, *Merkle hash tree* is used to store the previous hash values. The PoW is calculated by miner node using Eqs. (4)–(6) as follows.

$$P_1 = SHA - 512(C||MR(H)||previoushash) \tag{4}$$

$$P_2 = SHA - 512(P_1||t) \tag{5}$$

$$PoW = SHA - 512(P_2||N) \tag{6}$$

where, MR denotes Merkle root, t is a timestamp, and N is nonce (a random variable generated by miner nodes for finding PoW).

Each miner node uses different or same nonce value at the same time to compute the PoW. If the computed PoW matches with the DS, then the miner node gets rewarded otherwise, the computed value of PoW is stored in the ledger of the distributed nodes. This process is repeated to create the

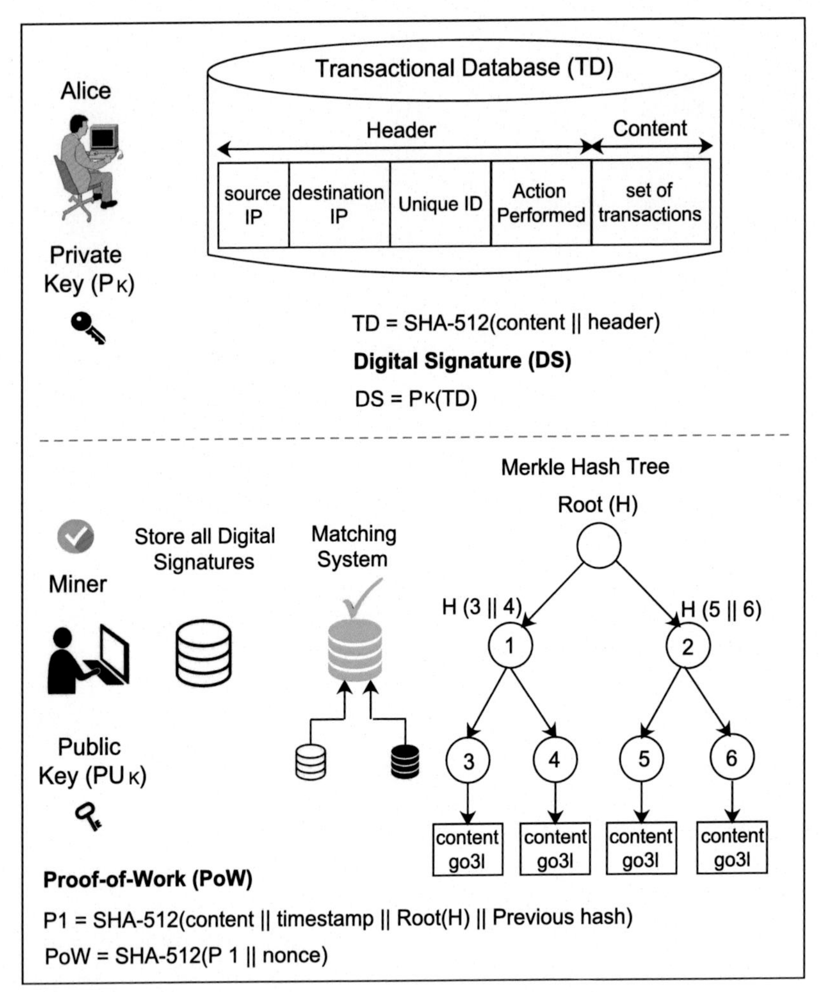

Fig. 14 Block verification process.

validated blocks, which are then chained together to form a blockchain. The entire process of block verification is as shown in Fig. 14.

4.2.3 C. Addition of blocks in the blockchain

The transactions of the block in the blockchain are verified by the miner nodes. Suppose, *Alice* has made a transaction of $100 to *Bob*. This transaction has passed all around the network called distributed nodes. The set of transactions in the network is called a transaction pool. All the transactions placed

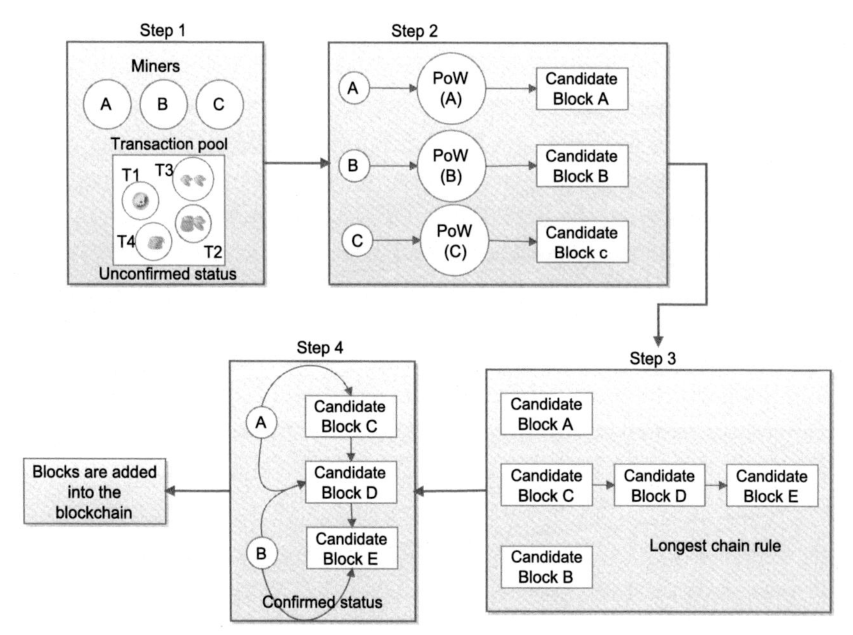

Fig. 15 Mining process.

in the transaction pool get unconfirmed status. So, the miner nodes are used for validating these transactions. They recorded all the detailed information in the global ledger of the distributed nodes in the blockchain. This process is known as "mining". The pictorial representation of this scenario is as shown in Fig. 15.

Following steps are used to explain the concept of mining.

1. There are three miners A, B, C who are trying to validate or confirm the transaction. From the transaction pool, miners select the transactions that have high transaction fees for them.
2. They create a candidate block by adding an unconfirmed transaction from the transaction pool. Then, they tried to solve the mathematical problem called PoW to add a block in the blockchain. We assume that all the three miners are able to solve the problem and have PoW. Then, we have three different candidate blocks with respect to each miner.
3. From the three candidate blocks, only one block is added to the blockchain-based on the longest chain rule. In this rule, the other miners try to create other valid blocks and keep on adding on the top of the earlier blocks. So, the other miners found that miner C has good and

enough power than other miners A and B to create a new block. So, they add the new block on the top of the candidate block of miner C. In this way, the longest chain is created.

4. After this, the miners A and B create another PoW for a new set of the transaction from the transaction base. Each confirmation is added into each block. Now, the unconfirmed transactions in the transaction pool become confirmed transactions and added it into the blocks of the blockchain.

The average confirmation time of each transaction is 10 min and 3–6 confirmations are required for a transaction to appear in the block of the blockchain. In the end, miners get transaction and block fees paid in the new bitcoin created an incentive to improve the transaction.

Reference

[1] F. Tschorsch, B. Scheuermann, Bitcoin and beyond: a technical survey on decentralized digital currencies, IEEE Commun. Surv. Tutorials 18 (3) (2016) 2084–2123.

About the authors

Shubhani Aggarwal is pursuing PhD from Thapar Institute of Engineering & Technology (Deemed to be University), Patiala, Punjab, India. She received the BTech degree in Computer Science and Engineering from Punjabi University, Patiala, Punjab, India, in 2015, and the ME degree in Computer Science from Panjab University, Chandigarh, India, in 2017. She has many research interests in the area of Blockchain, cryptography, Internet of Drones, and information security.

Neeraj Kumar received his PhD in CSE from Shri Mata Vaishno Devi University, Katra (Jammu and Kashmir), India in 2009, and was a postdoctoral research fellow in Coventry University, Coventry, UK. He is working as a Professor in the Department of Computer Science and Engineering, Thapar Institute of Engineering & Technology (Deemed to be University), Patiala (Punjab), India. He has published more than 400 technical research papers in top-cited journals such as IEEE TKDE, IEEE TIE, IEEE TDSC, IEEE TITS, IEEE TCE, IEEE TII, IEEE TVT, IEEE ITS, IEEE SG, IEEE Netw., IEEE Comm., IEEE WC, IEEE IoTJ, IEEE SJ, Computer Networks, Information sciences, FGCS, JNCA, JPDC, and ComCom. He has guided many research scholars leading to PhD and ME/MTech. His research is supported by funding from UGC, DST, CSIR, and TCS. His research areas are Network Management, IoT, Big Data Analytics, Deep Learning, and Cybersecurity. He is serving as editor of the following journals of repute: ACM Computing Survey, ACM·IEEE Transactions on Sustainable Computing, IEEE·IEEE Systems Journal, IEEE·IEEE Network Magazine, IEE·IEEE Communication Magazine, IEE·Journal of Networks and Computer Applications, Elsevier Computer Communication, Elsevier International Journal of Communication Systems, Wiley. Also, he has been a guest editor of various international journals of repute such as IEEE Access, IEEE ITS, Elsevier CEE, IEEE Communication Magazine, IEEE Network Magazine, Computer Networks, Elsevier, Future Generation Computer Systems, Elsevier, Journal of Medical Systems, Springer, Computer and Electrical Engineering, Elsevier, Mobile Information Systems, International Journal of Ad Hoc and Ubiquitous Computing, Telecommunication Systems, Springer, and Journal of Supercomputing, Springer. He has also edited/authored 10 books with international/national publishers like IET, Springer, Elsevier, CRC: Security and Privacy of Electronic Healthcare Records: Concepts, Paradigms and Solutions (ISBN-13: 978-1-78561-898-7), Machine Learning for Cognitive IoT, CRC Press, Blockchain, Big Data and IoT, Blockchain Technologies Across Industrial Vertical, Elsevier, Multimedia Big Data Computing for IoT Applications: Concepts, Paradigms and Solutions (ISBN: 978-981-13-8759-3), Proceedings of First International Conference on Computing,

Communications, and Cyber-Security (IC4S 2019) (ISBN 978-981-15-3369-3). One of the edited text-book entitled, "Multimedia Big Data Computing for IoT Applications: Concepts, Paradigms, and Solutions" published in Springer in 2019 is having 3.5 million downloads till June 6, 2020. It attracts attention of the researchers across the globe (https://www.springer.com/in/book/9789811387586). He has been a workshop chair at IEEE Globecom 2018 and IEEE ICC 2019 and TPC Chair and member for various international conferences such as IEEE MASS 2020 and IEEE MSN 2020. He is a senior member of the IEEE. He has more than 12,321 citations to his credit with current h-index of 60 (September 2020). He has won the best papers award from IEEE Systems Journal and ICC 2018, Kansas City in 2018. He has been listed in the highly cited researcher of 2019 list of Web of Science (WoS). In India, he is listed in top 10 position among highly cited researchers list. He is an adjunct professor at Asia University, Taiwan, King Abdul Aziz University, Jeddah, Saudi Arabia, and Charles Darwin University, Australia.

Core components of blockchain☆

Shubhani Aggarwal and Neeraj Kumar
Thapar Institute of Engineering & Technology, Patiala, Punjab, India

Contents

Abstract

In the blockchain network, there are two types of nodes such as miner nodes and normal nodes. The miner nodes are used for authentication, auditing, and validating the transactions and normal nodes are used to verify those transactions. Similarly, based upon several criteria, blockchain systems are classified as public and private blockchains. A public blockchain provides an open platform for people from various organizations and backgrounds to join, transact, and mine whereas private blockchain is used to facilitate private sharing and exchange of data among a group of individuals. Along with this, we have discussed the various blockchain platforms that can be used to execute and run smart contracts on the blockchain network.

☆ Introduction to blockchain.

Advances in Computers, Volume 121
ISSN 0065-2458
https://doi.org/10.1016/bs.adcom.2020.08.010

Chapter points

- In this chapter, we discuss the types of blockchain and classification of nodes used in the blockchain network.
- Here, we also discuss the blockchain platforms in detail used to execute the smart contracts.

1. Classification of nodes in a blockchain network

In a blockchain distributed network, there are two type of nodes, miner nodes and normal nodes which are described as follows.

- *Miner nodes*: A miner node is a central entity between devices and the network. It is used for authenticating, authorizing, and auditing all the transactions in the network. The miner node has its own microstorage where devices can store their data temporarily. It communicates with the controller node through remote switching units (RSUs), which are used to pass all the transactions in the network. The transactional data passed through the controller node is stored in its local storage and is authenticated by the miner node. Each miner node consists of various blocks and a ledger. A ledger is used to store the transaction information at any given instant of time. After the authentication of a transaction, the transactional data is appended into the blocks.
- *Normal nodes*: Normal nodes contain the full copy of the complete blockchain. They are used for verification, coordination, and validation of the transactions, which is authenticated by the miner nodes. The distributed blockchain network of nodes is as shown in Fig 1.

2. Classification of blockchain system

Blockchain can facilitate authentication and authorization without using any trusted authority. Due to its unique advantages and applications, the blockchain paradigm is on a fast track. The adoption and adaptation are on the higher side. Enterprises are rolling out newer blockchain use cases in order to make it more visible and viable. Resultantly there are a few blockchain types. There are certain use cases mandating for a couple of different types: public/permissionless and private/permissioned.

- *Permissionless blockchain*: With a public blockchain, anyone can join and leverage blockchain functionalities whereas private blockchain can be

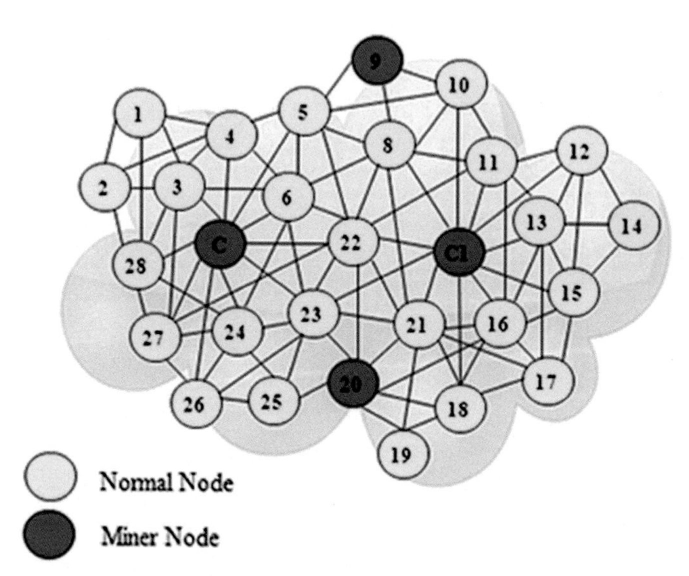

Fig. 1 Distributed network of blockchain.

accessed only by authorized users. Within a public blockchain, the much-required trust is being realized through game-theory and cryptography concepts. In the case of a public blockchain, anyone can simply download the appropriate application and join the decentralized blockchain to involve in transactions. It is not mandatory to have a previous relationship with the ledger, and there is no need of approval to join public blockchain.

- *Permissioned blockchain*: On the other hand, private/permissioned blockchain do not require such artificial incentives since all actors in the network are known to each other. New actors that want to join the network have to be approved by existing actors in the network. This enables more flexibility and efficiency of validating transactions. Private blockchains are typically used by a consortium of organizations that like to keep a shared ledger for settlement of transactions. Predominantly, financial services providers such as banks form a kind of a consortium to securely do transactions. Only, the members of the organizations of the consortium can see transactions.

The scenario of different types of blockchain is as shown in Fig. 2.

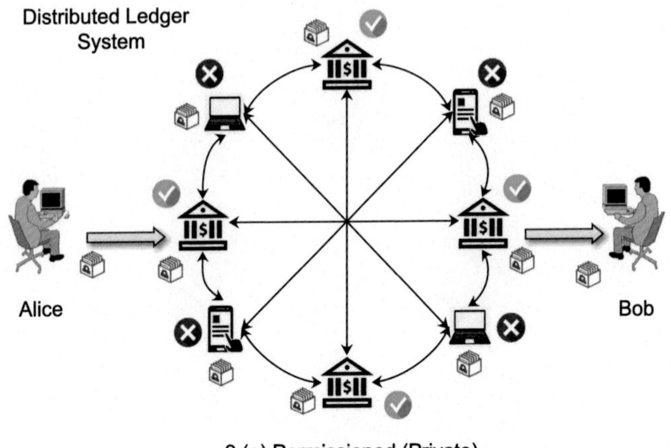

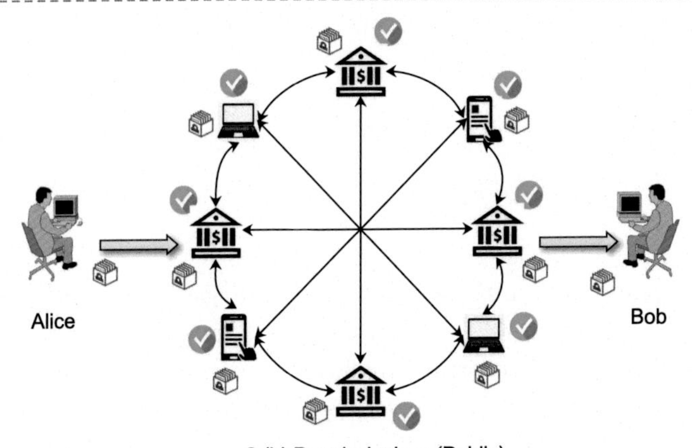

Fig. 2 Types of blockchain.

3. Blockchain platforms

Here, it is important to mention the various blockchain platforms that can be used by the enterprises. The detail description of the blockchain platforms is described as under and their difference is described in Tables 1 and 2.

3.1 Ethereum

It is an open-source and distributed blockchain-based platform suggested by Vitalk Buterin, a Russian–Canadian programmer. This platform is used to run smart contracts based on the customer's specifications. Ethereum virtual

Table 1 Blockchain platforms.

Platforms	Focus of industry	Type of ledger	Consensus algorithm	Use of smart contract	Governance
Ethereum	Cross-industry	Public	Proof-of-work	Yes	Ethereum developers
Hyperledger Fabric	Cross-industry	Private	Pluggable Framework	Yes	Linux Foundation
Hyperledger Sawtooth	Cross-industry	Private	Pluggable Framework	Yes	Linux Foundation
Hedera Hashgraph	Cross-industry	Private	Asynchronous Byzantine Fault Tolerance	Yes	Hedera Hashgraph Council
Ripple	Financial Services	Private	Probabilistic voting	No	Ripple Labs
Quorum	Cross-industry	Private	Majority voting	No	Ethereum developers and JP Morgan chase
Hyperledger Iroha	Cross-industry	Private	Chain-based Byzantine fault tolerance	Yes	Linux Foundation
Corda	Cross-industry	Private	Asynchronous Byzantine Fault Tolerance	Yes	Hedera Hashgraph Council
EOS	Cross-industry	Private	Delegated proof-of-stake	Yes	EOSIO Core Arbitration Forum (ECAF)
OpenChain	Digital Asset Management	Private	Partitioned consensus	Yes	Linux Foundation
Stellar	Financial Services	Both public and private	Stellar consensus protocol	Yes	Stellar Development Foundation
Dragonchain	Cross-industry	Public, private, and hybrid	Context-based verification with five levels of consensus	Yes	Dragonchain Foundation
NEO	Smart Economy	Private	Delegated Byzantine Fault Tolerance	Yes	NEO holders and NEO Foundation Support
MultiChain	Digital Asset Management	Private	Probabilistic voting	Yes	MultiChain Developers
IOTA	Digital Asset Management	Public	Proof-of-work	Yes	IOTA Foundation

Table 2 Blockchain platforms.

Platforms	Programming language	Cryptocurrency	Transaction cost for mining	Applications
Ethereum	Smart contract code written in Solidity	In-built (ether)	Yes	Banking, commodity trade finance, supply chain management, insurance, energy grid, oil and gas, real estate
Hyperledger Fabric	Chaincode written in GoLang, Java	No in-built (can be modeled in chaincode)	No	Supply chain for pharmaceuticals, trade financing, smart energy, supply chain management
Hyperledger Sawtooth	Python, C++, Go, Java, JavaScript, and Rust	Supports Ethereum via "seth"	No	Smart energy, supply chain management
Hedera Hashgraph	Smart contracts written in the Solidity	HBAR (in-built)	Yes (low)	Finance, real estate, gaming, and media and entertainment
Ripple	C++	XRP (in-built)	Yes (very-low)	Intra-bank applications, financial services
Quorum	Smart contract written in solidity	in-built (JPM Coin)	No	Banking, financial, insurance services
Hyperledger Iroha	C++	No in-built	No	Interbank settlement, central bank digital currencies, payment systems, national IDs, and logistics, among others

Corda	Smart Contract code written in Kotlin, Java	No in-built	No	Banking, financial services
EOS	C++	EOS tokens	No	Data hosting, usage management, and communication between the dApps
OpenChain	Javascript	No in-built	No	Digital asset management
Stellar	C++, Java, Python, GoLang, Javascript	In-built (lumen)	Yes (low)	Telecommunication, banking, financial services
Dragonchain	C++, Go, JavaScript, Java, Python, Ruby, Shell, C#, Hy, Rails, PHP	ERC-20 token runs on Ethereum	Yes	Currency agnosticism, and easy integration with business applications
NEO	C#	NEO & GAS tokens	Yes	Supply chain's deployment, hosting, and maintenance
MultiChain	Python, C#, JavaScript, PHP, Ruby	No in-built	No	Financial transactions, e-commerce
IOTA	JavaScript	mIOTA (in-built)	No	Internet-of-everything

Fig. 3 Ethereum.

machine (EVM) provides a run-time environment to smart contracts in ethereum. It is used to understand for the public (permissionless) blockchain platform and based on the PoW consensus mechanism which is slow in speed. A developer who builds an application using ethereum should pay charges in *Ether*, to execute transactions and to run an application on the network. This platform is launched by ethereum developers and contributed by ethereum enterprise alliance (EEA) [1]. The icon used to represent the ethereum is as shown in Fig. 3.

3.2 Hyperledger Fabric

Hyperledger Fabric is one of the projects of hyperledger that is designed for building blockchain-based applications and solutions. It basically uses as a modular architecture that allows network designers to alliance their important components such as membership services, agreement, and differentiate it from other blockchain solutions. It is used to understand private blockchain networks in which only known identities can participate in a system. The members of this platform should be authorized to take part in a blockchain. This project is launched by the Linux Foundation and contributed by digital assets and IBM [2]. The icon used to represent the hyperledger Fabric is as shown in Fig. 4.

3.3 Hyperledger Sawtooth

Hyperledger Sawtooth is an enterprise-grade used for designing to create, deploy, and execute the blockchain-based distributed ledgers that maintain

Fig. 4 Hyperledger Fabric.

Fig. 5 Hyperledger Sawtooth.

the digital records. It is used to understand private blockchain networks and based on the proof-of-elapsed time (PoET) consensus mechanism to integrate hardware security solutions. It is hyperledger's second open-source platform that gives a sense of confidence to blockchain companies. This project is launched by the Linux Foundation and contributed by digital assets and IBM [3]. The icon used to represent the Hyperledger Sawtooth is as shown in Fig. 5.

3.4 Hedera Hashgraph

Hedera Hashgraph is a new type of light, secure, fast, and fair platform that does not need the high computing power of the PoW consensus algorithm. The Hedera Hashgraph council is the governing body and the terms made in this platform ensures no single member or no small groups can have control and influence over the entire project. It is used to understand asynchronous byzantine fault tolerance (BFT) and has a strong level of security. It has the capability to handle thousands of transactions per second and authenticate one million signatures per second. It provides binding arbitration in which smart contracts use a list of public-key arbitrators which can be edited to correct the errors or to add new features [4]. The icon used to represent the Hedera Hashgraph is as shown in Fig. 6.

3.5 Ripple

Ripple was discovered in 2012 and is aimed at connecting financial services like digital assets exchanges, banks through the blockchain network called RippleNet. It also allows global payments like Ether or Bitcoin via a digital asset called ripple. It uses probabilistic voting to reach an agreement between the nodes. It is more scalable and faster than any other blockchain platforms. To test the potential of the ripple platform, some big brands like SBI

Fig. 6 Hedera Hashgraph.

Fig. 7 Ripple.

holdings, Deloitte, American Express, which integrates with this platform to make their payment system more fast, accurate and secure [5]. The icon used to represent the ripple is as shown in Fig. 7.

3.6 Quorum

The quorum platform was founded by J.P. Morgan. As similar to ethereum, it is an open-source and free to use in endurance. It is used for private blockchain networks and would not be open for everyone. It uses a vote-based consensus mechanism that enables to handle hundreds of transactions per second. It can handle the application requiring high throughput and high speed of the transaction. It also resolves the issue of confidentiality of records that other blockchain platforms are failed to solve this issue [6]. The icon used to represent the quorum is as shown in Fig. 8.

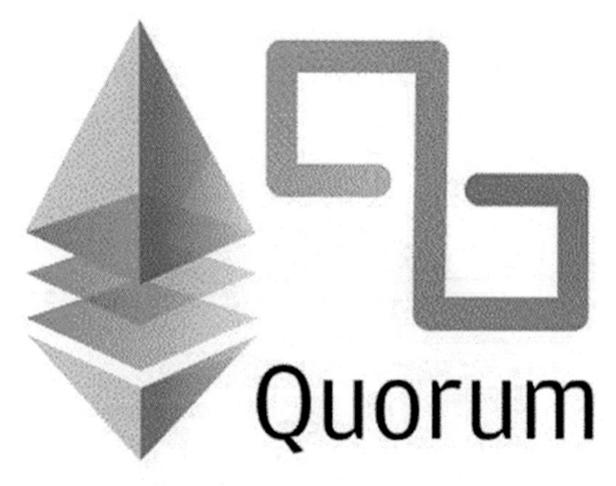

Fig. 8 Quorum.

Fig. 9 Hyperledger Iroha.

3.7 Hyperledger Iroha

Hyperledger Iroha is hosted by Linux Foundation and is used to build secure, fast, and trusted decentralized applications. This project is developed by the simultaneously working of both companies such as National Bank of Cambodia and Soramistu Cooperation Limited. It creates a healthy balance using transparency in the development process and attaining milestones. It is portable and supportive for mac operating system and Linux environment, which is used for supply chain and IoT use cases [7]. The icon used to represent the Hyperledger Iroha is as shown in Fig. 9.

3.8 R3.Corda

R3.Corda is an open-source platform developed in 2015. It is a cutting edge blockchain platform that enables financial institutions to transact with a smart contract directly. It does not have any cryptocurrency and is used for private blockchain networks that allows only authorized participants to access data. It increases privacy and provides access-control to digital records. It is used in various use cases like healthcare, supply chain, financial sectors, and government authorities [8]. The icon used to represent the R3. Corda is as shown in Fig. 10.

Fig. 10 R3.Corda.

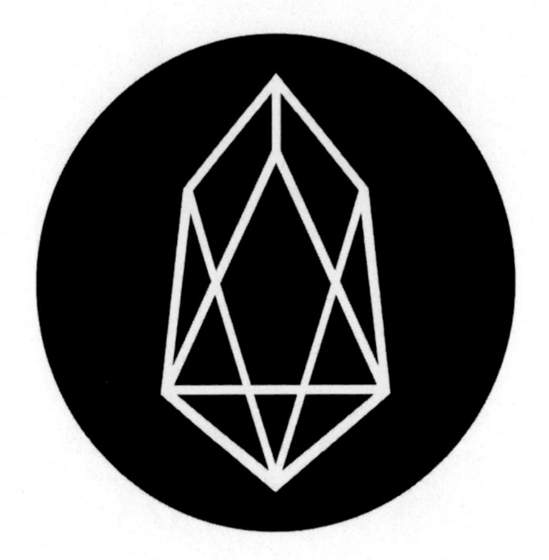

Fig. 11 EOS.

3.9 EOS

EOS was founded by a private "Block.One" company in June 2018 and is used as an open-source platform. It was launched for designing and development of decentralized applications. The goal of this platform is to provide decentralized storage, applications, and run smart contracts. It is free for all users who want to take advantage of decentralized applications. It is a part of private blockchain based on multithreading as well as proof-of-stake (PoS) consensus mechanism. The EOS has its own association called the EOS forum to discuss and to talk with the developers, investors, and their users based on esteem blockchain [9]. The icon used to represent the EOS is as shown in Fig. 11.

3.10 OpenChain

OpenChain was developed by Coinprism and is used as an open-source platform. It is used for those organizations who manages the digital assets and needs scalability and security. It is based on partitioned consensus where one instance will have one authority to validate the transaction. It is free of cost and validated by a service administrator that provides efficiency to the network [10]. The icon used to represent the OpenChain is as shown in Fig. 12.

Fig. 12 OpenChain.

Fig. 13 Stellar.

3.11 Stellar

Stellar is a distributed ledger technology that deals with exchanges of cryptocurrencies like Ripple. It is used to build banking tools, smart devices, and mobile wallets. It is based on stellar consensus protocol used to record financial transactions. The stellar protocol has better capabilities than PoW and PoS in terms of entry for a new participant in financial services. Some private banks and companies like ICICI bank, Ripple Fox are integrating with this platform to secure the transactions at the network [11]. The icon used to represent the Stellar is as shown in Fig. 13.

3.12 Dragonchain

Dragonchain used as a service-oriented blockchain platform that provides important resources to developers and enterprises at the time of application development. It was developed at Walt Disney company in 2014 and used as an open-source in 2016. It is used for both the blockchain networks such as public and private. It can support any programming language that offers flexibility in business-services and utilizing interchain capabilities with other blockchains. With five levels of agreement, it provides a large spectrum of trust and security [12]. The icon used to represent the dragonchain is as shown in Fig. 14.

3.13 NEO

NEO platform was founded by Da Hongfei (CEO) and Erik Zhang of blockchain Research and Development (R&D) company "OnChain" in Shangai. It was developed to design decentralized applications and generate GAS tokens using NEO tokens during the payment of transaction fees. It uses the delegated BFT consensus mechanism that provides better performance and scalability in comparison to other protocols. It maintains the X.509 standard compatible with digital identity to support point-to-point certificate issuance models with trust and privacy[13]. The icon used to represent the NEO is as shown in Fig. 15.

Fig. 14 Dragonchain.

Fig. 15 NEO.

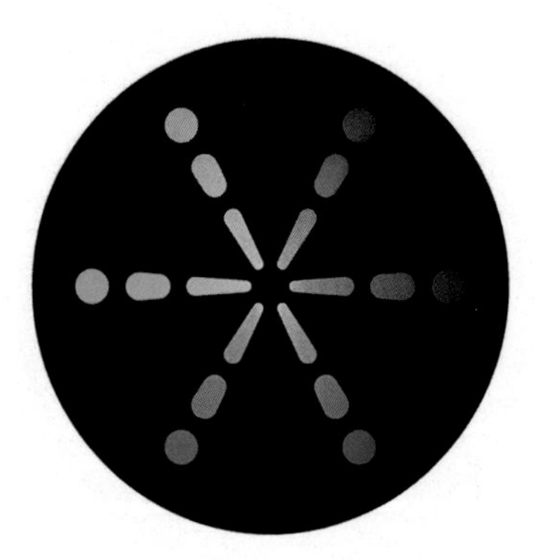

Fig. 16 MultiChain.

3.14 MultiChain

MultiChain is a free and open-source platform used to build and deploy DLT-based applications instantly. It can support any programming language that provides better performance and scalability in comparison to other protocols. Therefore, the users do not require to learn a new programming language. It is also used for encrypted records with an advanced level of data sharing and time-stamping [14]. The icon used to represent the MultiChain is as shown in Fig. 16.

3.15 IOTA

IOTA is an open and scalable distributed ledger that has to be designed for supporting frictionless data and value transfer. It is the first distributed ledger built for the "Internet of Everything"—a network for exchanging value and data between humans and machines with tamper-proof, feeless microtransactions, and low resource requirements. In IOTA, there are no blocks and miners but when sending an IOTA transaction then it validates the previous two transactions. This feature of IOTA allows us to overcome the cost and scalability limitations of blockchain. So, this platform acts to play a central role in the next industrial revolution, enabling economic relationships between machines and bridging the human and machine economies [15]. The icon used to represent the IOTA is as shown in Fig. 17.

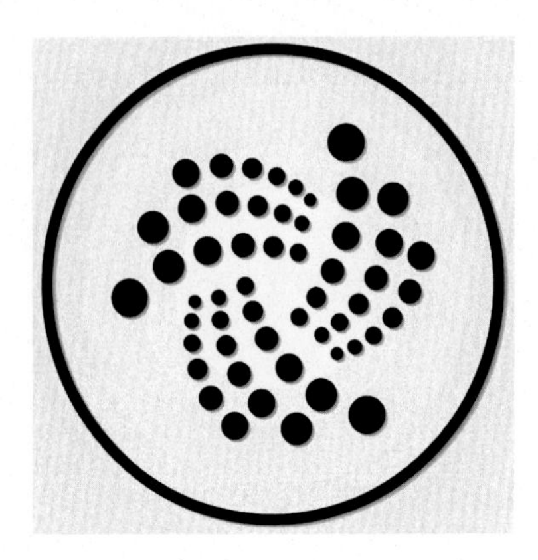

Fig. 17 IOTA.

References

[1] Ethereum, Available: https://ethereum.org/ (accessed 11 May 2020).
[2] Hyperledger Fabric, Available: https://www.hyperledger.org/use/fabric (accessed 11 May 2020).
[3] Hyperledger Sawtooth, Available: https://www.hyperledger.org/use/sawtooth (accessed 11 May 2020).
[4] Hedera Hashgraph, Available: https://www.hedera.com/ (accessed 11 May 2020).
[5] Ripple, Available: https://ripple.com/ (accessed 11 May 2020).
[6] Quorum, Available: https://www.goquorum.com/ (accessed 11 May 2020).
[7] Hyperledger Iroha, Available: https://www.hyperledger.org/use/iroha (accessed 11 May 2020).
[8] R3.Corda, Available: https://www.r3.com/corda-platform/ (accessed 11 May 2020).
[9] EOS, Available: https://eos.io/ (accessed 11 May 2020).
[10] OpenChain, Available: https://www.openchainproject.org/ (accessed 11 May 2020).
[11] Stellar, Available: https://www.stellar.org/ (accessed 11 May 2020).
[12] Dragonchain. Available: https://dragonchain.com/ (accessed 11 May 2020).
[13] NEO, Available: https://neo.org/ (accessed 11 May 2020).
[14] Multichain, Available: https://www.multichain.com/ (accessed 11 May 2020).
[15] IOTA, Available: https://www.iota.org/ (accessed 11 May 2020).

About the authors

Shubhani Aggarwal is pursuing PhD from Thapar Institute of Engineering & Technology (Deemed to be University), Patiala, Punjab, India. She received the BTech degree in Computer Science and Engineering from Punjabi University, Patiala, Punjab, India, in 2015, and the ME degree in Computer Science from Panjab University, Chandigarh, India, in 2017. She has many research interests in the area of Blockchain, cryptography, Internet of Drones, and information security.

Neeraj Kumar received his PhD in CSE from SMVD University, Katra (Jammu and Kashmir), India, and was a postdoctoral research fellow in Coventry University, Coventry, UK. He is working as an Associate Professor in the Department of Computer Science and Engineering, Thapar Institute of Engineering & Technology, Patiala (Punjab), India since 2014. Dr. Neeraj is an internationally renowned researcher in the areas of VANET & CPS Smart Grid & IoT Mobile Cloud computing & Big Data and Cryptography. He has published more than 150 technical research papers in leading journals and conferences from IEEE, Elsevier, Springer, John Wiley, and Taylor & Francis.

CHAPTER ELEVEN

Cryptographic consensus mechanisms☆

Shubhani Aggarwal and Neeraj Kumar
Thapar Institute of Engineering & Technology, Patiala, Punjab, India

Contents

Abstract

A consensus mechanism is a fault-tolerant mechanism used in a blockchain to reach an agreement on a single state of the network among distributed nodes. These are protocols that make sure all nodes are synchronized with each other and agree on transactions, which are legitimate and are added to the blockchain. Their function is to ensure the validity and authenticity of the transactions. Common consensus mechanisms that have been described in this chapter such as proof-of-work (PoW), proof-of-stake (PoS), delegated proof-of-stake (DPoS), practical Byzantine fault tolerance (PBFT), proof-of-capacity (PoC), proof-of-activity (PoA), proof-of-publication (PoP), proof-of-retrievability (PoR), proof-of-importance (PoI), proof-of-burn (PoB), proof-of-elapsed time (PoET), and proof-of-ownership (PoO).

☆ Introduction to blockchain.

Advances in Computers, Volume 121
ISSN 0065-2458
https://doi.org/10.1016/bs.adcom.2020.08.011

Chapter points

- In this chapter, we discuss the different type of consensus mechanisms used in the blockchain network.
- Here, we discuss synchronization and coordination of all the nodes on the blockchain network to reach on one agreement using consensus mechanisms.

Consensus mechanisms are used in blockchain to manage all the nodes that process transactions on the network. It makes sure that all the nodes on the network are synchronized with each other and agree on one consensus in which transaction is legitimate and then added to the blockchain. These mechanisms are a crucial part of the blockchain network. When everyone can take part in the blockchain and submit data to the network, then with the help of the consensus mechanism transactions are continuously checked and verified by all the nodes. Without an agreement, blockchain is at risk of various types of attacks like DoS, DDoS, sybil attack, etc. There are many types of blockchain consensus mechanisms, which are described in Sections 1–12.

1. Proof-of-work

PoW is the most common consensus mechanism used by the most popular cryptocurrency like Litecoin and Bitcoin. The PoW is known as mining and the participated nodes in the process are known as miners. In this, miners solve complex and difficult mathematical problems and puzzles with the help of high computation power and high processing time. The first miner who solves the puzzle to create a block gets a reward with cryptocurrency. A more detailed description of the PoW process is provided in this chapter.

2. Proof-of-stake

PoS is the second most common consensus mechanism alternative to PoW. It uses low-energy, less processing time, low cost, low computational power than PoW. In this consensus mechanism, it uses a randomized method to choose who gets to create a next new block in the chain. Instead of miners, validators are present in PoS. The users can stake their tokens to become a validator which means they lock their money for a certain period of time to create a new block. The user who has the biggest stake has the highest chance to become a validator and a chance to create a new block. This process also depends on that one user how long the coins have been staked. By using this consensus mechanism in the network, we

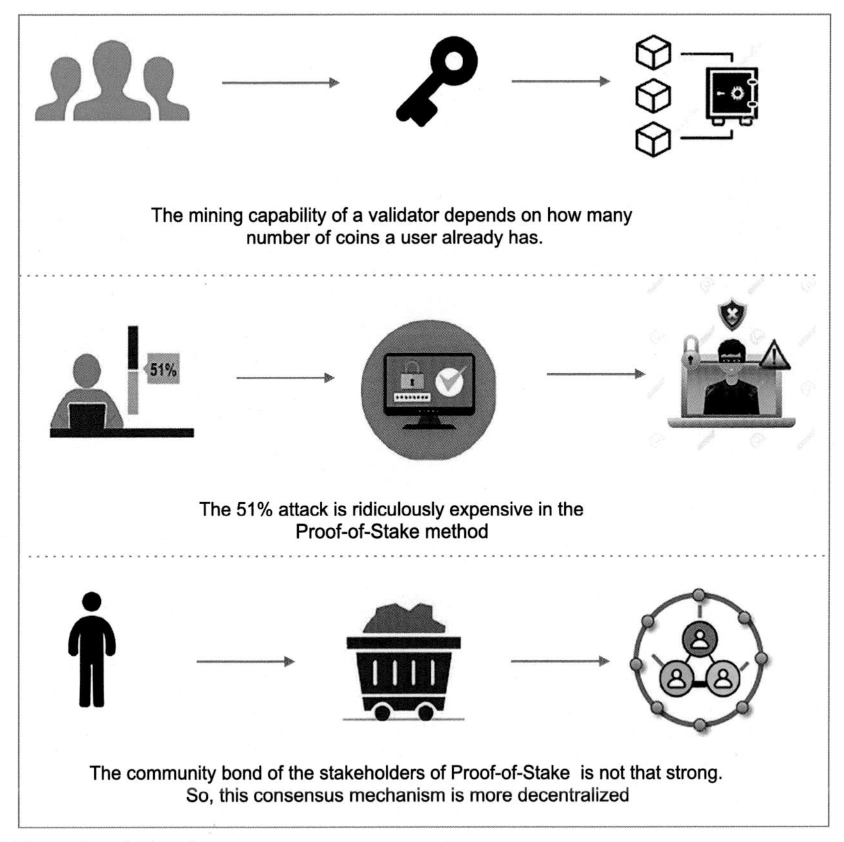

Fig. 1 Proof-of-stake.

can save the energy of other validators because only selected validators can create a block. It is a very useful consensus mechanism because in which when a validator does wrong things during the creation of block then they lose their stakes. Hence, the validator gets rewarded for honestly. The other validators who verify and validate the block take their transaction fees because they get no rewards, unlike PoW. It uses the Ethereum platform. The pictorial representation of the PoS consensus mechanism is as shown in Fig. 1.

The main difference between PoW and PoS is as shown in Fig. 2.

3. Delegated proof-of-stake

Delegated proof of stake (DPoS) is a very fast consensus mechanism and used for the implementation of EOS. Firstly, we understand the word

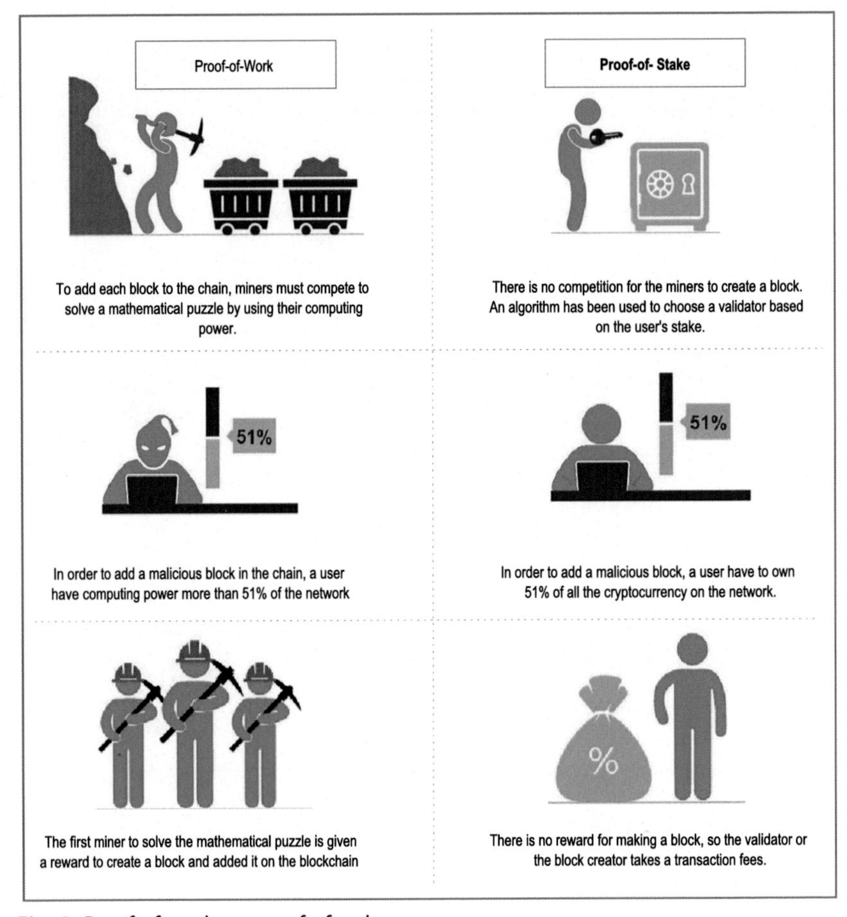

Fig. 2 Proof of work vs. proof of stake.

"*delegate*." It means a person or an organization that can produce blocks on the network. It receives the maximum number of votes from all the nodes of the network to create a block and gets rewarded. The delegates get rewarded either from the transaction fees or from a fixed amount of coins that are created during inflation. Secondly, the process of "*DPoS*" consensus mechanism. In this, the nodes of the network can stake their coins to vote for delegates. The weight of the vote depends upon the stakes. For example, If *A* stakes 5 coins for a delegate and *B* stake 1 coin then *A's* vote weight is 5 times more than the *B's* vote. The pictorial representation of DPoS is as shown in Fig. 3.

The DPoS is much more efficient at processing the transactions in comparison to another consensus such as PoW, PoS and is as shown in Fig. 4.

PROCESS

Staking Voting Forging

The nodes with the most votes are ranked and top N of these will become members of elected witness panel

NODES

Delegates Witnesses

People in the network allocate their tokens as votes for witnesses- the more tokens they have, the higher their voting weight

REWARDS

Transaction Fees Monthly Rewards

Nodes interested in becoming a witness make positive contributions to the network and actively engage in the community

Fig. 3 Delegated proof of stake.

Requires expensive computer calculations that is called mining

Requires coin holders chosen in a deterministic way that is called staking

Requires participant votes on a trusted representative that is called a delegate

Fig. 4 Proof of work vs proof of stake vs delegated proof of stake.

4. Practical Byzantine fault tolerance

Byzantine fault tolerance (BFT) is the resistance of a fault-tolerant distributed computer system against component failures. This is used by the NEO platform as a consensus mechanism. BFT is an analogy for the problem faced by a distributed computing system. The problems in BFT are described in Fig. 5.

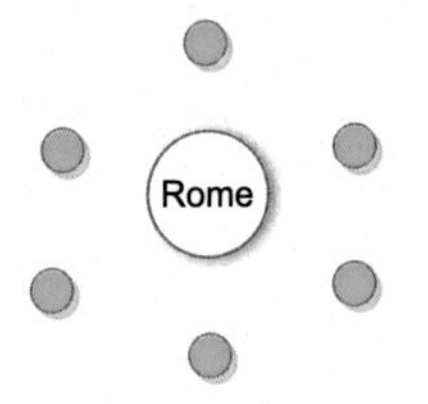

Step.1 Rome seized by six armies, each commanded by a General

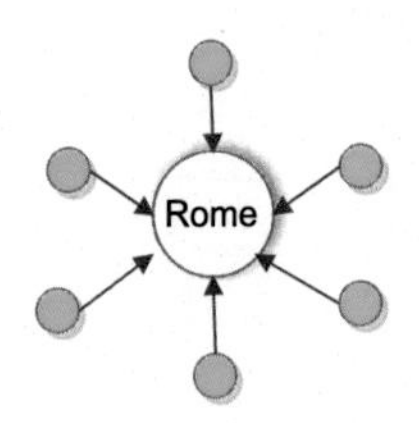

Step.2 In order to launch a successful attack, all armies have to do the same, otherwise they will be defeated by Rome

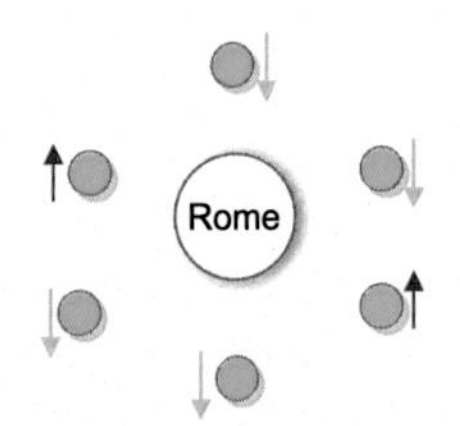

Step.3 The decision is either attack or retreat is put up to vote. Which option receives more than 50 % votes, that's the Generals will do.

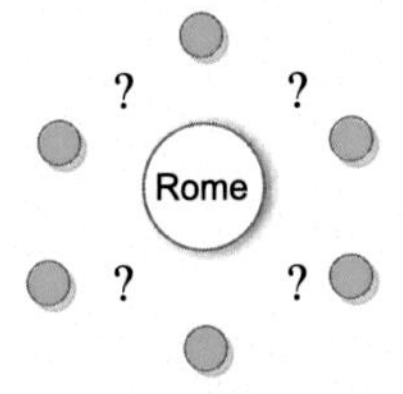

Problem.1 The Generals communicate via messenger who have passed the areas of Romans, risk happens or the message become corrupt

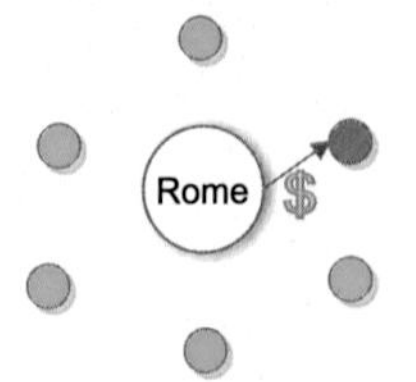

Problem.2 Each of the Generals could be bribed by the Romans:Traitorous Generals

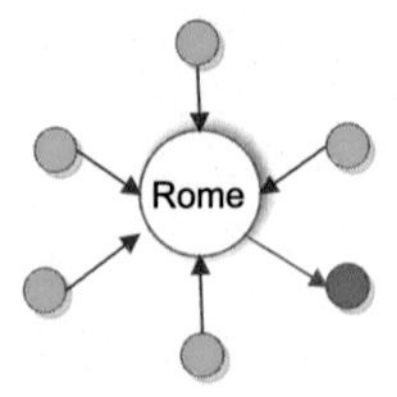

Problem.3 Any of the Generals can make wrong decision: Improper Functioning Generals

Fig. 5 Problems in Byzantine fault tolerance.

To solve the problem of a distributed computing system in BFT, practical Byzantine fault tolerance (PBFT) was developed. This consensus mechanism operated on the principle of BFT for verifying and the blocks using an election process comes after the validation process.

- Unlike DPoS, a *speaker* is chosen randomly from all the nodes then the remaining nodes on the network assume the role of delegates is as shown in Fig. 6.
- The*speaker* is responsible to construct a new block from the transaction. It verifies the transaction and also calculates the hash value. The pictorial representation is as shown in Fig. 7.
- Then, the block is sent to the delegates, who will validate the block and their transactions (scripts, data, claims, smart contracts). The pictorial representation is as shown in Fig. 8.
- The delegates validate the block by sharing and comparing their findings and all they reach to the same conclusion (more than 66.66% consensus). The pictorial representation is as shown in Figs. 9 and 10.

5. Proof-of-capacity

Proof of capacity (PoC) is a consensus mechanism used for plotting. In PoW, miners use computational power to choose a correct solution but in PoC, solutions are prestored in the memory hard-disks. The miners used this storage data to draw a plot. So, this process is called plotting. After the storage data has been plotted, miners can take part in the process of block creation. The more capacity, a miner have, the more solutions, a miner can store. So, in this way, the larger storage capacity of the miners has a high probability to create a new block using this mechanism.

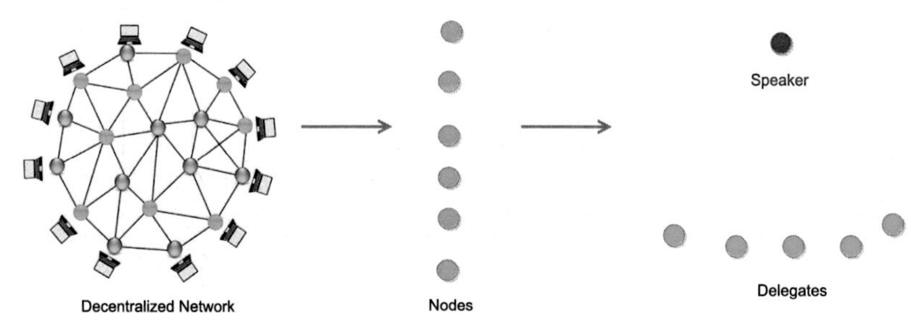

Fig. 6 Practical Byzantine fault tolerance-1.

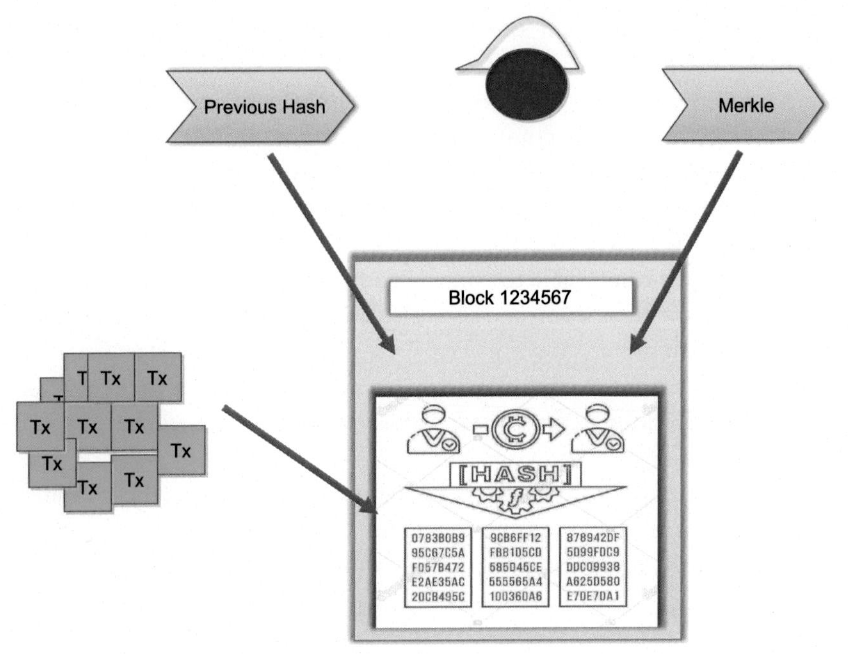

Fig. 7 Practical Byzantine fault tolerance-2.

6. Proof-of-elapsed time

Proof-of-elapsed time (PoET) is a consensus mechanism that chose miners in a random and fair manner. It also decides that who gets to produce a new block by choosing a miner. This consensus mechanism is based on the time that the miners have waited for the creation of the block. The process assigns a random and fair wait time to all the nodes on the network. The node on the network whose wait time finishes first gets to produce a new block. This mechanism works well for verification if a system has no multiple nodes and an assigned wait time is actually a random value.

7. Proof-of-activity

Proof-of-activity (PoA) consensus mechanism is much more work like PoW with reduced complexity in which the solution takes more time from a fraction of seconds to several minutes. It is a hybrid approach that combines PoW and PoS. Like PoW, miners solved the cryptographic puzzles and then, shifts to the PoS. The difference is that in which blocks contain templates instead of transactions that include header information and address

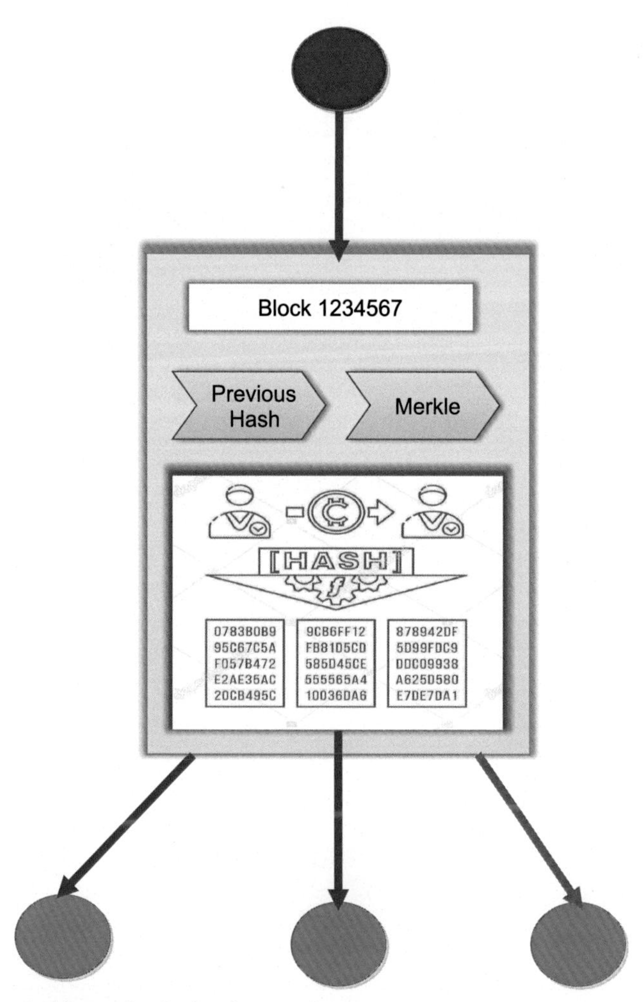

Fig. 8 Practical Byzantine fault tolerance-3.

of mining reward. In this, the blocks are verified by limiting the minimum possible time for the creation of a block that allows the maximum number of blocks added to the chain. Hence, preventing the network from spam transactions, i.e., emergence of floods.

8. Proof-of-publication

Proof-of-publication (PoP) is used in Bitcoin to check whether some particular information has been published at a certain time and date. This consensus mechanism entails the encoding of the secure hash of a certain plain text inside the bitcoin blockchain.

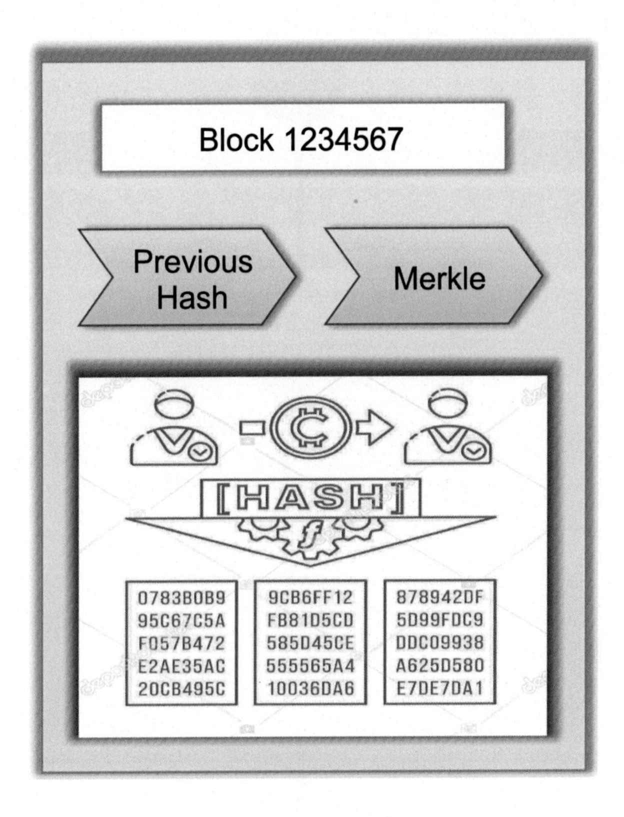

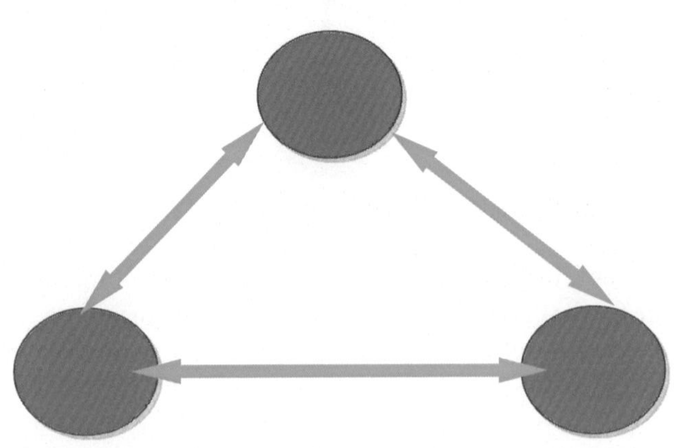

Fig. 9 Practical Byzantine fault tolerance-4.

9. Proof-of-retrievability

Proof-of-retrievability (PoR) is a consensus protocol wherein a server proves that a target file is fully downloaded and retrieved by a client from the server. The main advantage of PoR over other consensus mechanisms is

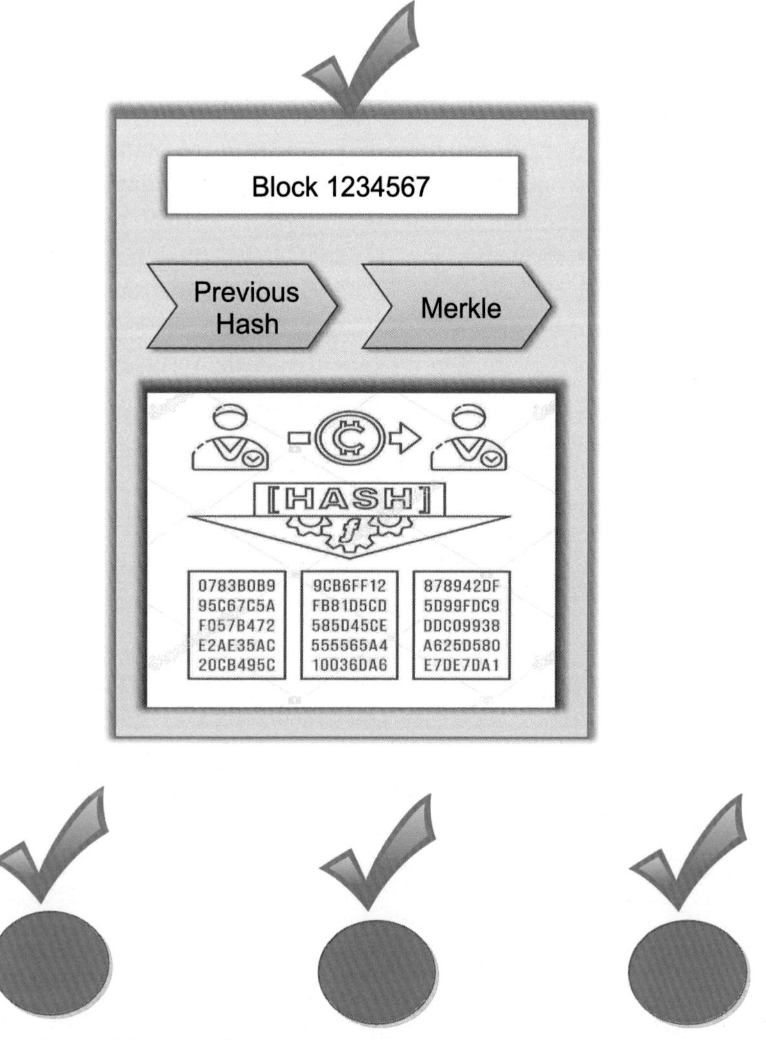

Fig. 10 Practical Byzantine fault tolerance-5.

efficiency. It is mainly deployed in an environment, where files are allocated across several systems in a redundant form.

10. Proof-of-importance

Proof-of-importance (PoI) is a consensus algorithm, introduced during the new economy movement (NEM), is used to check the entity responsible to verify the blockchain transactions.

11. Proof-of-ownership

Proof-of-ownership (PoO) is used to track the owners of some specific information at a certain time. This consensus mechanism can be used by entities, such as business organizations, to certify the integrity, date of publication, and ownership of their creations or contracts. It is implemented in CodeChain. As shown in Fig. 11, the buyer (prover) and the seller (verifier) can check the ownership of a asset. This allows safe P2P transactions. The buyer can check whether seller actually owns the pass before making the decision to buy. On the other hand, the receiver is guaranteed with an instant payment (due to the benefits of blockchain) as long as the seller is the actual owner.

12. Proof-of-burn

An alternative consensus protocol for PoS and PoW. In proof-of-burn (PoB) mechanism, the miners prove that they burn one cryptocurrency to create another currency, i.e., they are sent to a bitcoin address which is unsupendable. The significance of PoB depends on the burning tokens in an unrecoverable manner. As comparative to PoW and PoS, it is easily verifiable but hard to undo.

The difference among all consensus mechanisms are described in Table 1.

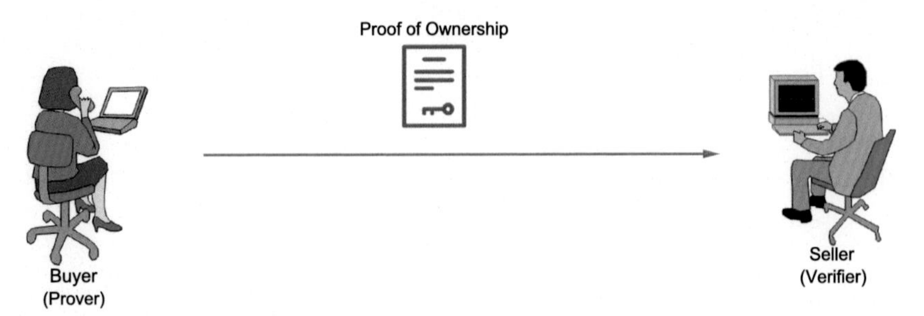

Fig. 11 Proof of ownership.

Table 1 Consensus mechanisms used in blockchain.

Consensus mechanism	Node identity	Language used	Execution environment	Energy efficient	Resource consumption	Cost	Through-put	Limitations
PoW	Public	Golang, C++, Solidity, Lisp Like Language (LLL)	Native, ethereum virtual machine (EVM)	No (high power)	High CPU	High	Low	Less secure, high power consumption
PoS	Public	Michelson	Native	Yes	Fast	Medium	Low	Consensus control to highest paid stakeholders
DPoS	Public	—	Native	Yes	Fast (faster than PoS)	Low	High	Limited token holders
BFT	Private	Any language	—	Yes	High CPU	Low	High	Semitrusted, complex with more nodes, less scalable
PBFT	Private	Golang, Java	Docker tool	Yes	High bandwidth	Low	High	Communication overhead is high for large nodes
PoC	Public	—	—	Yes	High memory	High	High	Chances of malicious vulnerable to mining tasks
PoET	Public	Python	Native	Yes	High	Low	Medium	Works only on dedicated hardware security

Continued

Table 1 Consensus mechanisms used in blockchain.—cont'd

Consensus mechanism	Node identity	Language used	Execution environment	Energy efficient	Resource consumption	Cost	Through-put	Limitations
PoA	Public	Solidity, Java, Python	EVM, Docker	No (but better than PoW)	High	High	High	Scalability and security is less
PoP	Private	Golang, C++, Solidity, Serpent, LLL	Native, EVM	Yes	Low	Low	High	Only used to check file publications
PoR	Public	Golang, C++, Solidity, Serpent, LLL	Native, EVM	Yes	Low	Low	Medium	Limited nodes usage
PoI	Public, Private	Java	—	Yes	Medium	Medium	Medium	Risk of nothing-at-stake issue
PoO	Public, Private	Any	C#	Yes	Medium	High	Medium	Expensive consensus
PoB	Public	Golang, C++, Solidity, Serpent, LLL	Native, EVM	No	Medium	Medium	Medium	costly for individual node, waste unnecessary resources

About the authors

Shubhani Aggarwal is pursuing PhD from Thapar Institute of Engineering & Technology (Deemed to be University), Patiala, Punjab, India. She received the BTech degree in Computer Science and Engineering from Punjabi University, Patiala, Punjab, India, in 2015, and the ME degree in Computer Science from Panjab University, Chandigarh, India, in 2017. She has many research interests in the area of Blockchain, cryptography, Internet of Drones, and information security.

Neeraj Kumar received his PhD in CSE from Shri Mata Vaishno Devi University, Katra (Jammu and Kashmir), India in 2009, and was a postdoctoral research fellow in Coventry University, Coventry, UK. He is working as a Professor in the Department of Computer Science and Engineering, Thapar Institute of Engineering & Technology (Deemed to be University), Patiala (Punjab), India. He has published more than 400 technical research papers in top-cited journals such as IEEE TKDE, IEEE TIE, IEEE TDSC, IEEE TITS, IEEE TCE, IEEE TII, IEEE TVT, IEEE ITS, IEEE SG, IEEE Netw., IEEE Comm., IEEE WC, IEEE IoTJ, IEEE SJ, Computer Networks, Information sciences, FGCS, JNCA, JPDC, and ComCom. He has guided many research scholars leading to PhD and ME/MTech. His research is supported by funding from UGC, DST, CSIR, and TCS. His research areas are Network Management, IoT, Big Data Analytics, Deep Learning, and Cybersecurity. He is serving as editor of the following journals of repute: ACM Computing Survey, ACM·IEEE Transactions on Sustainable Computing, IEEE·IEEE Systems Journal, IEEE·IEEE Network Magazine, IEE·IEEE Communication Magazine, IEE·Journal of Networks and Computer Applications, Elsevier Computer Communication, Elsevier

International Journal of Communication Systems, Wiley. Also, he has been a guest editor of various international journals of repute such as IEEE Access, IEEE ITS, Elsevier CEE, IEEE Communication Magazine, IEEE Network Magazine, Computer Networks, Elsevier, Future Generation Computer Systems, Elsevier, Journal of Medical Systems, Springer, Computer and Electrical Engineering, Elsevier, Mobile Information Systems, International Journal of Ad Hoc and Ubiquitous Computing, Telecommunication Systems, Springer, and Journal of Supercomputing, Springer. He has also edited/authored 10 books with international/national publishers like IET, Springer, Elsevier, CRC: Security and Privacy of Electronic Healthcare Records: Concepts, Paradigms and Solutions (ISBN-13: 978-1-78561-898-7), Machine Learning for Cognitive IoT, CRC Press, Blockchain, Big Data and IoT, Blockchain Technologies Across Industrial Vertical, Elsevier, Multimedia Big Data Computing for IoT Applications: Concepts, Paradigms and Solutions (ISBN: 978-981-13-8759-3), Proceedings of First International Conference on Computing, Communications, and Cyber-Security (IC4S 2019) (ISBN 978-981-15-3369-3). One of the edited text-book entitled, "Multimedia Big Data Computing for IoT Applications: Concepts, Paradigms, and Solutions" published in Springer in 2019 is having 3.5 million downloads till June 6, 2020. It attracts attention of the researchers across the globe (https://www.springer.com/in/book/9789811387586). He has been a workshop chair at IEEE Globecom 2018 and IEEE ICC 2019 and TPC Chair and member for various international conferences such as IEEE MASS 2020 and IEEE MSN 2020. He is a senior member of the IEEE. He has more than 12,321 citations to his credit with current h-index of 60 (September 2020). He has won the best papers award from IEEE Systems Journal and ICC 2018, Kansas City in 2018. He has been listed in the highly cited researcher of 2019 list of Web of Science (WoS). In India, he is listed in top 10 position among highly cited researchers list. He is an adjunct professor at Asia University, Taiwan, King Abdul Aziz University, Jeddah, Saudi Arabia, and Charles Darwin University, Australia.

Cryptocurrencies ☆

Shubhani Aggarwal and Neeraj Kumar
Thapar Institute of Engineering & Technology, Patiala, Punjab, India

Contents

☆ Introduction to blockchain.

Advances in Computers, Volume 121
ISSN 0065-2458
https://doi.org/10.1016/bs.adcom.2020.08.012

Abstract

A cryptocurrency is a digital or virtual currency, which is secured by cryptography that makes it impossible to double-spend on a distributed network. There are more than 1600 cryptocurrencies present in today's world that can be used for making payment transactions. Some cryptocurrencies like bitcoin, bitcoin cash, LTC, LINK, tether, etc. are based on decentralized networks like blockchain technology, which is a distributed ledger imposed by distinct nodes of the network. A blockchain behind cryptocurrencies is a public ledger, which is used to store the history of every transaction that cannot be tampered or changed. This makes the transactions secure and safer than the existing systems. But some blockchains do not use any cryptocurrency or tokens. So, it varies significantly that depends on the type of blockchain. In this chapter, we have described the functionality of some cryptocurrencies like bitcoin, ETH, XRP, LTC, USDT, BCH, LIBRA, XMR, EOS, BSV, and many others.

Chapter points

- In this chapter, we discuss the different type of cryptocurrencies that can be used in the blockchain network
- Here, we discuss the virtual or digital money that can be taken in the form of tokens or coins on a blockchain.

The crypto refers to cryptography in the cryptocurrency that allows for the creation and processing of digital payments and their transactions across decentralized systems. Around the world, the cryptocurrencies are free to be designed from government manipulation and control. The first cryptocurrency was developed by Satoshi Nakamoto in 2008, named Bitcoin. The other cryptocurrencies modeled after bitcoin are names as altcoins and represented as improved versions of bitcoin. In this chapter, we have discussed some of the important digital currencies used for digital payment transactions on a decentralized network [1].

1. Bitcoin

Bitcoin is a first digital cryptocurrency, a decentralized system that records transactions in a DLT called a blockchain. The miners in a bitcoin run high computing power computers to solve complex mathematical puzzles to confirm groups of transactions called blocks. Then, these blocks are added to the blockchain record and the miners are rewarded with a small number of bitcoins. The other nodes can buy or sell tokens through cryptocurrency in the bitcoin market. These bitcoin exchanges also work to defend themselves against potential theft through a trustless system. The icon represents to bitcoin cryptocurrency is as shown in Fig. 1.

2. Ethereum

The other most important cryptocurrency is ether used in Ethereum, which is a decentralized software platform that enables smart contracts and decentralized applications to be built and run without any downtime, fraud, control, or central authority [2]. The ether cryptocurrency was launched in 2015 [3]. This cryptocurrency is like a moving vehicle on the Ethereum platform to develop and run applications. As mentioned in the facts of

Fig. 1 Bitcoin cryptocurrency.

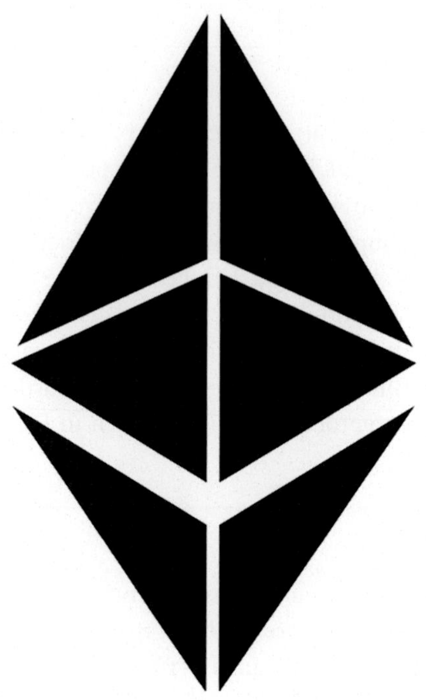

Fig. 2 ETH cryptocurrency.

the report [4], Ethereum (ETH) had a market cap of $15.6 billion and a per token value of $142.5. The icon represents to ETH cryptocurrency is as shown in Fig. 2.

3. Ripple

Ripple was launched in 2012, which is a real-time global settlement network that offers instant, certain, and low-cost international payments [5]. It also enables banks to settle cross-border payments in real-time, with end-to-end transparency, and at lower costs. The transaction confirmation method of ripple does not require any mining because Ripple's XRP tokens were premined before launch. So, there is no need for the creation of XRP over time, only the introduction and removal of XRP from the market supply according to the network. As it does not require any mining, it reduces the usage of computing power and minimizes network latency [6]. It is currently the third-largest cryptocurrency in the world by the overall market cap. As mentioned in the facts of the report [7], Ripple

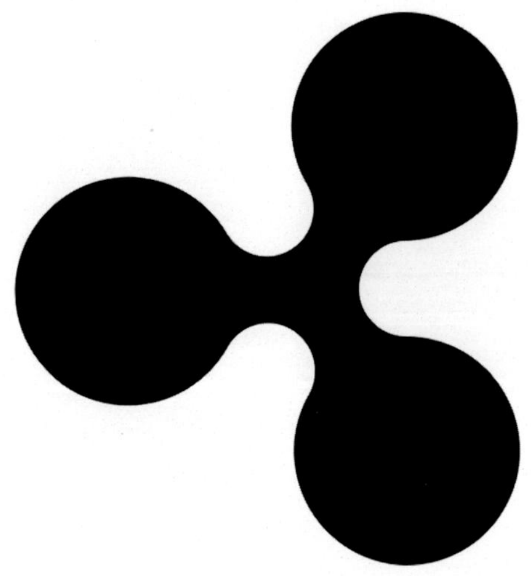

Fig. 3 XRP cryptocurrency.

had a market cap of $8.9 billion and a per token value of $0.202. The icon represents to XRP cryptocurrency is as shown in Fig. 3.

4. Litecoin

Litecoin (LTC) was launched in 2011 and was created by Charlie Lee, an MIT graduate, and former Google engineer [8]. This cryptocurrency has often been referred to as the silver to bitcoin's gold. It is based on an open-source global payment network, which is not controlled by any central authority and uses scrypt cryptography as a PoW. It is similar to bitcoin, which has a faster block generation rate and offers a faster transaction confirmation time. As mentioned in the facts of the report [9], LTC had a market cap of $3.01 billion and a per token value of $46.48. The icon represents to LTC cryptocurrency is as shown in Fig. 4.

5. Tether

Tether (USDT) was launched in 2014 and one of the most popular of a group called stablecoins [10]. Tether itself describes as a blockchain-based platform that facilitates the use of fiat currencies digitally. Effectively, this cryptocurrency allows individuals to utilize a blockchain network and related technologies to transact in traditional currencies while minimizing the volatility and complexity often associated with digital

Fig. 4 LTC cryptocurrency.

Fig. 5 USDT cryptocurrency.

currencies. As mentioned in the facts of the report [11], Tether was the fourth-largest cryptocurrency by market cap, with a total market cap of $9.1 billion and a per token value of $1.00. The icon represents to USDT cryptocurrency is as shown in Fig. 5.

6. Bitcoin Cash

Bitcoin Cash (BCH) was launched in 2017. It is one of the earliest and most successful hard forks of the original bitcoin. In the cryptocurrency world, a fork takes place as the result of arguments and controverts between developers and miners. The controvert which led to the creation of BCH had to do with the issue of scalability. But the bitcoin network has a strict limit of 1 MB on the size of blocks whereas BCH increases the block size from 1 to 8 MB for faster transaction times [12]. As mentioned in the facts of the report [13], BCH had a market cap of $4.6 billion and a value per token of $254.45. The icon represents to BCH cryptocurrency is as shown in Fig. 6.

7. Libra

Libra is one of the most-hyped cryptocurrency. A popular Facebook's incredible global reach and the potential for massive volumes of exchange across its platform, it has released its cryptocurrency in 2019 in the form of Libra [14]. Libra will be overseen in part by a new Facebook subsidiary, the financial services outfit Calibra [15]. The icon represents to Libra cryptocurrency is as shown in Fig. 7.

Fig. 6 Bitcoin cash cryptocurrency.

Fig. 7 LIBRA cryptocurrency.

8. Monero

Monero (XMR) is a reasonably private digital currency, which is fast, secure, and untraceable currency. It was launched in 2014 and is completely donation-based and community-driven [16]. The main focus of Monero is a strong focus on decentralization and scalability. It also enables privacy and security using a technique called ring signatures. By this, there is no contradiction to a cryptocurrency space that Monero has introduced important technological advances. As mentioned in the facts of the report [17], Monero had a market cap of $1.17 billion and a per token value of $67.05. The icon represents to XMR cryptocurrency is as shown in Fig. 8.

9. EOS

EOS was launched in June of 2018 and was created by cryptocurrency pioneer Dan Larimer. Like other cryptocurrencies, EOS is designed after Ethereum and offers a platform on which developers can build decentralized applications. It offers a DPoS mechanism that provides scalability and privacy to the platform. EOS consists of EOS.IO, similar to the operating system of a computer and acting as the blockchain network for the digital currency, as well as EOS coins. It is also revolutionary because of its lack of a mining process to produce coins. In this, block producers create blocks and are rewarded in EOS tokens based on their production rates. It will be

Fig. 8 XMR cryptocurrency.

Fig. 9 EOS cryptocurrency.

more democratic and decentralized than other cryptocurrencies [18]. As mentioned in the facts of the report [19], EOS had a market cap of $2.5 billion and a per token value of $2.77. The icon represents to EOS cryptocurrency is as shown in Fig. 9.

10. Bitcoin SV

Bitcoin SV (BSV) is a hard fork of Bitcoin Cash where "SV" stands for Satoshi Vision. It was launched in November of 2018. The developers of BSV recommended that this cryptocurrency restores Bitcoin protocol and allowing for new developments to increase stability and scalability. They have also prioritized security and fast transaction processing times [20]. As mentioned in the facts of the report [21], BSV had a market cap of $3.5 billion and a per token value of $193.05. The icon represents to BSV cryptocurrency is as shown in Fig. 10.

11. Binance Coin

Binance Coin (BNB) is the token of the Binance cryptocurrency exchange platform. It was launched in 2017. It allows Binance users to trade in dozens of different cryptocurrencies on the Binance platform efficiently and effectively. This platform facilitates transaction fees on the exchange and

Fig. 10 Bitcoin SV cryptocurrency.

Fig. 11 Binance Coin cryptocurrency.

can be used to pay for certain goods and services [22]. As mentioned in the facts of the report [23], BNB had a market cap of $2.7 billion and a per token value of $17.42. The icon represents to BNB cryptocurrency is as shown in Fig. 11.

12. Cardano

Cardano is a public blockchain-based decentralized platform, which is used to develop smart contracts to deliver more advanced features. This platform is based on a research-first driven approach, which uses the ADA cryptocurrency that can be used to send and receive digital funds. This digital currency represents future cash and can be used to make secure and fast payments through the use of cryptography [24]. As mentioned in the facts of the report [25], ADA had a market cap of $2.2 billion and a per token value of $0.08. The icon represents to ADA cryptocurrency is as shown in Fig. 12.

13. Tezos

Tezos (XTZ) is a blockchain network that links to a digital crypto-currency or token, which is called a tez. It is based on a PoS mechanism that does not rely on the mining process. Like Ethereum, it is designed to make use of smart contracts in decentralized applications. The reason for the popularity and high use of XTZ is to mitigate the problems of the lack of flexibility and scalability in Bitcoin [26]. As mentioned in the facts of the report [27], XTZ had a market cap of $2.15 billion and a per token value of $2.94. The icon represents to XTZ cryptocurrency is as shown in Fig. 13.

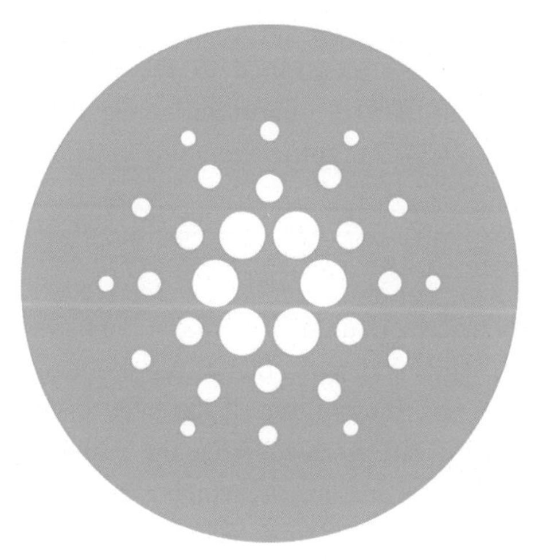

Fig. 12 ADA cryptocurrency.

Fig. 13 XTZ cryptocurrency.

14. Crypto.com Coin

Crypto.com is a cryptocurrency company that began its life by offering VISA-powered payment cards in 2017 [28]. It is a payment platform that offers attractive Visa cards, wallets, and portfolio building services. It is an easy-to-use payment application and uses CRO tokens as a cryptocurrency. The main features of crypto.com include trading, lending, borrowing, and staking. As mentioned in the facts of the report [29], CRO had a market cap of $1.8 billion and a per token value of $0.10. The icon represents to CRO cryptocurrency is as shown in Fig. 14.

15. Stellar

Stellar is an open network used for storing and transferring currency [30]. It is used to create, send, and trade digital representations of all forms of money, i.e., dollars, pesos, bitcoin, etc. It has been designed so that all the world's financial systems can work together on a single network. XLM is the currency symbol for Stellar Lumens, which is considered similar to XRP because of its focus on banking and financial system. As mentioned in the facts of the report [31], XLM had a market cap of $1.6 billion and a per token value of $0.079. The icon represents to XLM cryptocurrency is as shown in Fig. 15.

Fig. 14 CRO cryptocurrency.

Fig. 15 XLM cryptocurrency.

16. ChainLink

Chainlink was launched by SmartContract company in 2014 and is an Ethereum-based platform that serves as an intermediate stage between smart contracts and external data [32]. It is a version of the oracle. It provides reliable tamper-proof inputs and outputs for complex smart contracts on

Fig. 16 LINK cryptocurrency.

the blockchain network. As mentioned in the facts of the report [33], LINK had a market cap of $1.5 billion and a per token value of $4.38. The icon represents to LINK cryptocurrency is as shown in Fig. 16.

17. UNUS SED LEO

UNUS SED LEO (LEO) is a cryptocurrency created to expand the capabilities of all users of the iFinex platform and services that include Bitfinex exchange. This token is designed to cope with the crisis that arose after the exchange accused of illegally using Tether cryptocurrency funds [34]. As mentioned in the facts of the report [35], UNUS SED LEO had a market cap of $1.24 billion and a per token value of $1.24. The icon represents to LEO cryptocurrency is as shown in Fig. 17.

18. Tron

Tron was established in 2017 by a Singapore-based nonprofit organization called the Tron Foundation and headed by CEO Justin Sun [36]. It is a distributed ledger-based platform that aims to build a free and global digital

Fig. 17 LEO cryptocurrency.

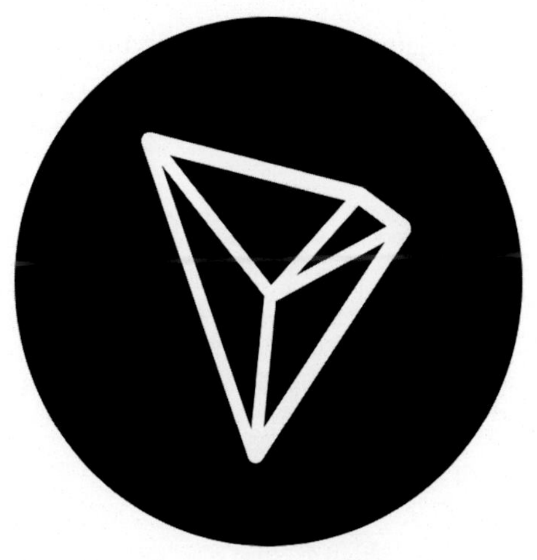

Fig. 18 TRX cryptocurrency.

system. It allows an easy and cost-efficient sharing of digital information and content. It uses its cryptocurrency called Tronix (TRX). This currency can be used by the consumers for paying the content they want to access. They either can be exchanged with other cryptocurrencies and can be used to pay for blockchain services. As mentioned in the facts of the report [37], TRON had a market cap of $1.19 billion and a per token value of $0.017. The icon represents to TRX cryptocurrency is as shown in Fig. 18.

19. Huobi Token

Huobi is a Singapore-based cryptocurrency exchange, which was created in 2013 by former Oracle engineer Leon Li. It supports over 280 cryptocurrency markets currently, with over $350 billion worth of crypto tokens traded daily, which makes it the third-largest cryptocurrency exchange in the world. The token or cryptocurrency used in Houbi is Huobi Token, which was launched in January 2018 to reward Huobi exchange users for using tokens. It can be exchanged with any cryptocurrency on the Huobi exchange [38]. As mentioned in the facts of the report [39], HT had a market cap of $969 million and a per token value of $4.40. The icon represents to HT cryptocurrency is as shown in Fig. 19.

20. NEO

NEO is a blockchain platform and cryptocurrency designed for digitizing assets using smart contracts and aiming to bring blockchain to a large scale. The NEO cryptocurrency was launched as Antshares (renamed

Fig. 19 HT cryptocurrency.

as NEO) by Da Hongfei in 2017 [40]. This platform has an active development team to develop decentralized applications. It uses a DBFT consensus mechanism that offers lower electricity costs and removes the possibility of a chain split. This makes to achieve its goal of digitalizing physical financial assets. As mentioned in the facts of the report [41], NEO had a market cap of $813 million and a per token value of $11.53. The icon represents to NEO cryptocurrency is as shown in Fig. 20.

21. BitTorrent

BitTorrent was launching its cryptocurrency BitTorrent token (BTT), which was issued by the Singapore-based BitTorrent Foundation in 2019 on the Tron network. BitTorrent, which pioneered P2P technology for sharing files on the Internet. It provides users with a way to spend and earn cryptocurrency while sharing files [42]. As mentioned in the facts of the report [43], BTT had a market cap of $63 million and a per token value of $0.0002. The icon represents to BTT cryptocurrency is as shown in Fig. 21.

22. BitShares

BitShares cryptocurrency (BTS) was launched in 2014 and is the creation of Steem and EOS cofounder along with Ethereum and Cardano cofounder. It is a decentralized exchange platform for trading

Fig. 20 NEO cryptocurrency.

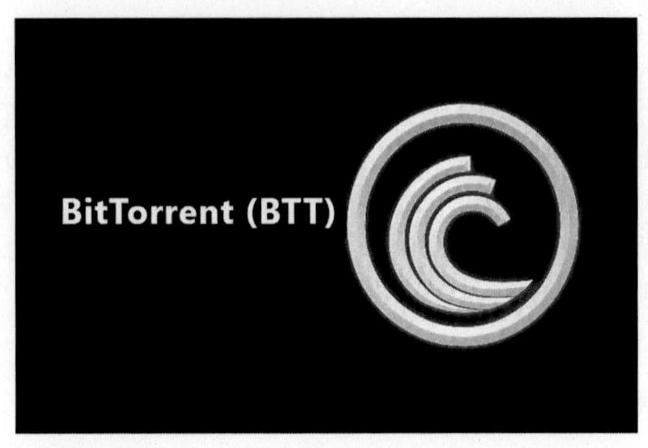

Fig. 21 BTT cryptocurrency.

Fig. 22 BTS cryptocurrency.

cryptocurrencies and based on the DPoS verification system. It is trustless, fast, and reliable to scale with use [44]. As mentioned in the facts of the report [45], BTS had a market cap of $59 million and a per token value of $0.02. The icon represents to BTS cryptocurrency is as shown in Fig. 22.

23. Ethereum Classic

Ethereum Classic (ETC) is an open-source, blockchain-based distributed cryptocurrency platform that runs smart contracts. It has emerged as a split version of the Ethereum's blockchain. The split has occurred on Ethereum in 2016 when $50 million worth of funds were stolen. This resulted in the two versions existing simultaneously. The newer one was called as ETH, and the older one was renamed as ETC [46]. As mentioned in the facts of the report [47], ETC had a market cap of $745 million and a per token value of $6.41. The icon represents to ETC cryptocurrency is as shown in Fig. 23.

24. USD Coin

USD Coin (USDC) was launched by the Centre Consortium in 2018. The technology and ruling framework were developed by the Centre, while Circle and Coinbase were the first commercial issuers of USDC. This cryptocoin has been used over the Internet and public blockchains. It opens up new opportunities for trading, lending, risk-hedging, and more [48]. As mentioned in the facts of the report [49], USDC had a market cap of $734 million and a per token value of $1.0. The icon represents to USDC cryptocurrency is as shown in Fig. 24.

Fig. 23 ETC cryptocurrency.

Fig. 24 USDC cryptocurrency.

25. Hedge Trade

Hedge Trade (HEDG) was launched in 2019 that aim is to revolutionize the social trading using blockchain technology. With the help of HedgeTrade, the more experienced users are provided with a platform that allows them to share their knowledge and acquire some HEDG tokens in the process of providing accurate information. At the same time, the less experienced users can purchase the trading predictions in the form of blueprints, which are further secured by the HEDG smart contracts. So, a piece of acquired knowledge from the blueprints should allow the users to engage in professional trading with the help of the market tools provided by the HEDG for pros [50]. As mentioned in the facts of the report [51], HEDG had a market cap of $500 million and a per token value of $1.74. The icon represents to HEDG cryptocurrency is as shown in Fig. 25.

26. Cosmos

Cosmos is a dual-layer network in which two blockchains exchange the token, data, and asset. Each blockchain on the network operates autonomous and independent [52]. ATOM is the original cryptocurrency token of the cosmos network, which operates based on BFT. This cryptocurrency is trading on various crypto exchanges such as Binance, Huobi Token, etc. As mentioned in the facts of the report [53], ATOM had a market cap of $546 million and a per token value of $2.87. The icon represents to ATOM cryptocurrency is as shown in Fig. 26.

Fig. 25 HEDG cryptocurrency.

Fig. 26 ATOM cryptocurrency.

Fig. 27 ZEC cryptocurrency.

27. ZCash

ZCash (ZEC) is an open-source cryptocurrency that provides anonymity and security to its users and its transactions on the decentralized network. It increases privacy and security by using zero-knowledge proofs that validate transactions without revealing the user's private information [54]. As mentioned in the facts of the report [55], ZEC had a market cap of $475 million and a per token value of $50.78. The icon represents to ZEC cryptocurrency is as shown in Fig. 27.

28. Hedera Hashgraph

HBAR is the original cryptocurrency of the Hedera Hashgraph, which is a public blockchain-based network. Developers use HBAR currency tokens to pay for network services and benefits such as running a smart contract, storing a file, or transferring cryptocurrency [56]. As mentioned in the facts of the report [57], HBAR had a market cap of $184 million and a per token value of $0.04. The icon represents to HBAR cryptocurrency is as shown in Fig. 28.

29. Bitcoin Gold

Bitcoin Gold was a hard fork open-source cryptocurrency and is branding itself a new version of bitcoin. It has been used to retain bitcoin's transaction history that means if the user-owned bitcoins before the fork, then the user owns the same number of Bitcoin Gold [58]. As mentioned in the facts of the report [59], BTG had a market cap of $150 million and a per token value of $8.57. The icon represents to BTG cryptocurrency is as shown in Fig. 29.

30. Bitcoin Diamond

Bitcoin Diamond (BCD) cryptocurrency is a fork of bitcoin, which occurs at the predetermined height of 495866 number of blocks and

Fig. 28 HBAR cryptocurrency.

Fig. 29 BTG cryptocurrency.

Fig. 30 BCD cryptocurrency.

therewith a new chain as the BCD will be generated [60]. As mentioned in the facts of the report [61], BCD had a market cap of $108 million and a per token value of $0.58. The icon represents to BCD cryptocurrency is as shown in Fig. 30.

31. Komodo

Komodo (KMD) is a cryptocurrency that provides more anonymity and security than Bitcoin. This komodo network relies on zero-knowledge

Fig. 31 KMD cryptocurrency.

proof that allows KMD transactions to be 100% private. The private development team of komodo stresses the concept of freedom, from the standpoint of both users and developers [62]. As mentioned in the facts of the report [63], KMD had a market cap of $88 million and a per token value of $0.73. The icon represents to KMD cryptocurrency is as shown in Fig. 31.

32. Nexo

Nexo is a blockchain-based lending platform, which uses cryptocurrency assets for private and secure loans [64]. Over 40 cryptocurrencies are supported and loans are made instantly, NEXO has gained a discount of 50%. As mentioned in the facts of the report [65], NEXO had a market cap of $72 million and a per token value of $0.12. The icon represents to NEXO cryptocurrency is as shown in Fig. 32.

33. Zcoin

Zcoin (XZC) cryptocurrency is a decentralized and digital currency, which mainly focus on the privacy and anonymity of cryptocurrency transactions. It uses zero-knowledge cryptographic proofs to provide confidential and secure transactions [66]. As mentioned in the facts of the report [67], XZC had a market cap of $46 million and a per token value of $4.5. The icon represents to XZC cryptocurrency is as shown in Fig. 33.

Fig. 32 NEXO cryptocurrency.

Fig. 33 XZC cryptocurrency.

34. Bytecoin

Bytecoin (BCN) is an open-source decentralized cryptocurrency based on CryptoNote technology. It has been designed to protect the user's privacy and works like Bitcoin's protocol. But, unlike Bitcoin, Bytecoin transactions hide all the connections of the sender and receiver [68]. As mentioned in the facts of the report [69], BCN had a market cap of $41 million and a per token value of $0.0002. The icon represents to BCN cryptocurrency is as shown in Fig. 34.

Fig. 34 BCN cryptocurrency.

35. Gnosis

Gnosis is an open-source decentralized infrastructure to predict markets that builts on Ethereum. The main aim is to build a framework for decentralized applications and allowing access from any party to its platform. It uses two cryptocurrencies like GNO and OWL that ensure its transactions are reliable and distributed [70]. As mentioned in the facts of the report [71], GNO had a market cap of $33 million and a per token value of$30.53. The icon represents to GNO cryptocurrency is as shown in Fig. 35.

36. Maker

Maker (MKR) is a cryptocurrency and a governance token. It is a digital token created on the Ethereum platform. The main purpose is to create a line of decentralized digital assets that would be tied to the value of real instruments such as-currency, gold, etc. It is one of the most potent and valuable coins in all of cryptocurrency [72]. As mentioned in the facts of the report [73], MKR had a market cap of $601 million and a per token value of $598.6. The icon represents to MKR cryptocurrency is as shown in Fig. 36.

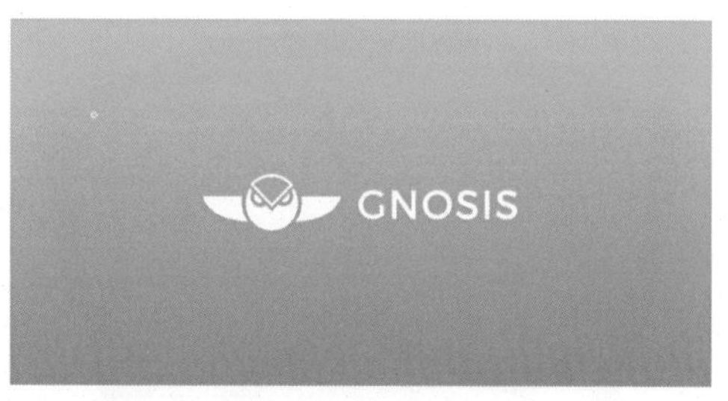

Fig. 35 GNO cryptocurrency.

Fig. 36 MKR cryptocurrency.

37. NEM

NEM stands for New Economy Movement and was launched in 2015 as a dual-layer blockchain. It is similar to Ethereum and smart contracts written in Java programming language. The cryptocurrency of NEM is XEM that uses a PoI algorithm for mining [74]. As mentioned in the facts of the report [75], XEM had a market cap of $395 million and a per token value of $0.04. The icon represents to XEM cryptocurrency is as shown in Fig. 37.

38. DOGE

Dogecoin is a P2P open-source cryptocurrency and was launched in 2013 with a Shibu Inus as its logo. It is a decentralized virtual currency, which

Fig. 37 XEM cryptocurrency.

Fig. 38 DOGE cryptocurrency.

is gaining wide popularity for secure and private online transactions. It uses scrypt-based cryptography and enables fast payments to anyone, anywhere across the globe [76]. As mentioned in the facts of the report [77], DOGE had a market cap of $312 million and a per token value of $0.002. The icon represents to DOGE cryptocurrency is as shown in Fig. 38.

39. THETA

THETA cryptocurrency is building a P2P mesh network, which aims to solve the current issues in video delivery networks using a blockchain. It

Fig. 39 THETA cryptocurrency.

has been designed to incentive the sharing of bandwidth and keep the delivery network at a continual high performance [78]. As mentioned in the facts of the report [79], THETA had a market cap of $206 million and a per token value of $0.23. The icon represents to THETA cryptocurrency is as shown in Fig. 39.

40. ICON

ICX is the cryptocurrency that is fundamental to the ICON network. It is a South Korean-based company, which is developed as a blockchain-based network. The company bills itself as an interconnected through the blockchain network that allow participants in a decentralized system [80]. As mentioned in the facts of the report [81], ICX had a market cap of $177 million and a per token value of $0.32. The icon represents to ICX cryptocurrency is as shown in Fig. 40.

41. QTUM

QTUM is a cryptocurrency that combines the assets of bitcoin and Ethereum. It uses a UTXO-based smart contract system with a PoS consensus model. It has been designed to use for large scale organizations [82]. As mentioned in the facts of the report [83], QTUM had a market cap of $168 million and a per token value of $1.74. The icon represents to QTUM cryptocurrency is as shown in Fig. 41.

Fig. 40 ICX cryptocurrency.

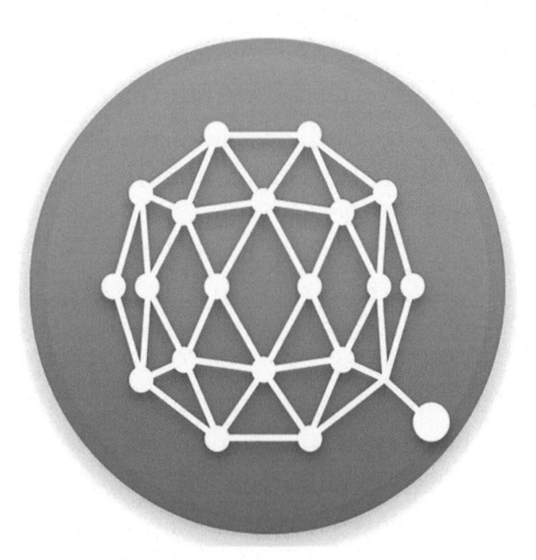

Fig. 41 QTUM cryptocurrency.

42. Siacoin

Siacoin (SC) is the cryptocurrency, which is to use the Sia platform. It is a blockchain-based cloud platform that provides a solution for decentralized storage. The nodes on the network can rent their hard drive space to other nodes and earn money in the form of SC [84]. As mentioned in the facts of the report [85], SC had a market cap of $114 million and a per token value of $0.002. The icon represents to SC cryptocurrency is as shown in Fig. 42.

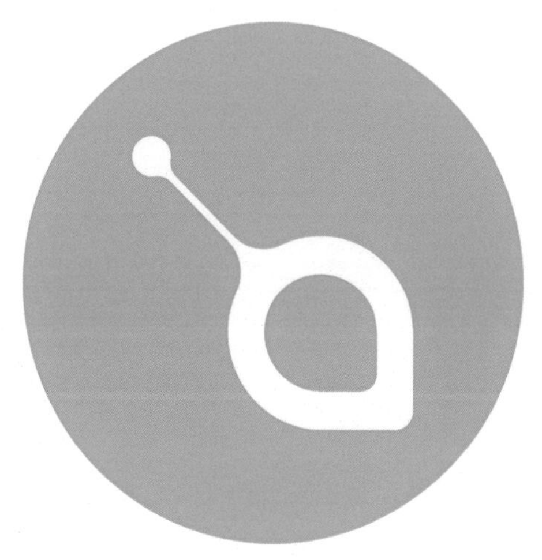

Fig. 42 SC cryptocurrency.

43. MonaCoin

MonaCoin was launched in 2013 by the Japanese internet forum as part of the MonaCoin Project. It is a decentralized P2P open-source platform-based payment network. The cryptocurrency of MonaCoin is MONA and has a total supply of 105 million coins. As mentioned in the facts of the report [86], MONA had a market cap of $115 million and a per token value of $1.76. The icon represents to MONA cryptocurrency is as shown in Fig. 43.

44. Terra

Terracoin was launched in 2012 and is one of the oldest cryptocurrency. The Foundation of Terracoin is used to helping the world via donations and the cryptocurrency of Terracoin is LUNA [87]. As mentioned in the facts of the report [88], LUNA had a market cap of $80 million and a per token value of $0.20. The icon represents to LUNA cryptocurrency is as shown in Fig. 44.

45. Flexacoin

Flexacoin (FXC) is a cryptocurrency token, which is used to coordinate payments on the Flexa network. FXC can be bought and sold for fiat

Fig. 43 MONA cryptocurrency.

Fig. 44 LUNA cryptocurrency.

currency and making them instant and secure. It has been designed to mitigate the friction between customers paying with cryptocurrency and merchants accepting fiat [89]. As mentioned in the facts of the report [90], FXC had a market cap of $81 million and a per token value of $0.002. The icon represents to FXC cryptocurrency is as shown in Fig. 45.

The comparison between the above mentioned cryptocurrencies is described in Table 1.

Fig. 45 FXC cryptocurrency.

Table 1 Comparison between the cryptocurrencies.

Cryptocurrency	Launch year	Symbol	Network	Block time	Transactions per second
Bitcoin	2009	BTC	Not Applicable	10 minutes	7
Ethereum	2015	ETH	Ethereum	15 s	20
Ripple	2012	XRP	RippleNet	Near instant	1500
Litecoin	2011	LTC	Not Applicable	2 min 30 s	56
Tether	2014	USDT	Ethereum	10 min	–
Bitcoin Cash	2017	BCH	Not Applicable	10 min	60
Libra	2019	LIBRA	Not Applicable	–	1000
Monero	2014	XMR	Monero	2 s	13
EOS	2018	EOS	EOS.IO	0.5 s	4000
Bitcoin SV	2018	BSV	Not Applicable	10 min 50 s	4127
Binance Coin	2017	BNB	Binance	1 min	–

Continued

Table 1 Comparison between the cryptocurrencies.—cont'd

Cryptocurrency	Launch year	Symbol	Network	Block time	Transactions per second
Cardano	2015	ADA	Cardano	20 s	6
Tezos	2018	XTZ	Tezos	60 s	15
Crypto.com Coin	2017	CRO	Crypto.com Coin	15 s	50,000
Stellar	2014	XLM	Stellar	5 s	1000
Chainlink	2014	LINK	Ethereum	5 min	3000
UNUS SED LEO	2019	LEO	iFinex	15 s	20
TRON	2017	TRX	Ethereum & EOS	10 s	–
Huobi Token	2018	HT	Houbi	–	500
NEO	2014	NEO	NEO	15 s	1000
BitTorrent	2019	BTT	TRON	3 s	3000
BitShares	2014	BTS	BitShares	–	1,00,000
Ethereum Classic	2016	ETC	Ethereum	13 s	20
USD Coin	2018	USDC	Circle	11 min	–
Hedge Trade	2019	HEDG	Hedge Trade	10 min	250
Cosmos	2014	ATOM	Cosmos	5 s	10,000
ZCash	2016	ZEC	Ethereum	75 s	23–26
Hedera Hashgraph	2017	HBAR	Hedera	3–5 s	10,000
Bitcoin Gold	2017	BTG	Not Applicable	10 min	–
Bitcoin Diamond	2017	BCD	Not Applicable	10 min	100
Komodo	2016	KMD	KOMODO	1 min	20,000
NEXO	2017	NEXO	NEXO	10 min	–
Zcoin	2016	XZC	Zcoin	5 min	14
Bytecoin	2012	BCN	CryptoNote	2 min	3–7

Table 1 Comparison between the cryptocurrencies.—cont'd

Cryptocurrency	Launch year	Symbol	Network	Block time	Transactions per second
Gnosis	2015	GNO	Ethereum	–	–
Maker	2017	MKR	Ethereum	–	–
NEM	2014	XEM	NEM	1 min	4000
DOGEcoin	2013	DOGE	DOGE	1 min	–
THETA	2019	THETA	THETA	6 s	8192
ICON	2017	ICX	ICON	15 min	3
QTUM	2017	QTUM	Ethereum	2 min	10,000
Siacoin	2013	SC	SIA	60 min	4000
MonaCoin	2014	MONA	MONA	1.5 min	–
Terra	2012	LUNA	TERRA	1 s	–
Flexacoin	2019	FXC	FLEXA	10 min	–

References

[1] A. Ghosh, S. Gupta, A. Dua, N. Kumar, Security of cryptocurrencies in blockchain technology: state-of-art, challenges and future prospects, J. Netw. Comput. Appl. (2020) 102635.
[2] Ethereum Blockchain APP Platform. 2008. Available: https://www.ethereum.org/ (accessed 27 July, 2019).
[3] Ethereum Blockchain APP Platform. 2018. Available: https://ethereum.org/what-is-ethereum/ (accessed 27 July, 2019).
[4] Securities and Exchange Commission, Report of Investigation Pursuant to Section 21 (a) of the Securities Exchange Act of 1934: The DAO, Washington, DC, July. vol. 25, 2017.
[5] Ripple, Available: https://ripple.com/ (accessed 11 May 2020).
[6] XRP Ledger, Available: https://xrpl.org/xrp-ledger-overview.html (accessed 11 May 2020).
[7] CoinMarketCap, XRP, Available: https://coinmarketcap.com/currencies/xrp/ (accessed 8 June 2020).
[8] Litecoin, Available: https://litecoin.org/ (accessed 11 May 2020).
[9] CoinMarketCap, Litecoin, Available: https://coinmarketcap.com/currencies/litecoin/historical-data/?start=20181002&end=20191002 (accessed 8 June 2020).
[10] Tether, Available: https://tether.to/ (accessed 10 May 2020).
[11] CoinMarketCap, Tether, Available: https://coinmarketcap.com/currencies/tether/ (accessed 8 June 2020).
[12] Bitcoin Cash, Available: https://www.bitcoin.com/get-started/what-is-bitcoin-cash/ (accessed 11 May 2020).
[13] CoinMarketCap, Bitcoin Cash, Available: https://coinmarketcap.com/currencies/bitcoin-cash/historical-data/?start=20181002&end=20191002 (accessed 8 June 2020).

[14] Libra Association, Libra white paper., (accessed June 19, 2019) (2019).
[15] Calibra, Available: https://about.fb.com/news/2019/06/coming-in-2020-calibra/ (accessed 8 June 2020).
[16] Monero, Available: https://web.getmonero.org/ (accessed 8 June 2020).
[17] CoinMarketCap, Monero, Available: https://coinmarketcap.com/currencies/monero/historical-data/?start=20181002&end=20191002 (accessed 8 June 2020).
[18] EOS, Available: https://eos.io/ (accessed 11 May 2020).
[19] CoinMarketCap, EOS, Available: https://coinmarketcap.com/currencies/eos/ (accessed 8 June 2020).
[20] Bitcoin SV, Available: https://bitcoinsv.io/ (accessed 8 June 2020).
[21] CoinMarketCap, Bitcoin SV, Available: https://coinmarketcap.com/currencies/bitcoin-sv/ (accessed 8 June 2020).
[22] Binance Coin, Available: https://www.binance.com/tr/buy-Binance-Coin (accessed 8 June 2020).
[23] CoinMarketCap, Binance Coin, Available: https://coinmarketcap.com/currencies/binance-coin/ (accessed 8 June 2020).
[24] Cardano, Available: https://www.cardano.org/ (accessed 8 June 2020).
[25] CoinMarketCap, Cardano, Available: https://coinmarketcap.com/currencies/cardano/ (accessed 8 June 2020).
[26] Tezos, Available: https://tezos.com/ (accessed 8 June 2020).
[27] CoinMarketCap, Tezos, Available: https://coinmarketcap.com/currencies/tezos/ (accessed 8 June 2020).
[28] Crypto.com.Coin, Available: https://crypto.com/en/chain.html (accessed 8 June 2020).
[29] CoinMarketCap, Crypto.com.Coin, Available: https://coinmarketcap.com/currencies/crypto-com-coin/ (accessed 8 June 2020).
[30] Stellar, Available: https://www.stellar.org// (accessed 11 May 2020).
[31] CoinMarketCap, Stellar (XLM), Available: https://coinmarketcap.com/currencies/stellar/ (accessed 8 June 2020).
[32] ChainLink, Available: https://chain.link/ (accessed 8 June 2020).
[33] CoinMarketCap, ChainLink, Available: https://coinmarketcap.com/currencies/chainlink/ (accessed 8 June 2020).
[34] LEO, Available: https://en.bitcoinwiki.org/wiki/LEOcoin (accessed 8 June 2020).
[35] CoinMarketCap, UNUS SED LEO, Available: (https://coinmarketcap.com/currencies/unus-sed-leo/ (accessed 8 June 2020).
[36] TRON, Available: https://tron.network/ (accessed 8 June 2020).
[37] CoinMarketCap, TRON, Available: https://coinmarketcap.com/currencies/tron/ (accessed 8 June 2020).
[38] Houbi Token, Available: https://www.huobi.com/en-us/ (accessed 8 June 2020).
[39] CoinMarketCap, Houbi Token, Available: https://coinmarketcap.com/currencies/huobi-token/ (accessed 8 June 2020).
[40] NEO, Available: https://neo.org/ (accessed 11 May 2020).
[41] CoinMarketCap, NEO, Available: https://coinmarketcap.com/currencies/neo/ (accessed 8 June 2020).
[42] BitTorrent, Available: https://www.bittorrent.com/token/btt/ (accessed 8 June 2020).
[43] CoinMarketCap, BitTorrent, Available: https://coinmarketcap.com/currencies/bittorrent/ (accessed 8 June 2020).
[44] BitShares, Available: https://bitshares.org/ (accessed 8 June 2020).
[45] CoinMarketCap, BitShares, Available: https://coinmarketcap.com/currencies/bitshares/ (accessed 8 June 2020).
[46] Ethereum Classic, Available: https://ethereumclassic.org/ (accessed 8 June 2020).
[47] CoinMarketCap, Ethereum Classic, Available: https://coinmarketcap.com/currencies/ethereum-classic/ (accessed 8 June 2020).

[48] USD Coin, Available: https://www.circle.com/en/usdc (accessed 8 June 2020).
[49] CoinMarketCap, USD Coin, Available: https://coinmarketcap.com/currencies/usd-coin/ (accessed 8 June 2020).
[50] Hedge Trade, Available: https://hedgetrade.com/ (accessed 8 June 2020).
[51] CoinMarketCap, Hedge Trade, Available: https://coinmarketcap.com/currencies/hedgetrade/ (accessed 8 June 2020).
[52] Cosmos, Available: https://cosmos.network/ (accessed 8 June 2020).
[53] CoinMarketCap, Cosmos, Available: https://coinmarketcap.com/currencies/cosmos/ (accessed 8 June 2020).
[54] Zcash, Available: https://z.cash/ (accessed 8 June 2020).
[55] CoinMarketCap, Zcash, Available: https://coinmarketcap.com/currencies/zcash/ (accessed 8 June 2020).
[56] Hedera Hashgraph, Available: https://www.hedera.com/cryptocurrency (accessed 8 June 2020).
[57] CoinMarketCap, Hedera Hashgraph, Available: https://coinmarketcap.com/currencies/hedera-hashgraph/ (accessed 8 June 2020).
[58] Bitcoin Gold, Available: https://bitcoingold.org/ (accessed 8 June 2020).
[59] CoinMarketCap, Bitcoin Gold, Available: https://coinmarketcap.com/currencies/bitcoin-gold/ (accessed 8 June 2020).
[60] Bitcoin Diamond, Available: https://www.bitcoindiamond.org/ (accessed 8 June 2020).
[61] CoinMarketCap, Bitcoin Diamond, Available: https://coinmarketcap.com/currencies/bitcoin-diamond/ (accessed 8 June 2020).
[62] Komodo, Available: https://komodoplatform.com/ (accessed 8 June 2020).
[63] CoinMarketCap, Komodo, Available: https://coinmarketcap.com/currencies/komodo/ (accessed 8 June 2020).
[64] NEXO, Available: https://nexo.io/ (accessed 8 June 2020).
[65] CoinMarketCap, NEXO, Available: https://coinmarketcap.com/currencies/nexo/ (accessed 8 June 2020).
[66] Zcoin, Available: https://zcoin.io/ (accessed 8 June 2020).
[67] CoinMarketCap, Zcoin, Available: https://coinmarketcap.com/currencies/zcoin/ (accessed 8 June 2020).
[68] Bytecoin, Available: https://bytecoin.org/ (accessed 8 June 2020).
[69] CoinMarketCap, Bytecoin, Available: https://coinmarketcap.com/currencies/bytecoin-bcn/ (accessed 8 June 2020).
[70] Gnosis, Available: https://gnosis.io/ (accessed 8 June 2020).
[71] CoinMarketCap, Gnosis, Available: https://coinmarketcap.com/currencies/gnosis-gno/ (accessed 8 June 2020).
[72] Maker, Available: https://makerdao.com/en/ (accessed 8 June 2020).
[73] CoinMarketCap, Maker, Available: https://coinmarketcap.com/currencies/maker/ (accessed 8 June 2020).
[74] NEM, Available: https://nem.io/ (accessed 8 June 2020).
[75] CoinMarketCap, NEM, Available: https://coinmarketcap.com/currencies/nem/ (accessed 8 June 2020).
[76] Dogecoin, Available: https://dogecoin.com/ (accessed 8 June 2020).
[77] CoinMarketCap, Dogecoin, Available: https://coinmarketcap.com/currencies/dogecoin/ (accessed 8 June 2020).
[78] THETA, Available: https://www.thetatoken.org/ (accessed 8 June 2020).
[79] CoinMarketCap, THETA, Available: https://coinmarketcap.com/currencies/theta/ (accessed 8 June 2020).
[80] ICON, Available: https://icon.foundation/?lang=en (accessed 8 June 2020).
[81] CoinMarketCap, ICON, Available: https://coinmarketcap.com/currencies/icon/ (accessed 8 June 2020).

[82] QTUM, Available: https://qtum.org/en (accessed 8 June 2020).
[83] CoinMarketCap, QTUM, Available: https://coinmarketcap.com/currencies/qtum/ (accessed 8 June 2020).
[84] Siacoin, Available: https://sia.tech/ (accessed 8 June 2020).
[85] CoinMarketCap, Siacoin, Available: https://coinmarketcap.com/currencies/siacoin/ (accessed 8 June 2020).
[86] CoinMarketCap, Monacoin, Available: https://coinmarketcap.com/currencies/monacoin/ (accessed 8 June 2020).
[87] TERRA, Available: https://terra.money/ (accessed 8 June 2020).
[88] CoinMarketCap, Terra, Available: https://coinmarketcap.com/currencies/terra-luna/ (accessed 8 June 2020).
[89] Flexacoin, Available: https://flexacoin.org/ (accessed 8 June 2020).
[90] CoinMarketCap, Flexacoin, Available: https://coinmarketcap.com/currencies/flexacoin/ (accessed 8 June 2020).

About the authors

Shubhani Aggarwal is pursuing PhD from Thapar Institute of Engineering & Technology (Deemed to be University), Patiala, Punjab, India. She received the BTech degree in Computer Science and Engineering from Punjabi University, Patiala, Punjab, India, in 2015, and the ME degree in Computer Science from Panjab University, Chandigarh, India, in 2017. She has many research interests in the area of Blockchain, cryptography, Internet of Drones, and information security.

Neeraj Kumar received his PhD in CSE from Shri Mata Vaishno Devi University, Katra (Jammu and Kashmir), India in 2009, and was a postdoctoral research fellow in Coventry University, Coventry, UK. He is working as a Professor in the Department of Computer Science and Engineering, Thapar Institute of Engineering & Technology (Deemed to be University), Patiala (Punjab), India. He has published more than 400 technical research papers in top-cited journals such as IEEE TKDE, IEEE TIE, IEEE TDSC,

IEEE TITS, IEEE TCE, IEEE TII, IEEE TVT, IEEE ITS, IEEE SG, IEEE Netw., IEEE Comm., IEEE WC, IEEE IoTJ, IEEE SJ, Computer Networks, Information sciences, FGCS, JNCA, JPDC, and ComCom. He has guided many research scholars leading to PhD and ME/MTech. His research is supported by funding from UGC, DST, CSIR, and TCS. His research areas are Network Management, IoT, Big Data Analytics, Deep Learning, and Cyber-security. He is serving as editor of the following journals of repute: ACM Computing Survey, ACM·IEEE Transactions on Sustainable Computing, IEEE·IEEE Systems Journal, IEEE·IEEE Network Magazine, IEE·IEEE Communication Magazine, IEE·Journal of Networks and Computer Applications, Elsevier·Computer Communication, Elsevier·International Journal of Communication Systems, Wiley. Also, he has been a guest editor of various International Journals of repute such as IEEE Access, IEEE ITS, Elsevier CEE, IEEE Communication Magazine, IEEE Network Magazine, Computer Networks, Elsevier, Future Generation Computer Systems, Elsevier, Journal of Medical Systems, Springer, Computer and Electrical Engineering, Elsevier, Mobile Information Systems, International Journal of Ad Hoc and Ubiquitous Computing, Telecommunication Systems, Springer, and Journal of Supercomputing, Springer. He has also edited/authored 10 books with international/national publishers like IET, Springer, Elsevier, CRC: Security and Privacy of Electronic Healthcare Records: Concepts, Paradigms and Solutions (ISBN-13: 978-1-78561-898-7), Machine Learning for Cognitive IoT, CRC Press, Blockchain, Big Data and IoT, Blockchain Technologies Across Industrial Vertical, Elsevier, Multimedia Big Data Computing for IoT Applications: Concepts, Paradigms and Solutions (ISBN: 978-981-13-8759-3), Proceedings of First International Conference on Computing, Communications, and Cyber-Security (IC4S 2019) (ISBN 978-981-15-3369-3). One of the edited textbook entitled, "Multimedia Big Data Computing for IoT Applications: Concepts, Paradigms, and Solutions" published in Springer in 2019 is having 3.5 million downloads till June 6, 2020. It attracts attention of the researchers across the globe (https://www.springer.com/in/book/9789811387586). He has been a workshop chair at IEEE Globecom 2018 and IEEE ICC 2019 and TPC Chair and member for various international conferences such as IEEE MASS 2020 and IEEE MSN 2020. He is a senior member of the IEEE. He has more than 12,321 citations to his credit with current h-index of 60 (September 2020). He has won the best papers award from IEEE Systems Journal and ICC 2018, Kansas City in 2018. He has been listed in the highly cited researcher of 2019 list of Web of Science (WoS). In India, he is listed in top 10 position among highly cited researchers list. He is an adjunct professor at Asia University, Taiwan, King Abdul Aziz University, Jeddah, Saudi Arabia, and Charles Darwin University, Australia.

Empowering digital twins with blockchain

Pethuru Raj
Site Reliability Engineering (SRE) Division, Reliance Jio Platforms Ltd. (JPL), Bangalore, India

Contents

Abstract

A digital twin is an exact digital/logical/cyber/virtual representation/replica of any tangible physical system or process. And the digital twin runs on a competent IT infrastructure (say, cloud centers). In essence, a digital twin is typically a software program that takes various real-world data about a ground-level physical system as prospective inputs and produces useful outputs in the form of insights. The outputs generally are the value-adding and decision-enabling predictions or simulations of how that physical system will act on those inputs. These help in quickly and easily realizing highly optimized and organized products with less cost and risk.

The manufacturing industry had embraced the digital twin technology long time back to be modern in their operations, outputs, and offerings. The distinct contributions of the digital twin paradigm, since then, have gone up significantly with the seamless synchronization with a number of pioneering technologies such as the Internet of Things (IoT), artificial intelligence (AI), big and streaming data analytics, data lakes, software-defined cloud environments, blockchain, etc. With the concept of cyber physical systems (CPS) is being adopted and adapted widely and wisely, complicated yet sophisticated electronics devices at the ground level are being blessed with their corresponding digital twins. The digital twins enable data scientists and system designers to optimize a number of things including process excellence, knowledge

Advances in Computers, Volume 121
ISSN 0065-2458
https://doi.org/10.1016/bs.adcom.2020.08.013

discovery and dissemination in time, better system design, robust verification and validation, etc. In the recent past, with the flourishing of the blockchain technology, the scope for digital twins has gone up remarkably. This unique combination is bound to produce additional competencies and fresh use cases for enterprises. This chapter is to explain how they integrate and initiate newer opportunities to be grabbed and gained for a better tomorrow.

1. Introduction

It is natural that business requirements and people expectations are varying and growing invariably. Breakthrough technologies individually and collectively have been fulfilling them with ease and elegance. With the explosion of IoT devices in our everyday environments, there is a need expressed widely to empower them with required intelligence in order to exhibit adaptive behavior in their actions and reactions. The mandate is to transition connected devices to be smart. The prime and swanky idea toward the smartness is to transition gleaned data into information and into knowledge. The changeover process gets accelerated through a few technological solutions such as machine and deep learning algorithms, integrated data analytics platforms, etc. That is, gathered data is subjected to a variety of deeper and decisive investigations in order to extract useful information and workable insights.

For making IoT devices to be smart, there came a number of ways and means vehemently expounded by many. One of them is to establish digital twins for IoT devices. Digital twins collect value-adding data from their corresponding physical twins continuously and crunch them in order to emit out useable insights out of data heaps. The acquired knowledge gets looped back to the physical devices to adequately nourish them in completing their assignments well. This continuous nurture helps them to behave intelligently according to the fast-changing situation.

However, with digital assets and data growing briskly everywhere, there is a need for utmost security and transparency to realize and reap the promised value for digital artifacts. Blockchain chips in here with its unique competencies and characteristics. Thus, the distinct combination of digital twin and blockchain technologies is being proclaimed and presented as the best-in-class method to eliminate any loophole and to boost the confidence of users on leveraging digital assets for their everyday transactions. In the subsequent sections, we are to discuss how this combination is going to be a great game-changer for myriads of industry houses.

2. Briefing the digital twin (DT) paradigm

As indicated above, digital twins are the digital versions of physical objects. In addition, digital twins comprise the details of the physical objects' surroundings or processes related to them. That means, digital twins consist of the relevant IT components for status updates and connectivity. They are also stuffed with defined data structures and user interfaces to facilitate knowledge visualization through 360-degree dashboards.

When you are using your smartphone for your communication, computing, and collaboration purposes, surfing the Internet using your laptop for knowledge gain, storing your data in cloud environments for affordability, using social networking tools to socialize, etc., you are actually creating a digital footprint of yourself. This digital footprint is a kind of digital twin. Similarly, you can think of digital blueprints for a building as its digital twin. A computer simulation of an aircraft engine can be as its digital twin.

Digital twin in the human context is all the data footprint that we as human create through our attributes, interactions, and online presence. These decision-enabling elements earnestly get combined with one another and their combined outputs are getting crunched using advanced and real-time AI and data analytics platforms, it is possible to predict our health condition and other leanings accurately.

- *Attributes* could be name, age, gender, address, ideology, language, culture, religion, ethnicity, education, salary, etc. These are actually categorized as base data about us as individuals, and these can go a long way in predicting whether we are prone to any specific health issue. Where we were born and we live lately are also contributing handsomely in arriving at correct prediction. All these data are meticulously gathered and stored digitally. Thus, digital twins are possible for humans also.
- *Interactions*—This is all about our everyday interaction with the world. The details of where and when we are traveling and our eating out come handy in precisely predicting our health status. Our health monitoring wearables can minutely capture a number of our body parameters such as fitness and blood sugar levels, heart-beating rates at different times, etc., and deposited the captured in our own digital twin. There are other electronic instruments, digital assistants, and smartphones to automatically monitor and measure our everyday movements, exercises, work stress, etc. All these are meticulously get aggregated and stocked in

our own digital twin. In conjunction with our unique attributes, digital twin can predict the health state prediction and prescribe of what to do.
- *Online presence*—We have a number of professional and social media applications that ultimately articulate our personality. Thus, human digital twins are also gaining prominence. Digital twin for humans can help predict if someone has a possibility to be down with any life-threatening disease. Also, it can help provide insights on changing lifestyle to stay healthy and live longer.

The digital twin of machines can predict the need for their timely maintenance. Similarly, digital twins help to come out with the next-generation machine design that is versatile, resilient, and robust. Precisely speaking, digital twins are digital replicas of a business, process, or product over its life cycle.

Digital twins are being meticulously built for physical twins. Software experts begins building futuristic digital twins leveraging their education, experience, and expertise on data science, statistics and mathematics, computer algorithms, etc. Digital twins' developers deeply research the physics that underlie the physical system being mimicked. That information greatly helps to visualize and develop a mathematical model that elegantly simulates the real-world physical system. The mathematical model gets turned into software package to run in the digital space. Maneuvering, managing, and maintaining digital twins are quite easy and fast. Thus, for adding new features to existing physical systems, digital twins contribute immensely in identifying and articulating the pros and cons of incorporating the features. Above all, before building complex and highly integrated systems from the ground up, their digital twins come handy in envisaging the impending risks and opportunities and in answering the difficult questions. All these acquired knowledge through digital twins goes a long way in setting up and sustaining complicated and sophisticated physical systems. Thus, digital twins are being projected as complexity-mitigation mechanism.

Physical systems like aircraft engines are being embedded with a number of multifaceted sensors, which continuously gather a lot of useful information of physical systems and pass it on to the digital twin in real time. This empowers the twin to crunch the received data quickly to offer insights (performance, health condition, security, failure prediction, etc.) to physical system designers and operators.

There is another tweak called as "Predictive twin." This can model the future state and behavior of the device. This prediction is derived based on historical data from other devices. Predict twins have the inherent capability to simulate breakdowns and other situations that need immediate attention.

Microsoft has come out with another term "Process Digital Twin." This is the next level compounding product digital twin benefits throughout the factory and supply chain. Process twins can highlight some advanced manufacturing scenarios that product digital twins cannot provide.

In summary, the digital twin typically can include a description of the devices, a 3D rendering, and details on all the sensors embedded in the device. Based on continuously generated sensor readings, the digital twins can simulate real-life options and operations of the device.

3. Digital twins: The industrial use cases

The aspects of digitization and digitalization are grasping the attention of enterprising businesses across the world. Digitization is all about empowering our everyday objects to be digitized. Digitalization is all about deriving digital intelligence out of digital data. The onset of integrated data analytics platforms, the faster maturity and stability of artificial intelligence (AI) algorithms, the realization of highly optimized and organized IT through cloudification, which represents IT industrialization, smartphones, and other I/O devices for enabling IT consumerization, virtualization, and containerization under the scope of IT compartmentalization, etc., have laid down a stimulating and sparkling foundation for the impending digital era. Cities are becoming digitally transformed cities. Similarly, other entitles, such as homes, hotels, hospitals, etc., are being transitioned into digitally transformed entities. There are a number of digital innovations and disruptions fulfilling the real digital transformation. Digital life has become the new normal.

The emerging concept of digital twins is one of the transformational technologies toward the realization of digital era. The noteworthy point of any digital twin is that it can be continuously updated with data from its physical counterpart. With trillions of sensors and billons of connected devices, digital twins will be readied and deployed for millions of physical things. An aircraft engine, a human heart, and an entire city can have its own digital twin that mirrors the same physical and biological traits as the real thing.

The implications are definitely profound in many ways. Real-time assessments and diagnostics are now much more precise. Trial and errors, chaos experiments, and course correction can be easily accomplished through digital twins rather than physical twins. Thus, digital twins induce and inspire faster, cheaper, and clever innovations. Digital twins are to change the innovation game by enabling three critical drivers.

- *Continuous evaluation*—Smart sensors being pasted on a product are capable of capturing and continuously updating the product's digital twin throughout its lifetime. This empowerment goes a long way for the betterment of the products. Predictive maintenance of products is being achieved. For example, advanced cars are being beneficially endowed through their own digital twin. A slew of multifaceted sensors are attached on vehicles to capture the operational value of each of the important components. That is, vehicles on the road can send data to their digital twin to do real-time processing to emit out actionable insights in time. In short, digital twin-enabled vehicles simply enjoy a number of distinct benefits. The fuel efficiency can go up whereas the vehicle performance is bound to rise up considerably.
- *Faster and cheaper prototyping*—The virtual version of any physical system comes handy in producing prototypes easily and quickly. Besides speeding up the innovation, the cost of producing physical products comes down sharply. Oklahoma State University developed a digital twin of an aerosol drug, which is intended to reach lung to annihilate tumor cells. By smartly varying parameters on the digital twin such as inhalation rate and particle size, scientists could clearly increase the number of particles reaching their target from 20% to 90%. There are a number of such promising case studies illustrating the strategic significance of building and running digital twins.
- *Uninhibited innovation*—As inscribed above, the digital twin paradigm has unquestionably amplified the innovation quotient. There are companies exploring the digital twin idea in experimenting and experiencing realistic innovations. Predicting and managing traffic congestion in particular locations are being made possible through the leverage of digital twins. SenSat, a company specializing in creating digital twins of cities, has created a digital twin of Cambridge, England. All the traffic snarls and hiccups get eliminated from the city streets.

As a digital and dynamic model, test scenarios can be performed on it to gain a deeper understanding and predict future events. In operation, as the physical asset undergoes changes, these modifications are being minutely captured and handed over to the user on a real-time basis. The overall functional role of a digital twin is three pronged—observation, optimization, and operation. The digital twin technology guarantees to transform the way industrial products and machineries are being instrumented, implemented, integrated, and operated across industries.

4. Digital twin industry use cases

Several business houses are keenly embracing this promising technology phenomenon in order to be ahead of their competitors in the knowledge-driven market condition. A few of them are given below:

- *Manufacturing* across industry verticals is the first and foremost one leveraging this unique technology in order to automate, accelerate, and augment several things in their day-to-day operations. New product designs are being facilitated while manufacturing processes are being continuously studied and optimized through actionable insights emitted by digital twins.
- *Automotive* industry is also seriously and sagaciously experimenting with the digital twin conundrum to develop next-generation vehicles. Increasingly cars are already fitted with advanced telemetry sensors, and their readings are continuously fed into digital twins in real time to bring in decisive and deeper automation especially in transitioning from connected cars to smart cars.
- *Healthcare* is another promising sector tinkering extensively with the digital twin paradigm. The mission-critical medical electronics devices and instruments are being literally empowered through their digital replicas, which are being hosted and run in cloud environments. As there is a need for large-scale IT resources for aiding big and streaming data analytics, cloud centers are being preferred to support and sustain the digital twin idea.

  There are band-aid-like sensors being used for tracking various body parameters. The captured health data is transmitted to digital twins, which can quickly crunch the received data and predict if there is any important information with the person's health. Digital twins can be used to predict different outcomes based on variable data. In manufacturing, with highly instrument devices, the digital twins can simulate and showcase how the devices have performed over time. With such capability, digital twins can easily predict the device performance in future. Similarly, the failure prediction also can be done. With the explosion of IoT devices and sensors/CPS, the popularity of digital twins goes up considerably.

Digital twins of windmills can support predictive maintenance. Sensor and camera data from mission-critical establishments such as nuclear installations, wind parks and power plants ensure continuous surveillance. Further on, digital twins can proactively and preemptively alert the concerned if there is any nefarious action in and around walled and essential zones. Specialized sensors are being used in cars in plenty these days. There is a statistical information saying that every single sophisticated car involves hundreds of microcontrollers. Increasingly infotainment systems inside any vehicle act as the IoT gateway for taking sensor data to the digital twin to be instantaneously processed to take quick corrective action. The IoT dream is being facilitated through state-of-the-art digital twins. For the connected world, the next logical step is to leverage digital twins adroitly to move toward the projected smart world.

In summary, a digital twin is a dynamic and digital representation of a physical system. A digital blueprint of a house or building can be its digital twin. Similarly, a computer-aided simulation of an automobile transmission system can be considered as a digital twin. In short, things are increasingly becoming complex. Not only visualizing and building but also operating complicated systems has become a tough and time-consuming affair. There are a number of complexity-mitigation techniques and tips. The proven technique of "Divide and conquer" has been the hallmark for software engineering. Modeling, process-centricity, and simulation are the other key methods enabling the construction of complex systems across the industry verticals. With the advancements in the AI space, empowering systems to self-learn and adapt has gained prominence. Digital twins are also being portrayed as one of the pure and sure ways to tackle the increasing complexity of systems and their functioning.

5. Digital twins: The benefits

Digital twins offer a real-time and decisive look at what's happening with physical assets. The knowledge gained could reduce maintenance costs and pains drastically. For an example, Chevron is rolling out the proven digital twin technology for its oil fields and refineries. Siemens uses digital twins to accurately model and prototype objects, which are not yet manufactured. The information extracted can reduce product defects and hence the time to market gets reduced sharply.

Digital twins enable companies to track the past, current, and future performance throughout the asset's lifecycle. The asset, for example, a vehicle or spare part, sends its performance data and distinct events directly to its corresponding digital twin, even as the asset moves from the manufacturer to the dealer and ultimately the new asset owner. Blockchain databases can be used to securely document everything related to the asset and IoT provides the real-time monitoring and updates. Microsoft and VISEO are partnering to use blockchain to connect each new vehicle's maintenance events to the vehicle's digital twin. Thus, digital twins are able to streamline additive manufacturing. With a series of innovations and disruptions in the digital space, the world is looking forward toward real digital transformation. The digital twins idea with continuous nourishment is to contribute for the digital transformation, the world is earnestly yearning for.

6. Blockchain as a powerful antidote

In the first part of this chapter, we have extensively discussed about the futuristic and fabulous implications of the digital twin paradigm across industry verticals. The perpetual and perplexing question is that if data breaches happen here. As we all know, wrong or corrupted data leads to wrong decisions. Thus, securing digital data and twins gains.

Data is extremely essential for the intended success of the digital twin paradigm. New data streams in the connected world typically comprise the participating physical objects' attributes, the interaction data with other physical objects, and future states. The data interchange happens between the physical and digital worlds. While digital twin instances (DTIs) are mirroring a physical entity and activity, the digital twin aggregates (DTA) is a composite of individual DTIs to achieve bigger and better things. With this combination and collaboration, a breach of a DTI will compromise one product. A breach of DTA will expose a set of participating products.

The emerging blockchain technology has the innate potential to surmount the safety and security lacunae to put digital twin projects on the track. As blockchain technology is blessed with the decentralization and immutability attributes, digital twins projects can innovate better and faster through the shrewd usage of blockchain.

The cryptographic characteristics of blockchain ensure safe and secure data transfer. By having a robust authentication of users and data sources, data immutability along with the safety of digital twins promises a lot for the digital world. As explained in the previous chapters, through one or other consensus mechanism incorporated in the blockchain technology, multiple stakeholders, collaborators and users, can be brought under one umbrella. The stakeholders have become the gate-keepers enforcing transparency and accountability. Thus, confidential, customer, and corporate information gets transmitted without any fear.

Through the astute combination of DTAs and blockchain, we can unquestionably expect the blockchain-enabled and path-breaking digital twins. From there, every business domain can visualize next-generation products, processes, operational models, etc.

Blockchain technology is a decentralized ledger with public records of all physical products and their transactions. The data can be viewed and verified publicly. However, modifications to it can be made only by authorized executives, and the ledger retains a history of every modification. In the previous chapters, we have discussed about the power of blockchain in surmounting many current problems and in laying down a breakthrough platform for envisaging and implementing newer use cases. Having understood the impacts, every noteworthy business domain is fast experimenting and evolving a strategic and sustainable blockchain plan. The blockchain technology is being empowered through a smart integration with other pioneering technologies such as the IoT, digital twin, AI, etc. On other hand, these technologies are being succulently and solidly empowered through the sagacious synchronization with blockchain. There are white papers, case studies, research publications, and best practices authored by technical experts, evangelists, and exponents clearly accentuating and eulogizing how the technology combination works wonders for several perpetual problems. In this and subsequent sections, we are to expound and espouse how different popular industry domains are tremendously benefiting out of this unique combination.

Eliminating counterfeits—On the supply chain side, now by combining digital twin and blockchain, businesses, and brands can easily protect their products from being wilfully counterfeited and increase the revenue remarkably. As widely known, one of the biggest concerns for worldwide businesses these days is skilled counterfeiting of their products. With the arrival of advanced technologies and tools, fraudsters can quickly create replicas for various expensive products and sell them to unsuspecting people at a cheaper price. This dangerous situation not only damages the brand name but also causes huge financial losses. Digital twin in conjunction with blockchain can bring forth a competent solution to thwart such frauds. They together guarantee the authenticity of products. As articulated above, digital twins facilitate the digitization of physical artifacts. As explained in the previous chapters, blockchain with its decentralization capability will bring in the much-needed transparency. This strengthen the security of any digital data.

7. The combination of digital twins and blockchain

The advent of the Blockchain technology has redefined the progress of digital twins. At macro level, blockchain enables the storage of information in a decentralized manner with no single point of control. Fresh transactions can be added, but the existing ones cannot be changed or corrupted. This clearly ensures transparency and data integrity.

Creating a digital twin on blockchain would mean that all information regarding a physical product can be saved immutably. Furthermore, the product's transaction records can be saved. This means prospective buyers can get all the information of the product from its origin and also the journey it took off. This will help create proofs of authenticity and identity.

Consider a premium watch. There exist countless replicas of this costly watch. Fake watches can be made along with fake documents for proving their authenticity. However, creating digital twins on blockchain decimates such fake products and documentations. Digital twins, in a way, act as *digital certificates for the digital age.* Digital certificates stored and issued with the help of Blockchain will ensure that these certificates can't be stolen, modified, and used wrongly.

Blockchain is becoming a popular digital technology that allows every transaction to be tracked in an inviolable way. The blockchain technology is gaining the inherent capability to redefine the concept of digital twins. It could turn out to be an indispensable tool to skilfully aid the application of digital twins in the ensuing IoT era. Sectors such as manufacturing, healthcare, and retail are set to benefit immeasurably from these technologies interlinking with one another. There is a tight coupling between the physical world and the digital world.

A digital twin built with traditional technologies must need a central intermediary for stocking digital data and for doing data analytics. Now, with the booming blockchain technology, the aspect of digital twins is bound to go through strategically sound transformation. The security and immutability being guaranteed by blockchain are to take digital twin to the next level in the days to come.

Keeping a digital twin (more specifically digital certificate) on blockchain helps businesses to retain information about their products perpetually and securely. Further on, the data regarding transactions for these products also can be saved in the blockchain database. This allows any perspective buyer to get the complete details including the real manufacturer, who are the other owners of the product, etc., about the product during purchase.

Precisely speaking, it helps establish the authenticity and provenance of the product. It is possible to craft a fake document to prove its authenticity. By creating digital twins/certificates on the blockchain, fake products can be eliminated from the sordid market. An immutable and tamper-proof digital certificate for each of the products is the way forward to nullify any kind of fraud.

Blockchain actually stores digital certificates that embed all the relevant data regarding the products registered by businesses. Blockchain ensures that the data embedded in the digital certificates cannot be copied, modified, and even deleted by others. Businesses can register their items on blockchain platforms. As enunciated above, every concrete product has its corresponding digital twin and in this case, it is digital certificate. The digital certificate gets updated at every stage of the product cycle right from the manufacturing stage. When the product is shipped, the information regarding the place, mode, and shipping destination can be updated with timestamps. When customers buy the product, they can verify the authenticity of the item by verifying the digital twin of the product on the blockchain. After the item is purchased, the ownership gets transferred to the new owner by the authorized salesperson. All the previous data stored is maintained as such. No modification is possible.

Without an iota of doubt, blockchain brings the much-needed reliability and credibility to businesses. Clearly, blockchain is being touted as an anticounterfeiting and antitheft technology. If businesses across the world implement this with a proper care, then the evil things of faking and falsification can be easily surmounted. The confidence of people on various products gets a strong boost. Buyers get all the right information about any product instantaneously. Unverified ownership will become a thing of the past and people get to know the complete details about every product. This inspiring combination is penetrating into every industry vertical ranging from manufacturing, healthcare, advertisement, and logistics.

Digital twin and blockchain in logistics—Supply chain processes can be unequivocally streamlined and secured using digital twin and blockchain together. Shipping containers embedded with a number of IoT sensors are connected to the blockchain platform in order to ensure a complete transparency in the complex logistics process. The products empowered through multiple sensors can have their own digital twins, which land on blockchain. This powerful entitlement helps to track and trace the products. This helps the stakeholders to have complete information about the product and the transportation process. The documentation for the products also can be stocked in blockchain database eliminating any kind of modification and forgery. The documents with product and payment details can be tracked digitally. The transparency aspect of the logistics sector can be fulfilled through blockchain and digital twin combination.

Digital twin and blockchain in utilities—Digital twin and blockchain can be used in the utility sector to help utility operators in efficiently fulfilling their consumers' requirements. As we know, the production and consumption of power energy varies sharply based on consumer behavior. Nowadays, electricity meters in our homes are being attached with IoT sensors to be transitioned into smart electricity meters. These connected meters can directly communicate the consumption data every few minutes to the smart energy grid. The energy consumption details of every individual consumer can be digitally updated on the blockchain to guarantee transparency. The corresponding digital twin for the smart energy grid can receive and analyze the energy consumption data to extricate actionable intelligence. This gives energy service providers (energy generation, transmission, and distribution) a clear understanding about the consumer behavior. Accordingly, power producers and suppliers can do everything in an optimal manner. On the other hand, consumers based on peak energy costs could adjust the power usage time to time to reduce expenditure on energy.

Digital twin and blockchain in healthcare—The use of digital twin and blockchain together can bring in drastic changes for the healthcare industry. As the whole world is severely stricken with Covid 19 virus, the smart combination of multiple technologies is the way forward to comprehensively overcome the grave challenges especially in curtailing the growing fatalities.

For a prime example, the digital twin of a patient's heart can be realized through the digital version of a pacemaker. This digital twin can specially empower cardiologists, who are carrying out cardiac resynchronization therapy, in precisely identifying the position of leads on the specific patient. All the lead placement positions pinpointed through the digital twin can be tried virtually to determine the best suitable position. This happens well before the real surgery begins. The best suitable position enables the surgeons to complete the operation with all the intended success. Blockchain can safely hold records of the medical history of the patient. The trustworthy and timely data can be accessed anytime by healthcare providers and professionals to ponder about the next course of action in case the person develops any untoward complications in the future.

Thus, digital twins and blockchain can blend together to bring in a bevy of transformational effects across industry verticals. By merging digital twin with blockchain technology, a secure and decentralized digital ecosystem can be formed and sustained. This solemnly assures businesses and end-users the authenticity of the products and services.

Aircraft industry—The manufacturers of different components of aircrafts are being strictly governed by unambiguously written technical standards. The ultimate aim is to certify and monitor the component production process. Further on, this industry domain increasingly uses additive manufacturing technologies to accomplish rapid prototyping of product components. Aided by highly optimized supply chain, the time to market comes down sharply while not compromising on the product quality. There are recommendations for producing and sustaining digital twins for additive manufacturing. It is recognized that the fusing of digital twin and blockchain can be a trend-setter for many industry sectors.

IBM Watson IoT, a cognitive system that learns from and infuses intelligence into the physical world, is fervently expanding its scope further by creating digital twins by using blockchain. Also, IBM is implementing blockchain to existing digital twins. It is believed that blockchain can substantially increase the cost-effectiveness of digital twins. Utilizing blockchain with digital twins facilitates digital identity and data tracking through blockchain traceability.

With a greater understanding, enterprises are keenly experimenting with the blockchain paradigm, which is fast penetrating and participating beneficially in many business domains. As we all know, Internet is the world's largest information superhighway. However, data transmission through the Internet, which is the most affordable and global-scale communication infrastructure, is battling a few security and privacy challenges. Blockchain is being projected as the promising savior here. Further on, blockchain-based infrastructure is being pronounced as the next-generation phenomenon for the secure exchange of value.

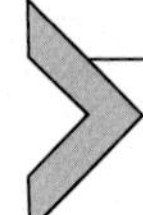

8. Blockchain and digital twins for enhanced digital value

With blockchain in place, the concept of digital value transfer will be implemented in a completely decentralized way. Crypto chips, which can be easily and quickly implanted on any physical asset, are enabling scores of physical assets getting directly linked to a blockchain platform of choice. The asset can be anything ranging from credit cards, personal devices, consumer electronics, drones, robots, and appliances, medical instruments, to industry machineries. The metadata of physical assets are being stored in blockchain databases. This guarantees that transactional and operational data actually originate from highly trustworthy sources and spread across with proven integrity. Private keys never leave the crypto chip and hence transaction data is fully encrypted while in transit until it gets formally recorded in an immutable ledger system.

Digital twins are increasingly stored on blockchains. Therefore, digital twins, while gaining the much-needed trust, are used efficiently to regulate business transactions involving multiple participants and parties. With trust enshrined through blockchain, we can safely expect that here will be many more innovations in the flourishing field of digital twin.

9. Digital twins for sharing economy

The realization of industrial IoT (IIoT) has led to smart manufacturing. The industry 4.0 vision gets accelerated through the astute usage of the IIoT competencies. Factory floors are skilfully automated through a variety of IoT sensors and actuators. IoT data, which is typically big in size, is being meticulously analyzed in order to get the useful insights in order to plan and execute appropriate tasks in order to steer businesses in the right direction. Industry machineries are not only connected but also made smart in their actions and reactions. IIoT in consonance with transformative technologies such as AI, blockchain, digital twin, etc., can be a fulsome hallmark for the future of our business powerhouses. However, industry players' efforts to monetize their industry IoT devices, which they own or use for production were without any grand success thus far. The rise of the sharing economy and its supporting technologies represents a paradigm shift in the access and consumption of IoT systems and services.

In traditional transactions, the transfer of ownership of a product to a buyer is the end of the story for the manufacturer or vendor in the value-creation process. In the new sharing economy, either the asset is a shared entity or the device/asset's output is portable. For example, a car is a physically portable asset that can be collaboratively consumed. But in the case of a windmill, it is the asset's output (wind energy) that gets shared across.

The dual concepts of digital twins and machine-to-blockchain networks are portrayed as key technology facilitators for intrinsically empowering industrial IoT devices to participate in the fast-evolving sharing economy. These concepts can support and sustain the distinct ideals of the sharing economy. As the digital era evolves, newer business models are emerging.

One promising model is a service-based business model. This can be deftly applied to individual IoT devices, and this can result in gaining an auxiliary revenue stream based on usage or outcomes. This new paradigm brings two important things. The first one is digital twins, which can virtually articulate the unique capabilities of the IoT device to any prospective buyer. The second is machine-to-blockchain, which facilitates smart and secure monetary transactions.

10. Conclusion

With the flourish of interactive IoT devices, a massive amount of multistructured data gets generated, transmitted, and stored in local and remote storage systems. There are powerful AI algorithms and integrated data analytics platforms for processing IoT data using batch and real-time processing methods to extract actionable insights. On the other hand, there are digital twins being built for IoT devices in order to deeply understand and articulate their structural properties and behaviors in different contexts. Existing and emerging IoT devices and their constant interactions with others in the vicinity and with remotely held applications and data sources/stores are being decisively investigated through their digital counterparts individually and collectively. Digital twins even simulate the working of physical products according to changing situations and hence a lot of useful knowledge gets delineated before the said products are physically manufactured. Thus, the scope of usage and benefits of digital twins are consistently on the rise. However, IoT devices can be remotely pierced through by brilliant hackers damaging the device networks to do irreparable damages to the IoT systems and applications. Similarly, IoT device data, when in transit and rest, can be breached to steal confidential information. Further on, digital twins can be made inaccessible, or corrupted data can be submitted to digital twins to make wrong decisions, etc.

With the faster maturity and stability of blockchain, the IT experts are exploring the possibility of establishing a linkage between digital twins and blockchain in order to ward off the security threats. The combination of these two technologies presents a positive future for the business and IT domains. How the blending of these two path-breaking digital technologies could bring forth a number of digital innovations and disruptions gets detailed in this chapter. The combined entity has laid down a strategically sound platform for envisaging and implementing a number of business, technology, and user cases.

Digital twins are also being continuously updated and upgraded by incorporating the cutting-edge technologies such as machine and deep learning (ML/DL) algorithms, computer vision (CV), and natural language processing (NLP) methods. Thus, with the cool synchronization with blockchain, the digital twin domain is to flourish in the years to come.

About the author

Pethuru Raj is working as the chief architect and vice-president in the Site Reliability Engineering (SRE) division of Reliance Jio Platforms Ltd., Bangalore. His previous stints are in IBM global Cloud center of Excellence (CoE), Wipro consulting services (WCS), and Robert Bosch Corporate Research (CR). In total, he has gained more than 19 years of IT industry experience and 8 years of research experience.

Finished the CSIR-sponsored Ph.D. degree at Anna University, Chennai and continued with the UGC-sponsored postdoctoral research in the Department of Computer Science and Automation, Indian Institute of Science (IISc), Bangalore. Thereafter, he was granted a couple of international research fellowships (JSPS and JST) to work as a research scientist for 3.5 years in two leading Japanese universities. Published more than 30 research papers in peer-reviewed journals such as IEEE, ACM, Springer-Verlag, Interscience, etc. Has authored and edited 20 books thus far and contributed 35 book chapters thus far for various technology books edited by highly acclaimed and accomplished professors and professionals.

Focuses on some of the emerging technologies such as the Internet of Things (IoT), Artificial Intelligence (AI), Big and Fast Data Analytics, Blockchain, Digital Twins, Cloud-native computing, Edge/Fog Clouds, Reliability Engineering, Microservices architecture (MSA), Event-driven Architecture (EDA), etc. His personal web site is at www.tinyurl.com/peterindia.

Phases of operation ☆

Shubhani Aggarwal and Neeraj Kumar
Thapar Institute of Engineering & Technology, Patiala, Punjab, India

Contents

Abstract

Data and records that can be generated from transactions get stored in the form of blocks. A blockchain transaction is a transfer of data or cryptocurrency between the nodes of the blockchain network. Each transaction generates a hash value and approved by the majority of nodes to add it to the block of the chain. Each block refers to a previous chain and combined to make the blockchain. Each node has its public ledger to store the history of the transactions. In this chapter, we have described the creation and verification of blockchain transactions. Then, we discuss the creation of block in the chain. After creation, the block verification and validation process using PoW has been described used to authenticate and authorize the transactions.

Chapter points

- In this chapter, we discuss the generation and verification of blockchain transactions.
- Here, we also discuss the block creation, block verification, and block validation on the blockchain network.

☆ Working model.

Advances in Computers, Volume 121
ISSN 0065-2458
https://doi.org/10.1016/bs.adcom.2020.08.014

1. Transaction generation and transaction verification

Blockchain technology is one of the latest innovations and it is the underlying technology behind bitcoin, i.e., digital cryptocurrencies. Apart from digital payments, blockchain technology has tremendous potential which is one of the reasons it has become so popular.

1.1 Transaction generation

Blockchain is a digital record of different transactions taking place between several bitcoin addresses. All of these transaction records are updated by the network across every node on the blockchain as the balance increases and decreases by any type of amount. The simplified bitcoin blockchain is as shown in Fig. 1.

Fig. 1 consists of blocks that contain one or more transactions, which are collected into a transaction part of the block. The first transaction of the block is called a *coinbase transaction or generation transaction* that should collect and spend the block rewards. To store all the transactions in one block, copies of each transaction are hashed, and these hashes are paired, hashed and paired again and then, hashed again until a single hash remains and returns the merkle root hash (MRH) tree as shown in Fig. 2A. If there are odd number of transactions then, the transaction without partner is hashed with a copy of itself as shown in Fig. 2B.

The calculated MRH is stored in the block header. Each block of the blockchain also stores the hash of the previous block's header that connects and binds the blocks together which is called block chaining. This method ensures that the transaction in the block cannot be tampered or changed without modifying the block that records it and all the following blocks. The transactions of the block are also chained together that spends the bitcoins to make a payment on the network. These bitcoin payments are

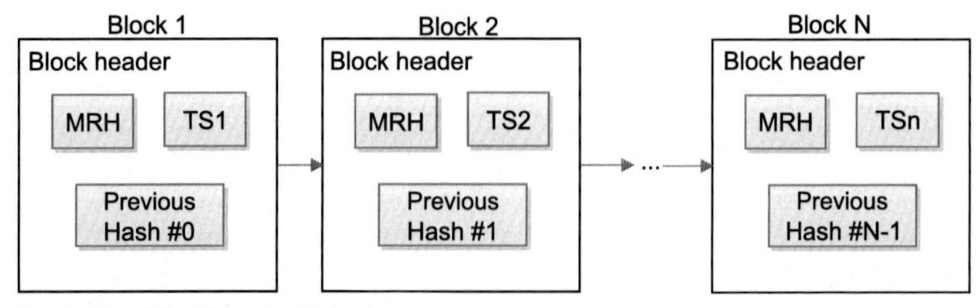

Fig. 1 Simplified bitcoin blockchain.

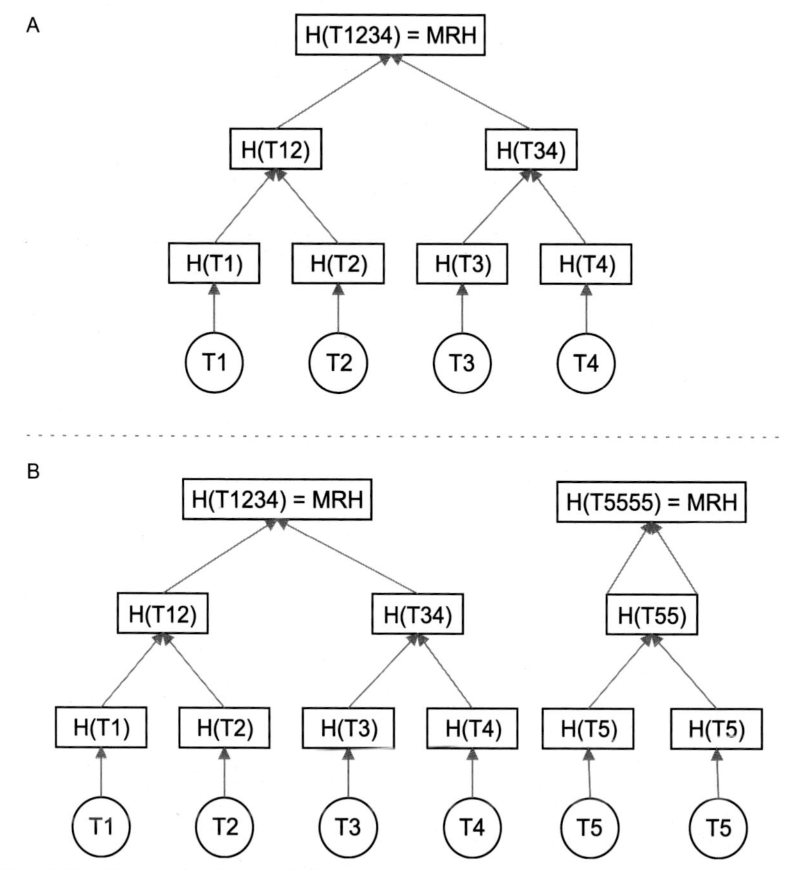

Fig. 2 Merkle root hash tree of the transactions.

previously received by any node in one or more earlier transactions, so the input of one transaction is the output of the previous transaction is as shown in Fig. 3.

As in Fig. 3 a single transaction can create a number of transaction outputs but each transaction output can only be used as once for input in the blockchain. The outputs of the transaction are joined together to form transaction identifier (TXIDs), that are the hashes of signed transactions. It can either be categorized as unspent transaction outputs (UTXOs) or spent transaction outputs (STXOs). The UTXOs of the coinbase transaction that is used as input cannot be spent for at least 100 blocks. To make a valid payment on the blockchain network, the blockchain node must use only UTXOs as input.

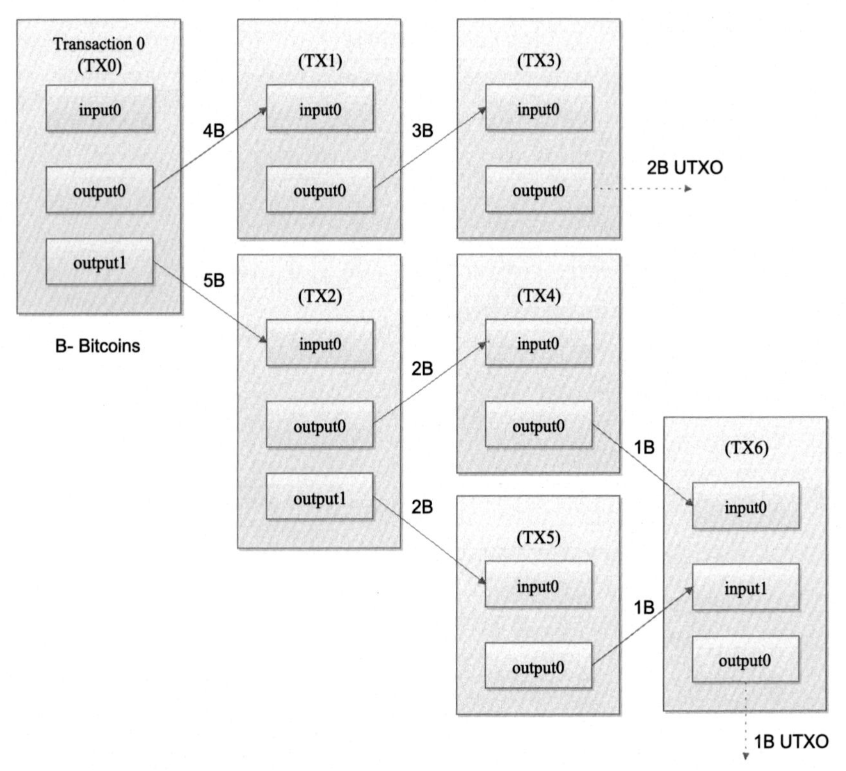

Fig. 3 Transactions-to-transactions payments.

1.2 Transaction verification

A method to verify the particular transactions are included in a block without the knowledge of complete block is called simplified payment verification (SPV). The MRH tree allows the users to verify the presence of transaction in the block. It is done by knowing the MRH and a list of intermediate hashes from a block header. For example, to verify the transaction (*T3*) in Fig. 2B. For this, a SPV client needs a copy of hashes of (*T4*) and (*T12*) transactions. The client does not need to know more hashes of any other transactions present in MRH. If all the five transactions (shown in Fig. 2B) in the block were used at the maximum size then, downloading the entire block requires over *5,00,000 bytes* whereas the downloading only two hashes with the block header requires only *100 bytes*.

1.3 Working of a transaction

For sending information to the blockchain requires access to both public and private keys. For example, when *Alice* wants to send or receive digital currencies from *Bob*, than *Alice* and *Bob* both must-have key pairs.

- *Public key*: It is a type of "address" to send or receive digital cryptocurrency.
- *Private key*: It is the key that authorizes the digital cryptocurrency transaction. This key is used as a "password" to the account.

Public keys are also called bitcoin addresses. It is a sequence of numbers and letters that work similarly to an email address. This is made public so that any user can share it with others to make a transaction. It is transparent and safe. Everyone can see what is inside but those who have private keys can only unlock it and access the funds. In the same manner, the private key is also a sequence of numbers and letters. It is used as a password and needs to be kept secret. For example, *Alice* wants to send some digital cryptocurrencies to *Bob*. To make the transaction, *Alice* uses his private key to sign on to generate the transaction. Then, this transaction is sent to the blockchain that includes following parameters.

- *Input*: It is the source transaction of coins which was previously sent to Alice's address.
- *Amount*: The number of digital cryptocurrencies to be sent from Alice to Bob.
- *Output*: Bob's public address.

The transaction is then broadcasted to the blockchain network, where miners verify *Alice's* key and allow him to access the input. This confirmation process is called mining and it requires high computation power and comes with big rewards. Miners are rewarded with digital cryptocurrencies after solving the mathematical problem. This is the process of new digital cryptocurrencies are created. All digital cryptocurrency transactions on the blockchain network are verified by miners. These miners do not mine the transactions but mined the blocks which are made from the transactions. Sometimes a transaction gets left out of a block and is put on hold until the next block is assembled. This can take up to a maximum of 10 min to mine again. In addition to this, the block size is limited to 1MB only that can further limit the number of transactions to enter the block.

2. Consensus execution

The full node of the blockchain network contains a number of blocks that are validated by that node and stores it on the network independently. When several nodes have the same number of blocks on the blockchain then, there is a need for a consensus mechanism for an agreement. The validation rules that follow control and maintain an agreement are called consensus rules. To maintain consensus, all the full nodes validate the new upcoming blocks by using consensus (validation rules). As time passes, consensus rules are changed and introducing new rules and features to prevent the blockchain from network attacks. When new rules are implemented then, it would take some period of time to implement properly on the network. So, the upgraded nodes follow the rules whereas nonupgraded nodes follow the old rules. This controversy can break the consensus between the nodes during the validation of the block. There are two ways in which the consensus can break are defined as under.

- *Case 1.* A new upcoming block follows the new rules which is accepted by the upgraded nodes and is rejected by the nonupgraded nodes. For example, a new feature has been used in an upcoming block and accepted by upgraded nodes whereas nonupgraded nodes reject that feature because it violates the old rules.
- *Case 2.* A block that violates the new consensus rules is rejected by the upgraded nodes but accepted by nonupgraded nodes. For example, when an old feature has been used in an upcoming new block and accepted by nonupgraded nodes whereas upgraded nodes reject that feature because it follows the old rules.

Mining is the process that creates valid blocks, which requires PoW in which miners are the devices or users that mine or verify the block by solving a mathematical puzzle. In the first case, the block is rejected by nonupgraded nodes. So, the mining process gets blockchain data from those nonupgraded nodes who refuse to build on the same chain as the mining process getting data from upgraded nodes. This process will create permanently different chains, one for nonupgraded nodes and one for upgraded nodes called a hard fork. The pictorial representation of hard fork is as shown in Fig. 4.

In the second case, the new upcoming block is rejected by upgraded nodes. Here it is possible to keep the blockchain from permanently diverging if upgraded nodes control a majority of the hash rate. This is because, in this case, nonupgraded nodes will accept the new upcoming block as valid and all the same old type of featured blocks as upgraded nodes. So, the

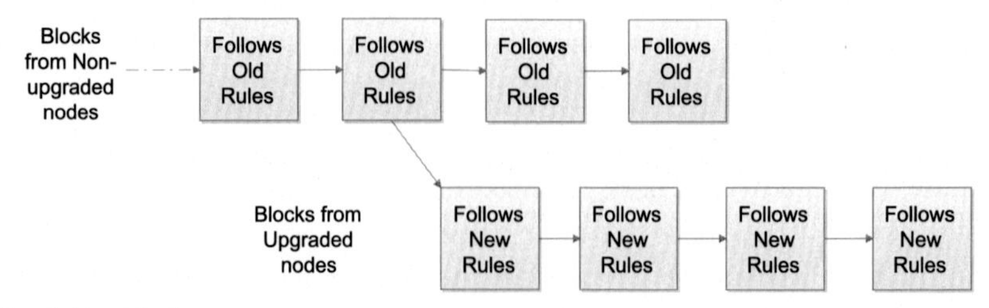

Fig. 4 Hard fork.

upgraded nodes can build a stronger chain that the nonupgraded nodes that will accept as the best valid blockchain. This process is called a soft fork. The pictorial representation of soft fork is as shown in Fig. 5.

2.1 Block height and forking

A fork is an actual divergence in blockchain in which consensus rules change that described to create either a soft fork or hard fork. The miner who successfully hashes a block header to a value below the target threshold can add the entire block to the blockchain. The block height is determined by the number of blocks between them and the first block (genesis block). When two or more miners produce a block at the same time then, multiple blocks have the same block height. This creates an apparent fork in the blockchain, as shown in Fig. 6. When the miners produce simultaneous blocks at the end of the blockchain then, each node of the network individually chooses which block to accept. A block height should not be used as a globally unique identifier of the block. These are also referenced by the hash of their header. The pictorial representation of block height and forking is as shown in Fig. 6.

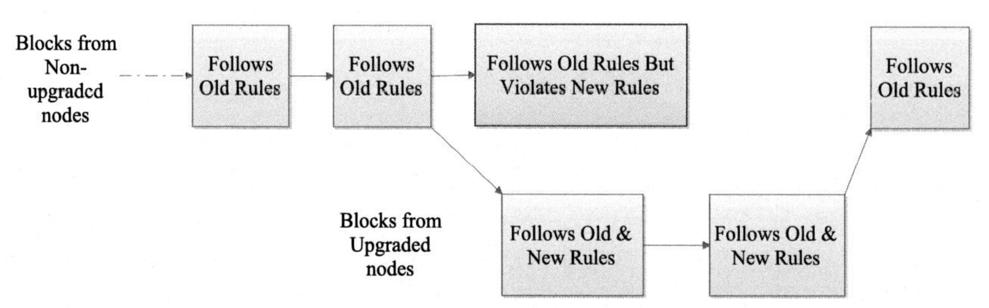

Fig. 5 Soft fork.

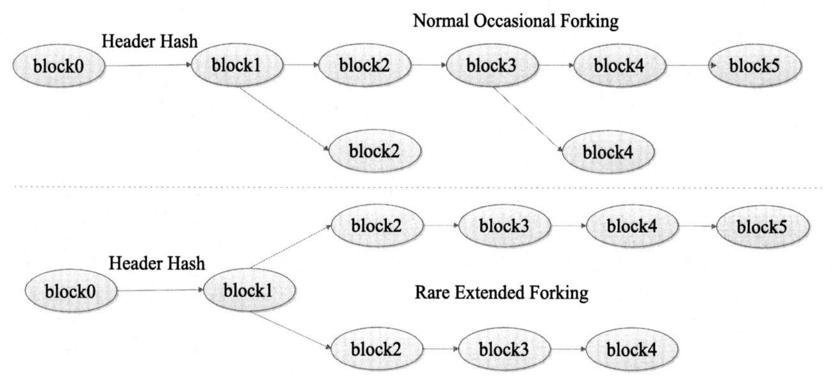

Fig. 6 Block height and forking.

In the end, the miner produces another block that attaches to only one of the competing simultaneously mined blocks. This makes that side of the fork stronger than the other side. The normal node always follows the most difficult chain to recreate and discard the stale blocks that belong to shorter forks.

2.2 Detecting forks

Sometimes, the nonupgraded nodes may use and distribute incorrect information that leads to financial loss. In particular, these nodes may relay and accept transactions that are considered invalid by upgraded nodes. These nodes may also refuse to relay blocks or transactions which have already been added to the best blockchain and provides incomplete information. The full nodes can also check block and transaction version numbers. If the block or transaction version numbers seen in several recent blocks are higher than the node uses, it can assume it does not use the current consensus rules. In addition, SPV clients can also detect a hard fork. It ensures that if there is a divergence then, the client can disconnect from nodes with weaker chains. So, bitcoin core includes code that detects a hard fork by looking at blockchain PoW.

3. Block verification and validation

The blockchain is a chain of data blocks. Each block can be thought of as a page in a ledger. The transactional data is permanently recorded in files called *blocks*. Each block consists of two main components, (i) the list of transactions and (ii) the Block Header. The list of transactions represents the number of transactions present in one block is approximately *500*. The Block Header consists of the *hash of the current block*, the *hash of the previous block*, *timestamp* of when the current block was hashed, the *target or difficulty* of the block, and the nonce.

Bitcoin uses the *SHA256* algorithm to "hash" the input data into a fixed size of a *256-bit* number. It means put any input data into the algorithm and it will spit out a fixed size of a *256-bit* number that uniquely identifies that input data. However, any changes (a single bit, spelling, capitalization) in the input data will also change the complete hash and will get a completely different number unrelated to the previous hash. There is no backward step to use the *SHA256* hash on data input. We can only check by taking the same piece of input data and get the same *SHA256* hash or *256-bit* number. If both the hash matches, it means the upcoming data is authentic otherwise not.

The concept behind a blockchain is illustrated by example. Suppose two blocks, *block A* and*block B*. Firstly, *block A* is added to the network. The miners collect all the transactions into *block A* and hash all of them to generate a fixed size of a *256-bit* number that uniquely identifies the *block A*, called it the hash *H(A)* of block A. Then, miners work to create the *block B* to add on the top of the *block A*. The miners collect another set of transactions and add *H(A)* to the *block B*. The miners then hash the new set of transactions + hash *H(A)* to get the hash *H(B)* of the *block B*. The pictorial representation of an example is as shown in Fig. 7.

If any malicious actor were to go back to block *A* and change the transactions of block *A* then, *H(A)* of the block *A* would completely change, which in turn would change the subsequent hash *H(B)* of block *B*. It requires a high amount of computing power to create a block and add it to the network. It would require even more computing power to change a previous block, say block *A*, add that to the network and then recreate block *B* with a new hash value and add it to the network all before the rest of the mining network has moved on to next blocks.

3.1 Mathematical problem

The hash of the transactions of the block cannot just add it to the block and then, add it to the network. In order to add a new block in the blockchain, the miner finds the correct hash by solving a mathematical puzzle to win a competition. There is no way to find any starting and backward point of the piece of data that gave the hash. Bitcoin protocol uses hash cryptography to create a difficult mathematical problem. This problem stipulates that the first miner to produce a hash with a certain amount of leading *0s* will be the

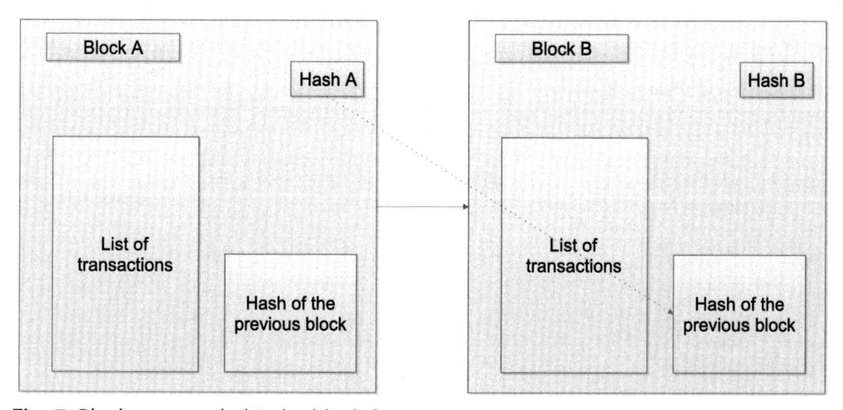

Fig. 7 Block concept behind a blockchain.

winner of that block and be able to add that block to the network. A hash is just a 256-bit string of numbers and letters (hexadecimal number) as represented in the following Eq. (1).

$$\begin{array}{c} 93ef6f358fbb095c608024968630522 \\ 90d5t32735b7fe5bdaac821de96a53a9a \end{array} \tag{1}$$

This number can start with any number *0–9*or letter *A–F*, so finding a hash that starts with say *17* leading *0s* would take a lot of work. The pictorial representation of the hash is as shown in Fig. 8.

3.2 The nonce

The nonce added to each block of the block header. It is a number that miners can continuously change until they find the correct nonce that solves the mathematical problem. The miners can continuously change the Nonce value until the *SHA256* hash function results in a hash with a certain amount of leading 0s reaches. When a miner broadcasts the block to the network then, this nonce value has been used by everyone to find the correct hash of the block with the correct number of leading *0s*.

3.3 Target or difficulty

To solve a mathematical puzzle on the network, more and more miners will add their computing power to the network and choose a Nonce value that gives them a hash with the correct number of leading 0s. The only way to win a reward over the competition is to use high computing power that have a high hash rate, i.e., hashes per second are called hashing power. As more hashing power attempts in 1 s to solve the puzzle, then, surely the puzzle will be solved faster and faster. This is the difficulty or target used in the blockchain. The bitcoin protocol has the goal to add a block once every 10 min over the long run. In order to maintain this order, the protocol has adjusts the Difficulty up or down depending on how quickly blocks

Hashes	
Hash	0000000000000000009719904c0880382f77c216beb4e9278459fa33d26a03c6
Previous Block	000000000000000000cd38a03ffbd813200988cee3abaddef4414fbb292fdcfc
Next Block(s)	
Merkle Root	6bcaa6ab913d86ec6018515aeba59fdc1fae95da3b36e6ac8da6509684fdd29b

Fig. 8 Hashes of the current block, previous block, and merkle root.

are added to the network. Every 2016 blocks (about 2 weeks), the protocol looks back at the last 2016 blocks and measures how long it took to solve them and then adjusts the Difficulty accordingly. The difficulty means how many leading numbers of 0s that miners will need in their hash to claim a valid block.

3.4 Block rewards and fees

The incentive for dedicating computing resources to the network and continuously expending energy to verify transactions is the block reward and transaction fees. For every successful block that adds it to the blockchain, the miner wins the bitcoin reward as compensation to find the appropriate Nonce and their corresponding hash. In the beginning, the block reward was 50 bitcoins per block, but every 4 years, the block reward halves until there are no more block rewards.

The second incentive is the transaction fees associated with each transaction. The bitcoin protocol has coded that each block can only fit a maximum of 1 MB of transaction data. The more the transaction fee paid to the miner to fit into a block, the more miners will prioritize adding the transaction to the next block so that increasing the overall block reward.

3.5 Bockchain working

Blockchain allows for the secure management of a public and shared ledger, where the history of the transactions are verified and stored on a network without any central authority. This technology can come in different configurations, ranging from governing to public, open-source networks to private networks that require explicit permissions to work on the network. The step-wise description of block verification and validation is as under and as shown in Fig. 9

1. *Transaction*: In the first step, two parties exchange the data that represents money, contracts, medical records, customer details, or any other types of information that can be described in a digital form.
2. *Verification*: The transaction is either verified or put into a secure record and placed in a queue of pending transactions. The verification part of the transaction is done by the nodes (computers or servers) of the network based on a set of rules.
3. *Structure*: Each block is identified by a hash value, created by using a hash algorithm agreed upon by the network. A block contains a block header, a reference to the previous block's hash, merkle root hash, and a list of

1. Transaction

Two parties exchange data that represents money, contracts, customer details, medical records or any other things described in digital form.

2. Verification

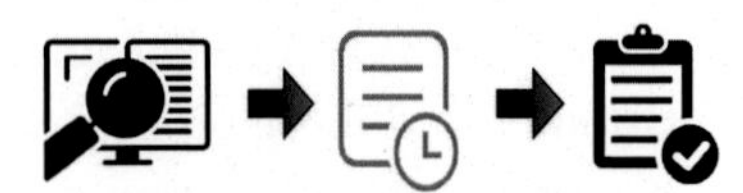

Depending on the network's parameters, the transaction is either verified instantly or arranged into records and placed in a queue of pending transactions.

3. Structure

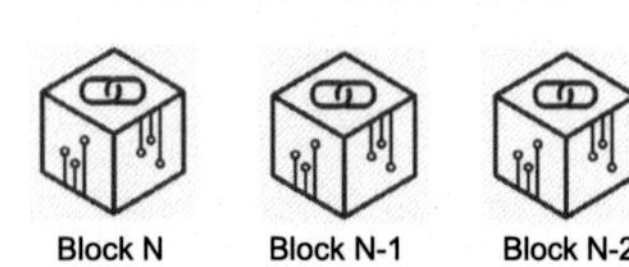

Each block is identified by hash(256-bit) created using an algorithm agreed on consensus by the network. The sequence of linked hashes creates a secure and interdependent chain.

4. Validation

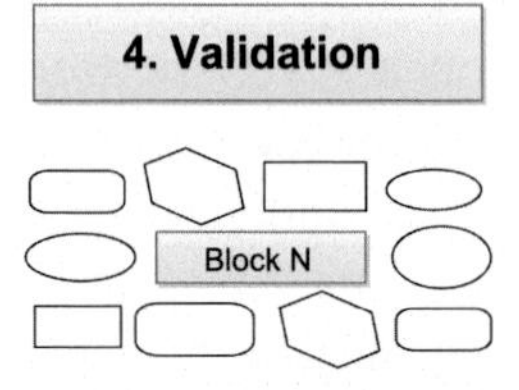

First blocks must be validated to be added to the blockchain. The most accepted form of validation is PoW: solution to a mathematical puzzle derived from the block's header

5. Mining

Block N

Miners try to solve the puzzle by changes to one variable until the solution satisfies. This is called PoW (potential solutions) that proves the appropriate level of computing power.

6. Built-in-defense

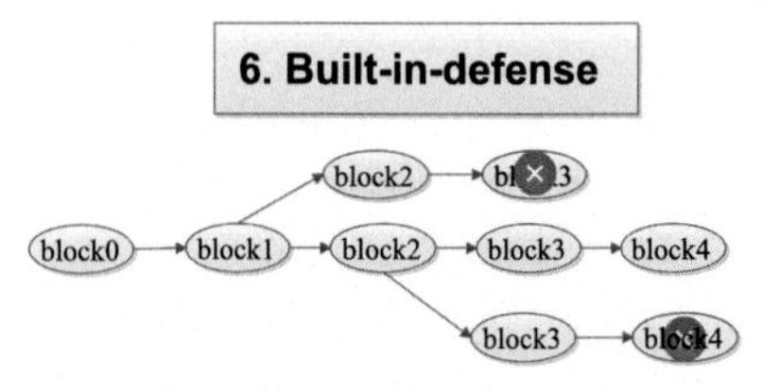

If a malicious miner tries to alter the block by changing the hash function then, all the following blocks would also change. The other nodes would detect the changes and reject the block by majority of the nodes. Hence, preventing corruption.

7. The Blockchain

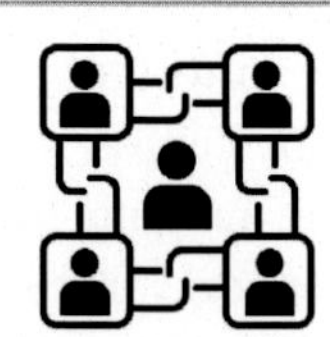

When a block is validated, the miners that solved the puzzle are rewarded and block is broadcasted on the network. Each node adds the block to the majority chain, immutable and auditable blockchain

Fig. 9 Block verification and validation.

transactions. The sequence of linked hashes of the block creates a secure and interdependent blockchain.

4. *Validation*: Before added the block to the blockchain, it must be validated. The validation part of the block is done mostly by a consensus mechanism PoW. It is a solution to a mathematical puzzle derived from the block's header that requires a high amount of computation power.

5. *Mining*: The miners try to solve a mathematical puzzle of the block by making incremental changes to a nonce value until the solution satisfies a network-wide target. They have used a high amount of computation power and resources to get a potential solution. The more they have computation power, the more chance to solve a mathematical puzzle and to win a reward.
6. *Built-in-defense*: If a malicious miner tries to submit an altered or changed block to the blockchain then, the hash function of that block, and all following blocks, would also change. The other nodes would detect these changes and reject the block from the majority votes of the nodes on the blockchain, preventing corruption.
7. *The blockchain*: When a block is fully verified and validated by the miners then, the miner who solved the puzzle is rewarded by the bitcoins. And the block is distributed and broadcasted on the network. Each node of the network adds the block to the majority chain called the network's immutable and auditable blockchain.

About the authors

Shubhani Aggarwal is pursuing PhD from Thapar Institute of Engineering & Technology (Deemed to be University), Patiala, Punjab, India. She received the BTech degree in Computer Science and Engineering from Punjabi University, Patiala, Punjab, India, in 2015, and the ME degree in Computer Science from Panjab University, Chandigarh, India, in 2017. She has many research interests in the area of Blockchain, cryptography, Internet of Drones, and information security.

Neeraj Kumar received his PhD in CSE from Shri Mata Vaishno Devi University, Katra (Jammu and Kashmir), India in 2009, and was a postdoctoral research fellow in Coventry University, Coventry, UK. He is working as a Professor in the Department of Computer Science and Engineering, Thapar Institute of Engineering & Technology (Deemed to be University), Patiala (Punjab), India. He has published more than 400 technical research papers in top-cited journals such as IEEE TKDE, IEEE TIE, IEEE TDSC, IEEE TITS, IEEE TCE, IEEE TII, IEEE TVT, IEEE ITS, IEEE SG, IEEE Netw., IEEE Comm., IEEE WC, IEEE IoTJ, IEEE SJ, Computer Networks, Information sciences, FGCS, JNCA, JPDC, and ComCom. He has guided many research scholars leading to PhD and ME/MTech. His research is supported by funding from UGC, DST, CSIR, and TCS. His research areas are Network Management, IoT, Big Data Analytics, Deep Learning, and Cybersecurity. He is serving as editor of the following journals of repute: ACM Computing Survey, ACM·IEEE Transactions on Sustainable Computing, IEEE·IEEE Systems Journal, IEEE·IEEE Network Magazine, IEE·IEEE Communication Magazine, IEE·Journal of Networks and Computer Applications, Elsevier Computer Communication, Elsevier International Journal of Communication Systems, Wiley. Also, he has been a guest editor of various international journals of repute such as IEEE Access, IEEE ITS, Elsevier CEE, IEEE Communication Magazine, IEEE Network Magazine, Computer Networks, Elsevier, Future Generation Computer Systems, Elsevier, Journal of Medical Systems, Springer, Computer and Electrical Engineering, Elsevier, Mobile Information Systems, International Journal of Ad Hoc and Ubiquitous Computing, Telecommunication Systems, Springer, and Journal of Supercomputing, Springer. He has also edited/ authored 10 books with international/national publishers like IET, Springer, Elsevier, CRC: Security and Privacy of Electronic Healthcare Records: Concepts, Paradigms and Solutions (ISBN-13: 978-1-78561-898-7), Machine Learning for Cognitive IoT, CRC Press, Blockchain, Big Data and IoT, Blockchain Technologies Across Industrial Vertical, Elsevier, Multimedia Big Data Computing for IoT Applications: Concepts, Paradigms and Solutions (ISBN: 978-981-13-8759-3), Proceedings of First International Conference on Computing, Communications, and Cyber-Security (IC4S 2019) (ISBN 978-981-15-3369-3). One of the edited text-book entitled, "Multimedia Big Data Computing for IoT Applications: Concepts, Paradigms, and Solutions" published in Springer in 2019 is having 3.5 million downloads till June 6, 2020. It attracts attention of the researchers across the globe (https://www.springer.com/in/book/9789811387586). He has been a workshop chair at IEEE Globecom 2018 and IEEE ICC 2019 and TPC Chair and member for various international conferences such as IEEE MASS 2020 and IEEE MSN 2020. He is a senior member of the IEEE. He has more than 12,321 citations to his credit with current h-index of 60 (September 2020). He has won the best papers award from IEEE Systems Journal and ICC 2018, Kansas City in 2018. He has been listed in the highly cited researcher of 2019 list of Web of Science (WoS). In India, he is listed in top 10 position among highly cited researchers list. He is an adjunct professor at Asia University, Taiwan, King Abdul Aziz University, Jeddah, Saudi Arabia, and Charles Darwin University, Australia.

CHAPTER FIFTEEN

Blockchain 2.0: Smart contracts☆

Shubhani Aggarwal and Neeraj Kumar
Thapar Institute of Engineering & Technology, Patiala, Punjab, India

Contents

Abstract

A smart contract is an agreement among the nodes on the blockchain network. They run on a blockchain and stored it on a public ledger. It is a computer protocol used to digitally facilitate, verify, or enforce the negotiation of a contract. The network transactions are run in a smart contract, which is processed and executed by the blockchain automatically. So, whenever a transaction happens between the nodes, a function is invoked that calls the smart contract, and the processing starts. Hence, the transaction can be submitted to any node on the blockchain, which broadcasts it to the entire network so that all the nodes will see the transaction. With this, we have been discussed the Ethereum platform used for writing smart contracts and introduce the solidity programming language.

☆ Working model.

Advances in Computers, Volume 121
ISSN 0065-2458
https://doi.org/10.1016/bs.adcom.2020.08.015

Chapter points

- In this chapter, we firstly discuss the role smart contracts in blockchain technology and mentioned their uses in real-time applications.
- Then, we discuss the Ethereum platform used for writing smart contracts with solidity programming language.

1. Smart contracts

Blockchain technology is a powerful and decentralized platform—capable of going much further than bitcoin transactions. The concept of blockchain 2.0 was the extended version of blockchain 1.0 for more interesting application known as *Smart Contracts* as shown in Fig. 1. It can be defined as an automated computerized protocol which is used to digitally facilitate, verify, or apply the agreement for the performance of a legal contract. It avoids the central authority or intermediary and directly validates the contract in a faster, cheaper, and more secure way over a distributed platform. Let us understand by taking an example of two persons who are coming in contact with each other for some legal contracts. They may contact each other by blockchain technology where smart contracts are used to control and manage these type of legal advisory contracts without any lawyer. So, with the help of blockchain technology using smart contracts, there is no need for intermediaries to make a legal contract with anyone at any-time.

The term smart contract was developed by *Nick Szabo* in 1994, a computerized scientist and a cryptographer. Szabo claimed that "*smart contracts can be realized with the help of DLT called blockchain*". Hence, blockchain is developing technology used for the realization of smart contracts. In this context, smart contracts can be converted into computer codes that can be stored and replicated at the network and supervised by the network nodes present in the blockchain. It is a computer protocol and self-enforcing segment which is controlled and managed by a P2P network. It helps in

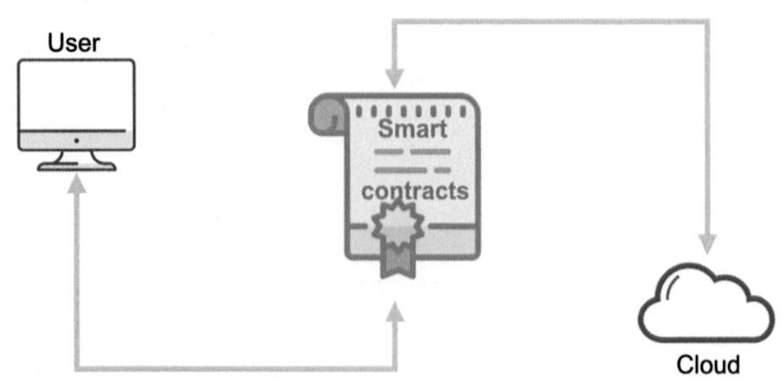

Fig. 1 Smart contracts.

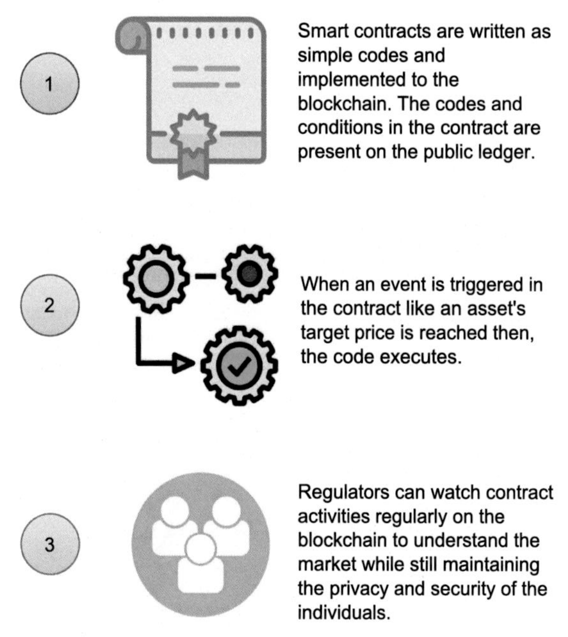

Fig. 2 Step-wise process of smart contracts.

exchanging share, currency, property, or money in a transparent and secure way while escaping the services of a central authority. The step-wise process of smart contracts is as shown in Fig. 2.

The best way to understand smart contracts is to compare it with a vending machine. Traditionally, a person would go to a notary or the lawyer for their documents, pay them, and wait while a person gets the documents back. But, with the help of smart contracts, a person simply drops a currency coin into a vending machine (like a lawyer) and drops the documents (driving license, Aadhaar card, etc.) into the account. These smart contracts not only define the rules and regulations of an agreement. These also implemented and controlled the agreement codes automatically and processed them.

1.1 Properties of smart contracts

The manually writing codes may have some errors to execute and spend chunks of time to complete but smart contracts use software codes that automatically run tasks and to complete it within a span of time. The important properties that smart contract gives us in a real-time environment are as follows and as shown in Fig. 3.

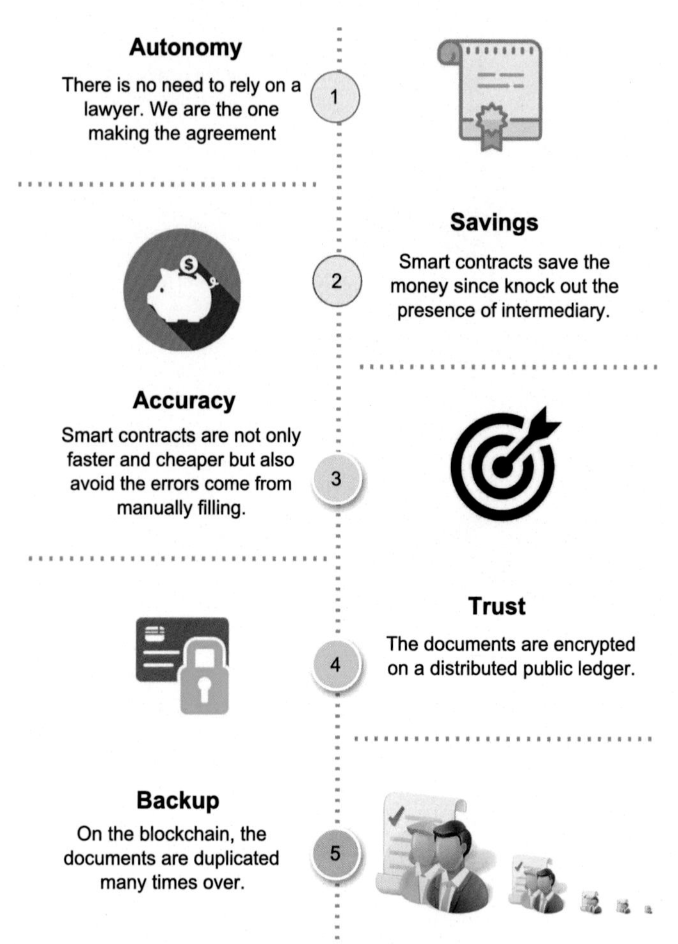

Fig. 3 Properties of smart contracts.

- *Autonomy*: By using smart contracts, there is no need for a broker or a lawyer. In this, a person itself who is making the agreement. It protects the system from third-party validator and execution of code is managed and controlled by the blockchain network rather than by one or more individuals.
- *Savings*: Smart contracts save our currency or money because it knocks out the presence of an intermediary. A person itself is the only one who has to pay a notary to witness their transactions.
- *Accuracy*: Smart contracts are not only cheaper or faster but also bypass the errors that come from manually writing the codes.

- *Trust*: In blockchain technology, documents or files are shared in an encrypted form. So, there is no way to lost or to hack a file from DLT.
- *Backup*: On the blockchain technology, each and every node has a copy of data in their public ledgers. So, there is a number of duplicate copies of documents or files present at the blockchain network in the form of backup.

1.2 Use of smart contracts

Jerry Cuomo, Vice President at *IBM* for blockchain technology, believes that smart contracts can be used all across the world in various applications from financial to healthcare to insurance. The brief description of smart contracts in real-time use cases is as follows.

- *Government*: The corruption-free system can be built using smart contracts on blockchain technology that ensures smooth governance. The traditional voting system based on a central authority which may lead to a single point of failure. So, smart contracts would play an important role in voting system by providing infinite security using blockchain. The ledger protected votes would need to be decoded and required high computation power to access. But, no one has this much amount of computing power. So, there is no way to hack or to alter the system. By this technology, people can also cast their votes online that reduces time and infrastructure cost.
- *Management*: The blockchain technology provides not only DLT as a source of trust but also cut down the possible confusion in work-flow and communication for accuracy, security, transparency, and automated system. It also manages and cuts down all the dissimilarities that occur due to independent processing and may lead to costly arguments and high delays.
- *Supply chain*: Supply chains are prevented by paper-based systems, where documents have to transfer by network channels for confirmation that increases to loss and fraud. But, blockchain technology abolishes all these issues by providing security, the accessible digital version to all the network nodes and do the tasks and payments automatically.
- *Automobile*: Smart contracts play an important role in self-autonomous or self-parking vehicles where they could detect the faults in vehicle collisions. Using smart contracts, an insurance automobile company may charge different rates based on the parking system. They may also charge according to the conditions, under which the customer can park their vehicles.

- *Real estate*: By using smart contracts, people can earn more money. The DLT cuts down the cost of an intermediary. In blockchain technology, all the payments are done through bitcoin and encode the contract on public ledgers. So, everyone can see and accomplish the fulfilments automatic. Hence, brokers, money lenders, and anyone associated with real-estate can profit from smart contracts.
- *Healthcare*: In the healthcare system, blockchain provides security and confidentiality to the medical-records. These reports and receipt of surgeries can be encoded and stored on the public ledgers of blockchain with a private key. These public ledgers would be accessed only by some particular individuals. These ledgers can also be used for management in the healthcare systems such as testing results, surgery reports, regulation check-up, supervising drugs, and managing medical supplies.

Let us understand the smart contract with the help of an example. Assume, *Alice* rent an apartment from *Bob*. *Alice* can do all the payments through blockchain technology and get a receipt in a virtual contract. *Bob* gave the digital key to *Alice* on a specified date and time. If *Alice* does not receive a key on a particular date and time then, the blockchain releases a refund to *Alice*. So, in this way, the system works on if-then announce and is noticed by the hundreds of people present in the blockchain network. Similarly, If *Alice* give the key to *Bob*. Then, it is sure that *Bob* has to be paid. If *Bob* sends a particular amount of bitcoin cryptocurrency to *Alice* then, only *Bob* can receive a key. The system and the document are automatically canceled after the completion of contract time. The smart contract code cannot be run by one or more individuals since all the individuals are participated and alerted simultaneously.

2. Advantages of smart contracts

Fig. 4 shows the advantages of smart contracts and their brief description is as follows.

1. *Record keeping*: All the contracts are stored in chronological order inside the blockchain. Therefore, no one can tamper or change the contracts because it is secured cryptographically using hash functions with full privacy.
2. *No intermediary*: There is no involvement of third-party validator in smart contracts. It allows transparency and creates direct relationships between the customers and the dealers.

Fig. 4 Advantages of smart contracts.

3. *Fraud detection and reduction*: Smart contracts stored inside the blockchain where no one can change or alter the contract. If anyone wants to change it forcefully then, they require a huge amount of computing power. Also, the changes in the contract can be detected by network nodes. Further, they will get marked as invalid and not stored in the blockchain network.
4. *Resistance to failure*: There is not only a single node present in control of digital services. So, if one node of the blockchain detaches itself from the network, the contract remains the same because of decentralization.
5. *Enhanced trust*: Smart contracts create trust between the supports and the product team and the agreements between them are automatically run by writing the contract inside the blockchain. These are immutable, unbreakable, and undeniable in nature.
6. *Saves money and resources*: Smart contracts eliminate the need for intermediaries (lawyers, notaries, witnesses) that leads to reduce costs. It also eliminates paperwork that leads to paper-saving and money-saving.

2.1 Smart contracts used in a decentralized system

Traditionally, the concept of smart contract used in a centralized platform which is explained with the help of *crowdfunding* example. There are number of crowdfunding companies like Kickstarter. In this context, the crowdfunding system works on the principle that if anyone wants to open their business or to start the project, but they do not have any sufficient amount of money. Then, they submit their project to the crowdfunding companies in which multiple supporters are present who can support the particular project by giving them with small funds.

The crowdfunding company is ensuring that the fund is going to the contracted project. So, when the project completes some event then, the project executers getting that fund from the supporters or if the project is not completed successfully then, the fund sent back to the supporter's account. The step-wise process of crowdfunding in a centralized system as shown in Fig. 5.

2.2 Crowdfunding platform in a centralized system

In this type of crowdfunding platform, there is a need for a trusty relationship between the product team and the supporters. The product team expects that their fund gets paid on the basis of project progress and whenever some

Fig. 5 Step-wise process of crowdfunding in a centralized system.

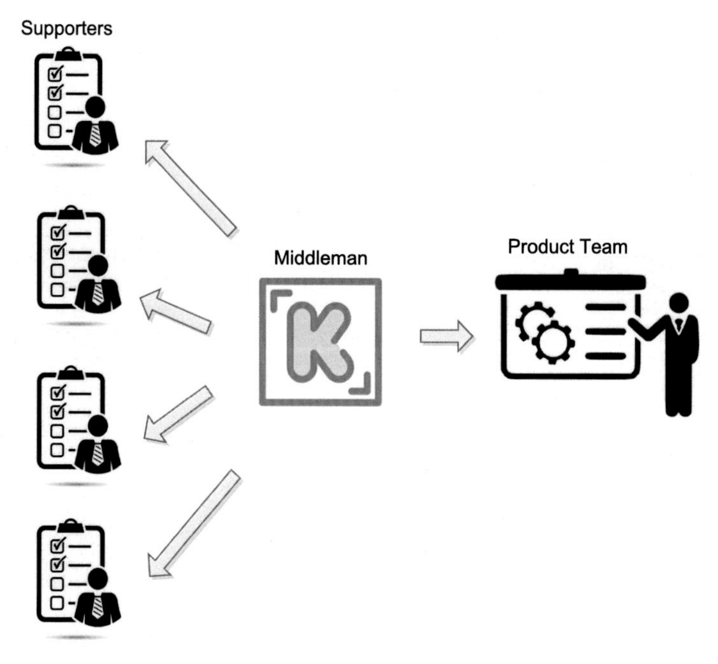

Fig. 6 Crowdfunding platform in a centralized system.

event completed then, they will get paid by the supporters. Similarly, the supporters expect that their fund is going to the right project or if the project gets crashed in between then, they will get back their fund from the product team.

Here, the crowdfunding platform is working as a middleman as shown in Fig. 6. They take significant charges from the product team as well as from the supporters. So, in a centralized platform, a need of trust and provides a significant amount of charge to middleman because it maintains the risk factors such as- project might not get completed, supporters may claim not to support the project further.

2.3 Crowdfunding platform in a decentralized system

In this type of crowdfunding platform, there is a presence of smart contracts between the product team and the supporters that further put on the blockchain for verification and validation. Fig. 7 shows the contract between the supporters and the product team. The code is written and is made available to all the stakeholders, the product team, and the supporters. If the code is present inside the blockchain then, anyone who has been participating in

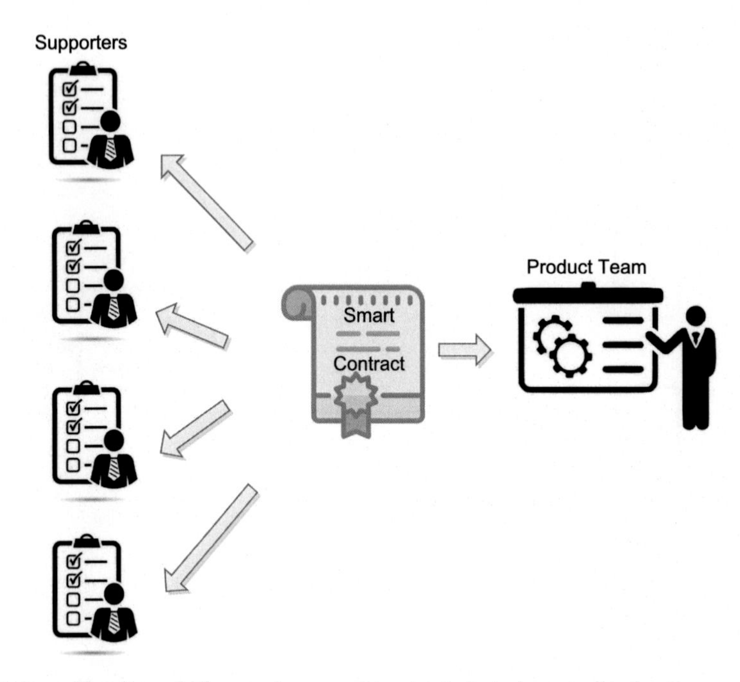

Fig. 7 Crowdfunding platform using smart contracts in a decentralized system.

the network can verify the contract. In this way, the centralized platform can change into a decentralized platform for crowdfunding.

Fig. 8 shows whenever the contract gets executed an event after 10 days of contractual time. Then, the fund is transferred from the supporters to the product team as it is automatically run based on the code written inside the contract.

Similarly, Fig. 9 shows whenever the contract gets failed or the product team says they are not able to make sufficient progress in the project. Then, the fund goes back to the supporters from the product team.

So, in this way, a code is written inside the contract that further put on the blockchain rather than put some transactions on the blockchain. Then, that contract is verified automatically by all the network nodes. So, it is impossible for any node to make changes in the contract or to deny the contract.

2.4 Need of smart contracts

By writing the contract on the blockchain, smart contracts can automate run many solutions that make things simple and more efficient to the project. Fig. 10 shows the difference between the smart contracts and traditional contracts.

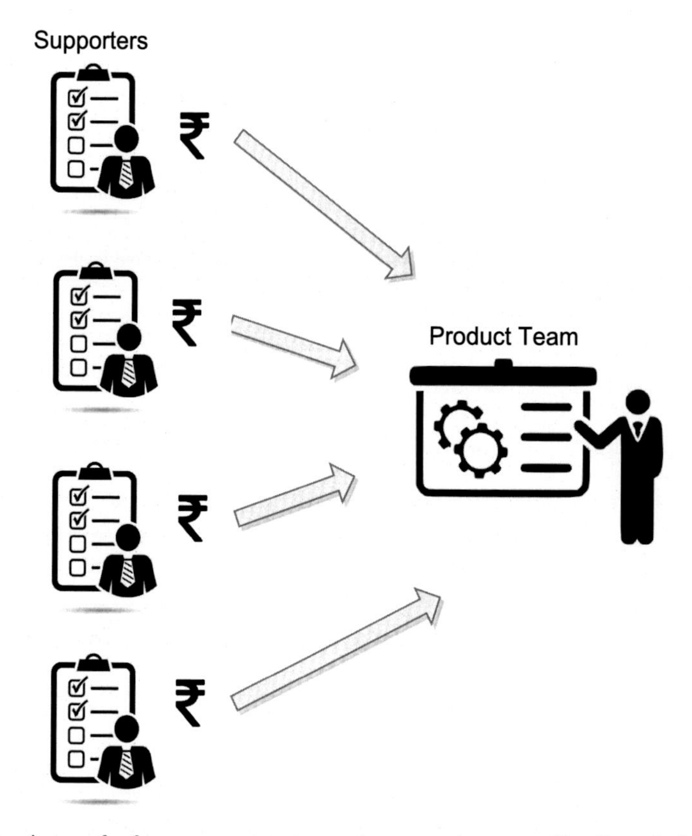

Fig. 8 Funds transfer from supporters to product team in a crowdfunding platform.

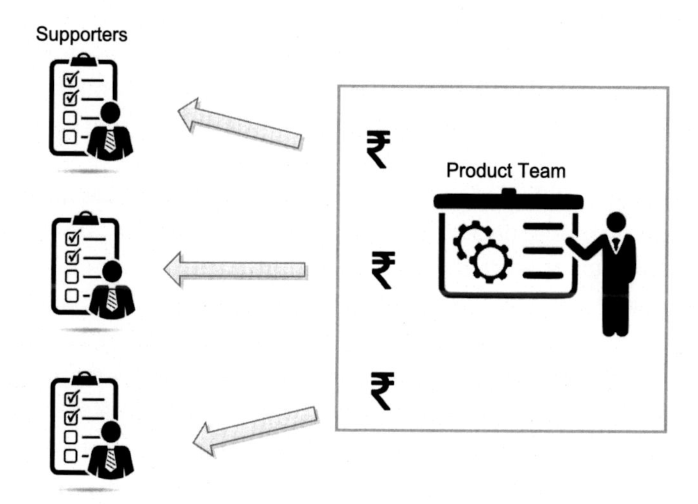

Fig. 9 Funds transfer from product team to supporters in a crowdfunding platform.

Name	Traditional Contracts	Smart Contracts
Timings	1-3 days	Minutes
Payment option	Manual remittance	Automatic remittance
Bond	Escrow necessary	Escrow may not be necessary
Cost	Expensive	Fraction of cost
Signature	Physical presence	Virtual presence (digital signature)
Intermediary	Lawyers necessary	Lawyers may not be necessary

Fig. 10 Difference between traditional contracts and smart contracts.

3. Blockchain platforms using smart contracts

The following are the blockchain platforms used for writing smart contracts.

1. *Bitcoins*: Bitcoin uses script language to process bitcoin transactions. This language has limited capabilities for processing the documents.
2. *Ethereum*: It is one of the popular blockchain platforms for writing smart contracts. It runs the code in any programming language and is accessible anywhere in the world.
3. *Hyperledger fabric*: The hyperledger fabric platform used for private blockchain where chaincode is coded programmatically on the network. It is executed and validated by chaincode validators during the consensus process.
4. *NXT*: The NXT platform used for public blockchain that includes limited number of templates for writing the smart contracts. It is used when the user cannot write the code on our own.
5. *Side chains*: This blockchain platform enhances the privacy protection of smart contracts. It also increases the performance of the blockchain by adding capabilities such as- secure handles and real-world property registry.

From foreknown blockchain platforms, Ethereum is the most popular platform used for writing smart contracts. It is a global and decentralized platform for new type of applications. It can accept a code in any programming language that control transactions and build applications anywhere in the world. Buterin publishes the white paper on Ethereum [1]. Then, the development of Ethereum platform was announced across the globe. It is an open-source platform based on blockchain technology that enables developers to build and use decentralized applications for better security and privacy.

4. Ethereum smart contracts using solidity

For Ethereum smart contracts, solidity and serpent are the two primary languages that can be used and is describes as follows.

- Solidity is a contract oriented high-level language with syntax similar to the javascript and is designed to target EVM.
- Serpern is a high-level language similar to python to write Ethereum contracts.

But, solidity is the preferred language for the development of Ethereum-based applications.

4.1 Defining a smart contract using solidity

Here, we introduce solidity for writing smart contracts. The basic steps followed for writing the smart contract using solidity are as follows.

1. *Build a smart contract*: A simple purchase order smart contract developed with the help of solidity.
2. *Structure of smart contract*: The smart contract is a collection of two main entities, i.e., data and function. Data maintains the current state of the contract and function is a logic that applies to the transition state of the contract. Each smart contract follows a standard structure. It begins with the following statements.
   - *Pragama directive*: The *pragama* keyword is used to enable certain features of the compiler as shown in Listing 1. The compiler does not compile the smart contract statements with a compiler version not earlier than 0.4.0 and not after the compiler version 0.6.0. This statement does not introduce any unintended behavior when a new version of compiler launched.

Listing 1 Pragama directive

```
1  pragma solidity >=0.4.0 <=0.6.0
2
```

- **Contract declaration:** The contract declaration in the smart contract declared by using the keyword *contract*. Listing 2 declares an empty contract and identified by *Purchase order*.

Listing 2 Contract declaration

```
1 contract Purchase Order
2 {
3 }
4
```

- *Store relevant data to the contract*: Every contract may require some data to store in smart contracts. Storing data provides a level of flexibility. The moving away from hard-coded values of data to user-provided values is an important feature. The variable used in the smart contract allows to store, label, retrieve, and manipulate the data. The variables used in the smart contract are of two types and described as follows.

  (a) Value type: These types of variables are passed by value, i.e., they copied at the time of function arguments. For example, integers and booleans.

  (b) Reference type: These type of variables are passed by reference. These have to be managed and controlled carefully at these are not fit into 256-bit.
- *Add data to the smart contract*: Here, add some data to the smart contract. We introduce a variable *Product_quantity* of unsigned integer of storage 256-bits and is referred as *uint256*. Its minimum value can be assigned as 0 and maximum value as 2^{256-1}. It is shown in Listing 3.

Listing 3 Add data to the smart contract

```
1 contract Purchase Order
2 {
3 uint256 Product_quantity;
4 }
5
```

- *Define the constructor*: The constructor is called when the contract is used. It initializes with some integer values as shown in Listing 4.

Listing 4 Define the constructor

```
constructor()
public{
Product_quantity = 100;
}
```

In this Listing 4, the *public*keyword refers that anyone can access this information, it is not a restricted function.

- *Add functions to the smart contract*: The addition of function makes the contract more interactive. The function can be declared as *function <function_name><access_modified><state_mutator><return_value>*.

  (a) Get function: To read the function, read_function or get_function is added in the contract is as shown in Listing 5.

Listing 5 Get function

```
function get_quantity()
public view returns(uint256)
{
return Product_quantity;
}

```

Here, <function_name> is get_quantity() (where no arguments are passed), <access_modified> is public (anyone can access the function), <state_mutator> is view (signifies only read the contract, does not change the state of the contract), and <return_value> is returns(uint256).

(a) Set function: Read the data is necessary but if requires to update or to write the data as well. Then, use the set_function in the contract that takes value from the user as input parameter and update the variable with that value as shown in Listing 6.

Listing 6 Set function

```
function update_quantity(uint256 value)
public{
{
 Product_quantity = Product_quantity + value;
}
```

Here, <function_name> is update_quantity(uint256 value), <access_modified> is public, <state_mutator> is not required, and <return_value> is returns a variable to type uint256.

3. *Defining the smart contract*: Combine all foreknown steps to make the smart contract. Listing 7 defines the purchase order smart contract.

Listing 7 Defining the smart contract

```
pragma solidity >=0.4.0 <=0.6.0;

contract PurchaseOrder{
uint256 product_quantity; //state variable

/*Called with the contract is deployed and initializes
  the value*/
constructor() public{
product_quantity = 100;
}

// Get Function
function get_quantity() public view returns(uint256){
return product_quantity;
}

// Set Function
function update_quantity(uint256 value) public {
product_quantity = product_quantity + value;
}
}

```

4.2 Deploy and run smart contract

Run the program shown in Listing 7 at remix online integrated development environment (IDE) [2]. It is a toolset to start the development of a smart contract and to test the smart contract without any installation on a local computer. The following steps are used to deploy and run smart contracts at remix online IDE.

1. Click on the plus icon to add a new file named "Purchase_order.sol".
2. Write the code of the contract in Purchase_order.sol as shown in Fig. 11.
3. Click on the icon "Solidity compiler" present on the left side to compute the contract. Under this, set the compiler version is 0.5.8 and then, compile the Purchase_order.sol as shown in Fig. 12. Once the smart contract compilation is done successful. The, click on the compilation details that

Fig. 11 Code of the contract.

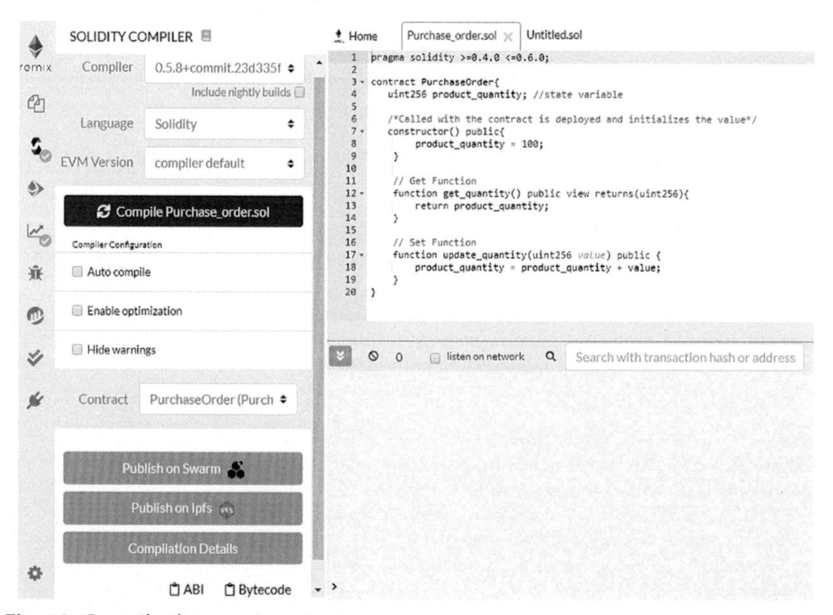

Fig. 12 Compile the smart contract.

provides two key formation, i.e., application binary interface (ABI) that details all the methods explored in the contract and bytecode is EVM operation code that converts smart logic gets into bytecode on compilation.

4. In order to test the smart contract, click on the "deploy and run transactions" icon on the left side. It provides the deployment option environment (JavaScript virtual machine, injected web3 provide, web3 provider), accounts (based on environment selection), gas limit (maximum amount to spend on transactions), and value (required to send across while using smart contract) as shown in Fig. 13.

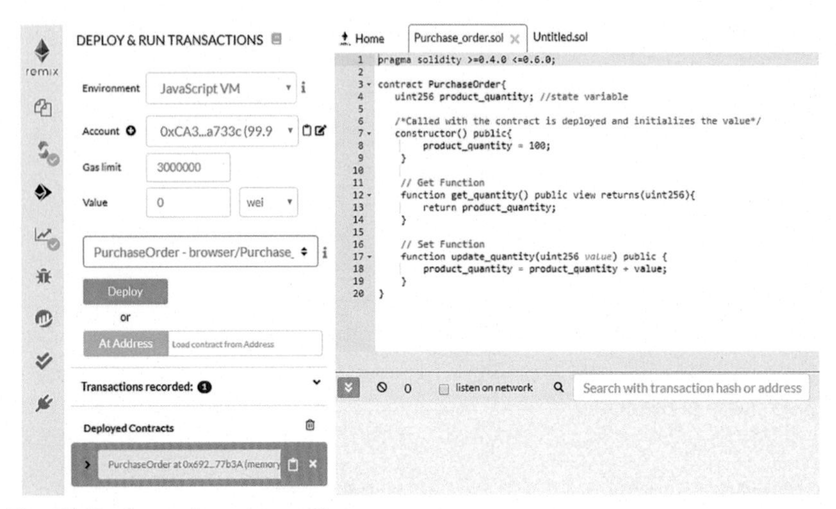

Fig. 13 Deploy and run transactions.

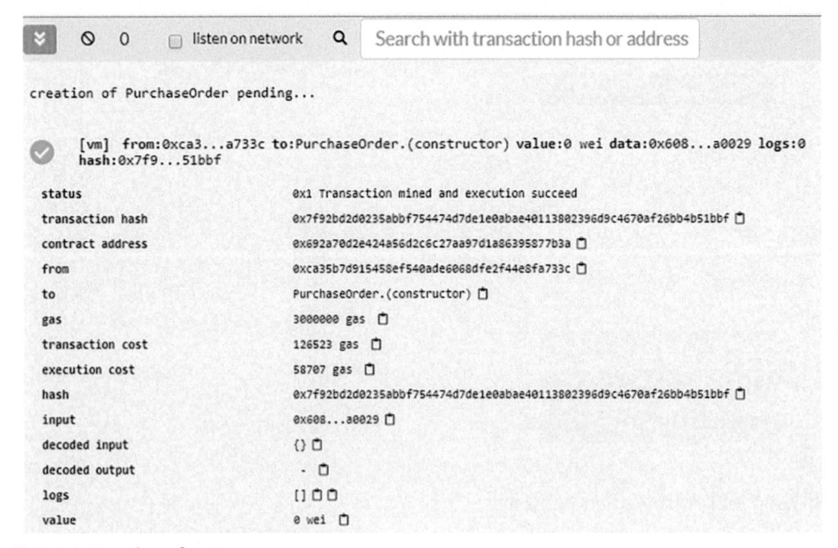

Fig. 14 Results of smart contract.

5. The results of the smart contract with transaction hash, contract address, transaction cost, and execution cost are as shown in Fig. 14.
6. After this, run the set and get functions introduced in the smart contract. During get function, the "*get_quantity*" method retrieves the value 100 as shown in Fig. 15. Similarly, during get function, the

Fig. 15 Results of get function.

```
0    listen on network    Search with transaction hash or address

[vm] from:0xca3...a733c to:PurchaseOrder.update_quantity(uint256) 0x692...77b3a value:0 wei
data:0x802...00032 logs:0 hash:0xf11...82b8b
status              0x1 Transaction mined and execution succeed
transaction hash    0xf11f0966372283184f411e8067b674ffb9de5cf30091275f72b4fc8d54b82b8b
from                0xca35b7d915458ef540ade6068dfe2f44e8fa733c
to                  PurchaseOrder.update_quantity(uint256) 0x692a70d2e424a56d2c6c27aa97d1a86395877b3a
gas                 3000000 gas
transaction cost    26890 gas
execution cost      5426 gas
hash                0xf11f0966372283184f411e8067b674ffb9de5cf30091275f72b4fc8d54b82b8b
input               0x802...00032
decoded input       {
                            "uint256 value": {
                                    "_hex": "0x32"
                            }
                    }
decoded output      {}
logs                []
value               0 wei
```

Fig. 16 Results of set function.

"*update_quantity*" method retrieves the value (user input, 50) is as shown in Fig. 16. The set function causes a transaction to happen but get function does not.

4.3 Examples of smart contract

The more examples of smart contract are as follows.

1. A simple example of smart contract for storage of data with get and set functions is as shown in Listing 8.

Listing 8 Storage example

```
1 pragma solidity >=0.4.0 <0.7.0;
2 contract SimpleStorage
3 {
4 uint storedData;
5 function set(uint x) public
6 {
7 storedData = x;
8 }
9 function get() public view returns (uint)
10 {
11 return storedData;
12 }
13 }
```

2. The following smart contract implements the simplest form of a crypto-currency as shown in Listing 9. In this contract, anyone can send coins without any registration process. There is only a need for an Ethereum key-pair.

Listing 9 Subcurrency example

```
1 pragma solidity >=0.5.0 <0.7.0;
2 contract Coin
3 {
4 address public minter;
5 mapping (address => uint) public balances;
6 event Sent(address from, address to, uint amount);
7 constructor() public
8 {
9 minter = msg.sender;
10 }
11 function mint(address receiver, uint amount) public
12 {
13 require(msg.sender == minter);
14 require(amount < 1e60);
15 balances[receiver] += amount;
16 }
17 function send(address receiver, uint amount) public
18 {
19 require(amount <= balances[msg.sender], "Insufficient
      balance.");
20 balances[msg.sender] -= amount;
21 balances[receiver] += amount;
22 emit Sent(msg.sender, receiver, amount);
23 }
24 }
```

References

[1] V. Buterin, Ethereum: a next-generation smart contract and decentralized application platform, 3 (37) (2014). White Paper. https://github.com/ethereum/wiki/wiki/%5BEnglish%5D-White-Paper.
[2] 2019. Available: https://remix.ethereum.org (accessed 4 October 2019) Remix Online IDE.

About the authors

Shubhani Aggarwal is pursuing a PhD from Thapar Institute of Engineering and Technology (Deemed to be University), Patiala, Punjab, India. She received the BTech degree in Computer Science and Engineering from Punjabi University, Patiala, Punjab India, in 2015, and the ME degree in Computer Science from Panjab University, Chandigarh, India, in 2017. She has many research interests in the area of Blockchain, cryptography, Internet of Drones, and information security.

Neeraj Kumar received his PhD in CSE from Shri Mata Vaishno Devi University, Katra (Jammu and Kashmir), India in 2009, and was a postdoctoral research fellow in Coventry University, Coventry, United Kingdom. He is working as a professor in the Department of Computer Science and Engineering, Thapar Institute of Engineering and Technology (Deemed to be University), Patiala, Punjab, India. He has published more than 400 technical research papers in top-cited journals such as IEEE TKDE, IEEE TIE, IEEE TDSC, IEEE TITS, IEEE TCE, IEEE TII, IEEE TVT, IEEE ITS, IEEE SG, IEEE Network, IEEE Communications, IEEE WC, IEEE IoTJ, IEEE SJ, Computer Networks, Information sciences, FGCS, JNCA, JPDC, and ComCom. He has guided many research scholars leading to PhD and ME/MTech. His research is supported by funding from UGC, DST, CSIR, and TCS. His research areas are Network management, IoT, Big Data Analytics, Deep learning, and cyber-security. He is serving as an editor of the following journals of repute: ACM Computing Survey, ACM·IEEE Transactions on Sustainable Computing, IEEE·IEEE Systems Journal, IEEE·IEEE Network Magazine, IEE·IEEE Communication Magazine, IEE·Journal of Networks and Computer Applications, Elsevier · Computer Communication, Elsevier·International Journal of Communication Systems, Wiley. Also, he has been a guest editor of various International Journals of repute such as IEEE Access, IEEE ITS, Elsevier CEE, IEEE Communication Magazine, IEEE Network Magazine, Computer Networks, Elsevier, Future Generation Computer Systems, Elsevier, Journal of Medical Systems, Springer, Computer and Electrical Engineering, Elsevier, Mobile Information Systems, International Journal of Ad hoc, and Ubiquitous Computing, Telecommunication Systems, Springer, and Journal of Supercomputing, Springer. He has also edited/authored 10 books with International/National Publishers like IET, Springer, Elsevier, CRC, Security and Privacy of Electronic Healthcare Records: Concepts, paradigms and solutions (ISBN-13: 978-1-78561-898-7), Machine Learning for Cognitive IoT, CRC Press, Blockchain, Big Data and IoT, Blockchain Technologies across industrial vertical, Elsevier, Multimedia Big Data Computing for IoT Applications: Concepts, Paradigms and Solutions (ISBN: 978-981-13-8759-3), Proceedings of First International Conference on Computing, Communications, and Cyber-Security (IC4S 2019) (ISBN 978-981-15-3369-3). One of the edited text-book entitled "Multimedia Big Data Computing for IoT Applications: Concepts, Paradigms, and Solutions" published in Springer in 2019 is having 3.5 million downloads till June 06, 2020. It attracts the attention of the researchers across the globe (https://www.springer.com/in/book/9789811387586). He has been a workshop chair at IEEE Globecom 2018 and IEEE ICC 2019 and TPC Chair and member for various International conferences such as IEEE MASS 2020 and IEEE MSN 2020. He is a senior member of the IEEE. He has more than 12,321 citations to his credit with current h-index of 60 (September 2020). He has won the best papers award from IEEE Systems Journal and ICC 2018, Kansas-city in 2018. He has been listed in the highly cited researcher of 2019 list of web of science (WoS). In India, he is listed in top 10 position among highly cited researchers list. He is an adjunct professor at Asia University, Taiwan, King Abdul Aziz University, Jeddah, Saudi Arabia and Charles Darwin University, Australia.

Hyperledger☆

Shubhani Aggarwal and Neeraj Kumar
Thapar Institute of Engineering & Technology, Patiala, Punjab, India

Contents

Abstract

Hyperledger is one of the biggest projects in the blockchain industry, which is comprised of a set of open source tools and subprojects. It includes leaders in different sectors who are aiming to build a robust, business-driven blockchain framework. It is an umbrella project, which aims to provide enterprise solutions and universal guidelines for blockchain implementation. The key projects within the Hyperledger framework are Hyperledger Fabric, Hyperledger Iroha, Hyperledger Indy, Hyperledger Sawtooth, Hyperledger Cello, Hyperledger Explorer, Hyperledger Composer. In this chapter, we have discussed the architecture, goals, and need for hyperledger projects in real-time applications.

Chapter points

- In this chapter, we discuss the hyperledger platform used for private transactions in an organization.
- Here, we discuss the most popular hyperledger fabric project and its architecture in detail.

☆ Working model.

Advances in Computers, Volume 121
ISSN 0065-2458
https://doi.org/10.1016/bs.adcom.2020.08.016

1. Hyperledger: An industrial approach to blockchain

Hyperledger is not a company, not a cryptocurrency, not a blockchain, not an IBM blockchain coin but it is one of the fastest-growing open source collaborative effort created to advance cross-industry blockchain technologies. It is hosted and governed by the Linux Foundation, including leaders in finance, banking, IoT, supply chain, manufacturing, and technology. It does not support bitcoin or any other cryptocurrency. But the platform of hyperledger is inspired by the blockchain technology. This platform has the capability to build a new generation of transactional applications that establishes trust, accountability, and transparency at their core while automating and streamlining business processes. So, with the help of the hyperledger platform, Linux Foundation plans to generate an environment in which members of the enterprise and companies meet and coordinate to build blockchain framework all over the world in real-time applications. Now, hyperledger has an impressive list of more than 100 members that includes wide range of well-known industries like Airbus and Daimler, IT-companies like IBM, Fujitsu, SAP, Huawei, Nokia, Intel and Samsung, financial institutions like Deutsche Börse, American Express, J.P. Morgan, BBVA, BNP Paribas and Well Fargo, as well as Blockchain startups like Blockstream, Netki, Lykke, Factom, and so on. A few members of hyperledger is as shown in Fig. 1. For its members, the hyperledger platform does not provide technical and software knowledge to companies only but also provides several contacts to companies and developers. The different type of hyperledger platforms defined by hyperledger organization in terms of domain-specific, libraries, tools, and distributed ledgers is as shown in Fig. 2.

The hyperledger project itself will never build its own cryptocurrency. This decision of hyperledger strongly helps to build industrial applications of blockchain technology and separating it from the other platforms that evolving cryptocurrency while using blockchain. It provides a secure platform and testing of interactions between business applications and private blockchain networks as shown in Fig. 3. It is used to develop business solutions described in Fig. 4.

1.1 Need of hyperledger

During testing and finding accurate results in private, developers realized blockchain networks, where every node wants to validate each and every transaction and run consensus simultaneously that makes the network less scalable. Also, the private and confidential transactions cannot be executed on public blockchains. So, there is a need for a platform that ensures the

Fig. 1 Hyperledger members.

integrity, privacy, and confidentiality of the transaction within a group of members called *Hyperledger*.

Let us understand by an example of *Alice* and *Bob*. Suppose, *Bob*, residing in *India*, wanted to buy some gifts from *Alice*, who resides in the *US*. As they know each other and were old friends, *Alice* decides to sell the gifts to *Bob* at a big discount. But the thing is that *Alice* sells her products to various kinds of markets and they buy the products from her at standard rates. To make a transaction between *Alice* and *Bob* on public blockchain, a lot of nodes have to verify the transaction in terms of quality assurance, logistic verification, payment verification and many more. But *Alice* does not want anyone except *Bob* to know about the special deal between them. So, both wanted to get a private network on the blockchain that completes their transaction successfully secure and confidential. Hence, they are generated a private

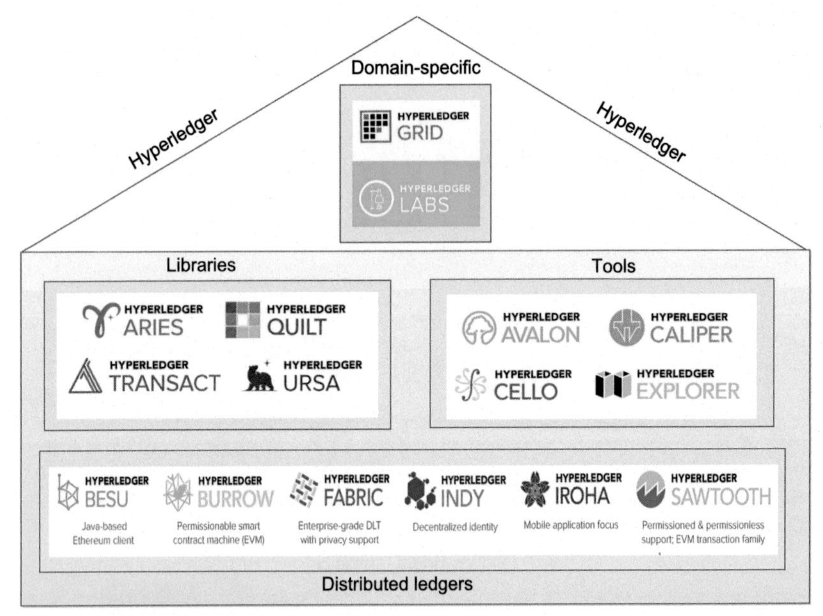

Fig. 2 Hyperledger platform.

Fig. 3 Testing between applications and blockchain networks.

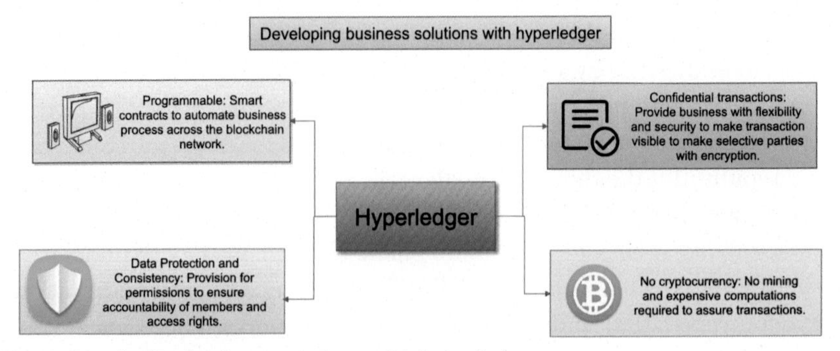

Fig. 4 Developing business solutions with hyperledger.

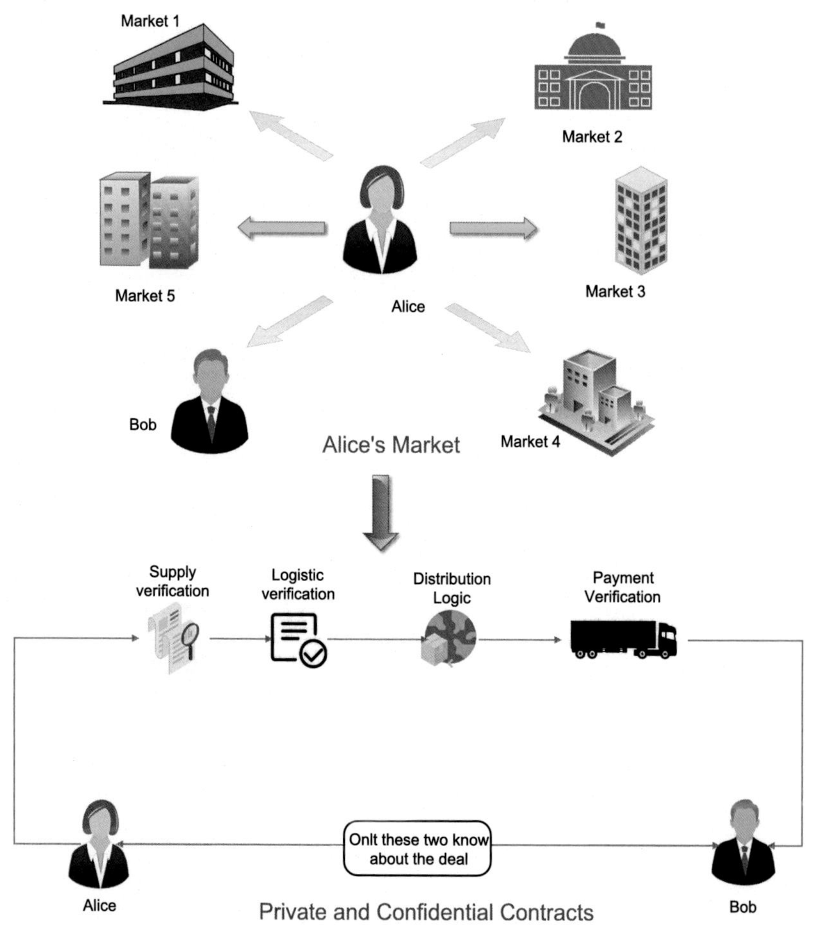

Fig. 5 Private and confidential contracts via hyperledger.

network using a hyperledger platform and complete the transaction with privacy and confidentiality. The pictorial representation of an example is as shown in Fig. 5.

1.2 Working of hyperledger transaction

On a hyperledger platform, the nodes directly communicate with each other regarding the deal and only their ledgers get updated about that deal. The third-party who involves in the deal only gets to know the exact amount of the transaction that has been transmitted during the transaction. Suppose, *Alice* and *Bob* were executing the deal transaction on a hyperledger platform as shown in Fig. 6. Then, *Alice* would look up *Bob* through an app which in return queries a membership service. After the membership has been validated, the two peers are connected and results are generated.

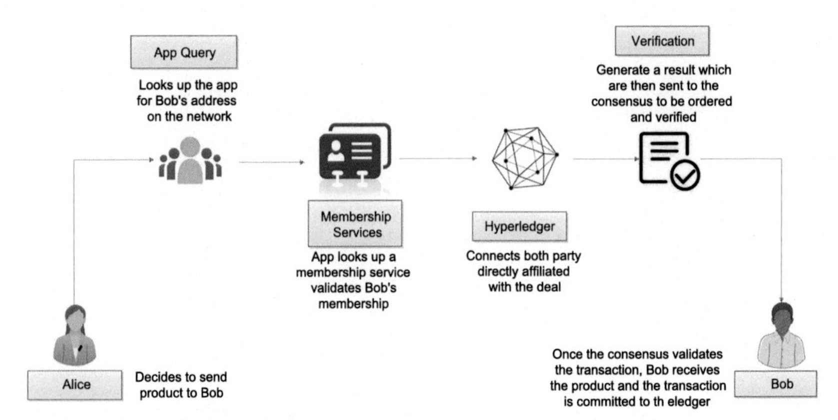

Fig. 6 Working of a hyperledger transaction.

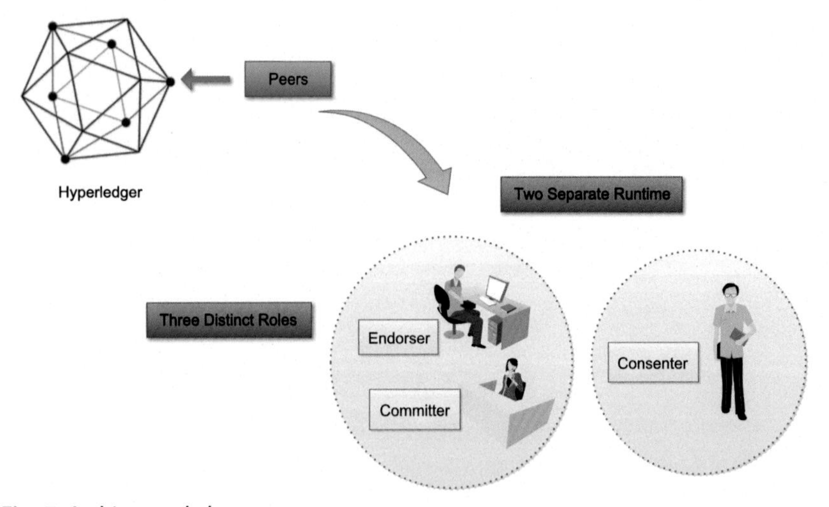

Fig. 7 Architectural changes.

These generated transactions are now sent to a consensus cloud for verification and validation. Then, only they are committed to their respective ledgers. The peers of this platform have been divided into two separate runtimes and three distinct roles are described as under. The pictorial representation of the nodal changed in the hyperledger is as shown in Fig. 7.

1. *Committer*: These peers only update the validated transactions and write into the respective ledgers that return from the consensus on the blockchain network.
2. *Endorser*: These peers are used to prevent non-deterministic and unreliable transactions that are simulating on a particular network. All

endorsers act as committers, while committers may or may not act as endorsers that depend on the network restrictions.

3. *Consenters*: These peers are use to take responsibility to run the consensus mechanism on the network. These peers are run on a different run-time, unlike endorsers and committers which run on the same run-time. These are also used to validate the transactions and deciding which ledger the transaction be committed to.

1.3 Hyperledger projects

There are several numbers of hyperledger projects promote a wide range of business blockchain technologies, frameworks, libraries, interfaces, and applications. The pictorial representation of hyperledger blockchain-based reference model is as shown in Fig. 8. Presently, hyperledger is the host of the mentioned following projects.

- *Hyperledger Sawtooth*: Hyperledger Sawtooth is an enterprise-grade used for designing to create, deploy, and execute the blockchain-based distributed ledgers that maintain the digital records. This project is a modular blockchain suite developed by Intel, which uses a PoET consensus mechanism.
- *Hyperledger Iroha*: Hyperledger Iroha is hosted by Linux Foundation and is used to build secure, fast, and trusted decentralized applications. It is

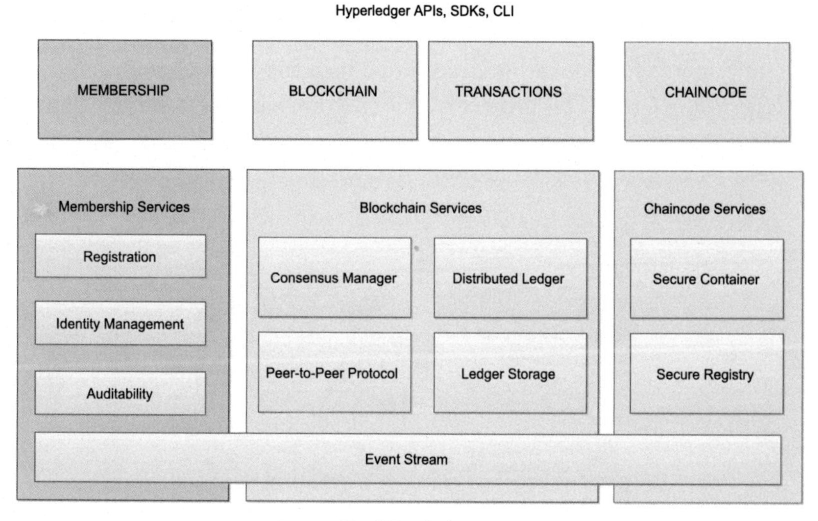

Fig. 8 Hyperledger reference architecture.

used to create an easy to incorporate the framework for a blockchain. This project is developed by the simultaneously working of both the companies like National Bank of Cambodia and Soramistu Cooperation Limited.

- *Hyperledger Burrow*: This project is used to develop a permissible smart contract along with the specification of Ethereum.
- *Hyperledger Fabric*: The most popular hyperledger framework is hyperledger fabric is as shown in Fig. 9. It is one of the important blockchain projects that helps to manage the transactions very efficiently and in a fast way (1000 transactions per second). It is quite different from other hyperledger projects as it is private and permissioned that has been used for private organizations. It is mandatory for all the members of a network to log in via a valid membership service provider. It also utilizes a PKI to create cryptographic certificates. Hyperledger Fabric has a greater strength of flexibility in comparison to other projects in permission and privacy.
- *Hyperledger Composer*: This project is used for building blockchain business networks. The framework is hyperledger composer is as shown in Fig. 10.
- *Hyperledger Explorer*: This project is designed to create a user-friendly Web application. It can view, invoke, use or query the blocks, transactions, and associated data, network information (name, status, list of nodes), chain codes, as well as any other relevant information stored in the public ledgers of the blockchain.
- *Hyperledger Indy*: This project is a collection of tools, libraries, and other components for digital identities present in the blockchain.
- *Hyperledger Cello*: This project is a blockchain-based as-a-service deployment model.

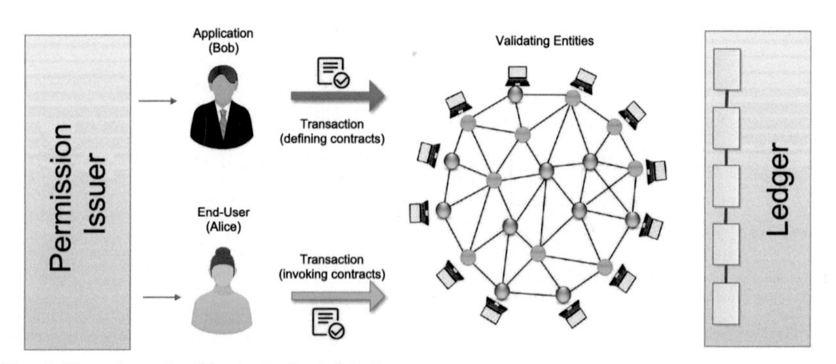

Fig. 9 Framework of hyperledger fabric.

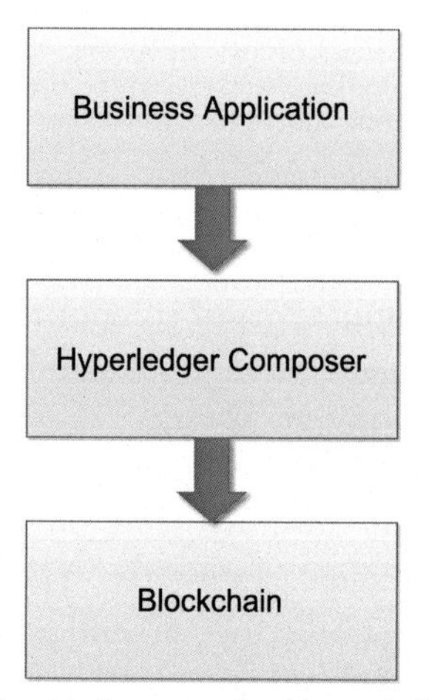

Fig. 10 Framework of hyperledger composer.

1.4 Goals of hyperledger

This hyperledger platform has many numbers of specific goals. That's why private companies and organizations are more concerned about participating in the projects of hyperledger and developing the technologies by using it in their business processes. The goals of hyperledger are described as under.

- This platform is used to create enterprise-grade, open source, distributed ledger frameworks and code bases that support business transactions.
- This platform is used to provide neutral, private, and permissioned blockchain that supports technical and governance business.
- This platform is used to build technical communities and organizations to develop a private blockchain that is deployed by authenticated members only.
- The other platforms of blockchain projects using cryptocurrencies but the projects under hyperledger determines high-scaling industry applications having nonmonetary.
- This platform provides private and confidential transactions that will not done on the public blockchain.

2. Hyperledger fabric

Hyperledger Fabric is a platform used to provide solutions for distributed ledgers with a high degree of flexibility, scalability, confidentiality, and resiliency. The modular and adaptable design satisfies a broad range of industry use cases. In this section, we focused on the architecture and the transaction flow of the hyperledger fabric in a blockchain network that would help us to build a business application.

2.1 Architecture of the hyperledger fabric

The architecture of hyperledger fabric consists of four types of components and their description is described as follows. The pictorial representation of the architecture of hyperledger fabric is as shown in Fig. 11.

1. *Membership services*: The component membership service is used in the hyperledger fabric for providing the identity to the nodes of the blockchain network. It is simply done by digital certificates. These certificates are used by the nodes to ensure that only the authenticated node enters into the blockchain network. Then, these nodes have all the rights to access, sign, and validate the transactions and submit it to the blockchain.
2. *Certificate authorities*: The component certificate authority in the hyperledger fabric is used to provide a certificate to all the nodes of the network. It is present anywhere in the network, i.e., inside (internally) or outside (externally) the network. These certificates depend on public key infrastructure that ensures public and private key-pair to all the nodes of the network. Then, these nodes would use the public and private key-pair to make a transaction on the blockchain network. It supports clustering for high availability characteristics, LDAP for user authentication, and HSM for security is as shown in Fig. 12.

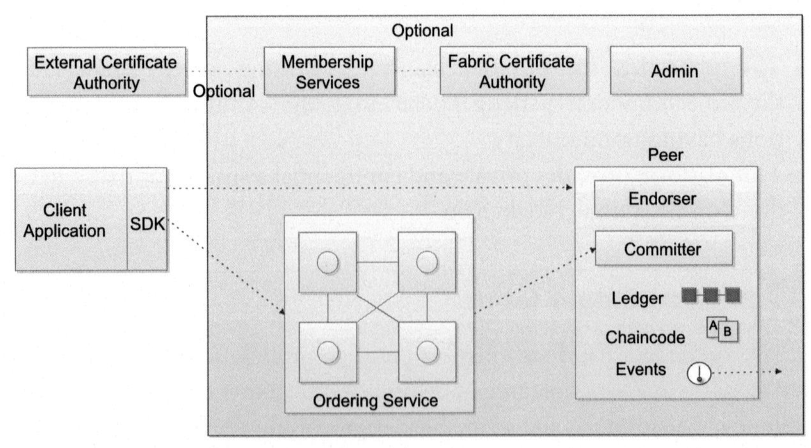

Fig. 11 Architecture of the hyperledger fabric.

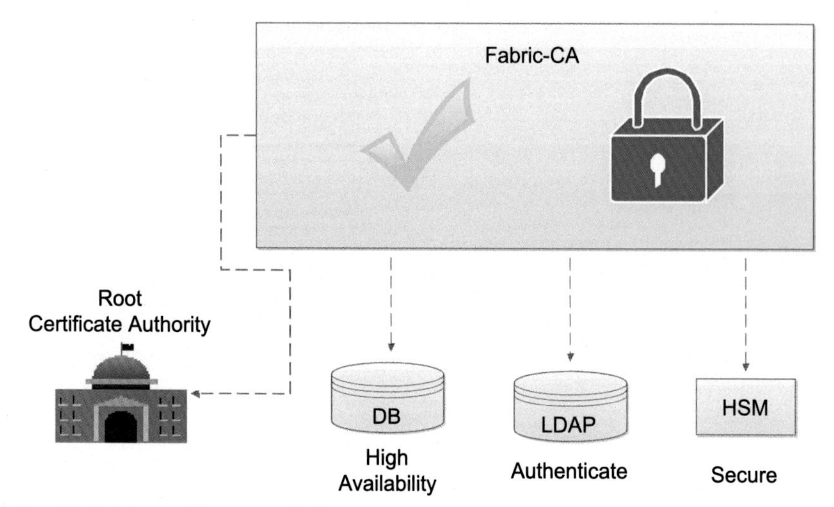

Fig. 12 Fabric certificate authority.

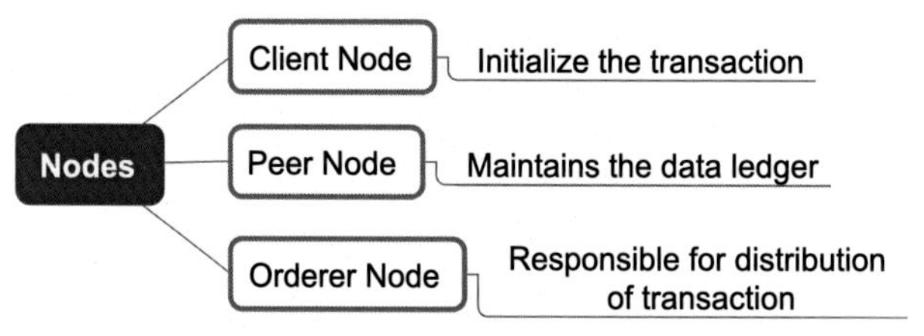

Fig. 13 Nodes in the hyperledger fabric network.

3. *Nodes*: In the permissionless blockchain, each node has equal rights to join, access and validate the transaction on the network. But in the permissioned blockchain, all the nodes are not equal as in the public or permissionless blockchain. In permissioned like hyperledger platform consists of different types of nodes such as client node, peer node, and orderer node and each node has its role to play described in Fig. 13.
4. *Peers*: The component peers are an important part of the blockchain network. Each organization and the business application would have one or more peers in the blockchain network. They play a different role according to the task assigned during the setup of the network. The different types of peers and their roles are described in Fig. 14. Each fabric peer connected to one or more channel and each channel has its ledger to maintain the data. The chaincodes are stored in the dockers and shared across the channels is as shown in Fig. 15.

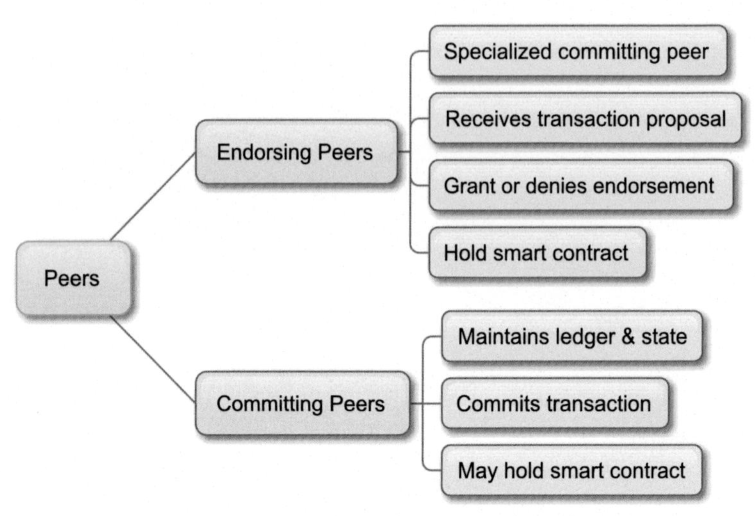

Fig. 14 Peers in the hyperledger fabric network.

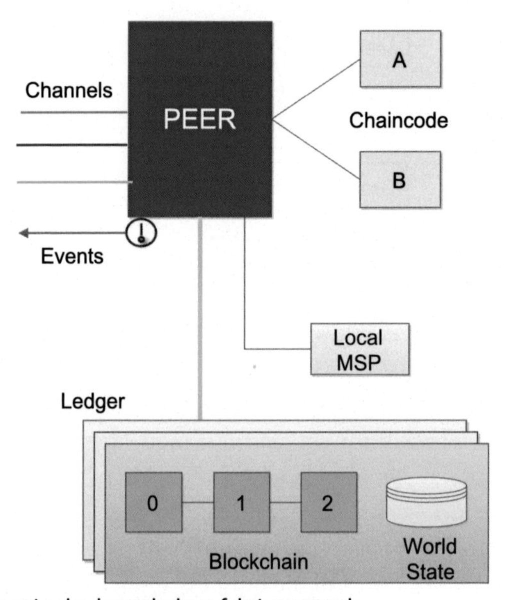

Fig. 15 Fabric peer in the hyperledger fabric network.

The more details about the parts of the architecture of hyperledger fabric is described as follows.

1. *Ordering-service*: The orderer node is used in the architecture for the distribution of data. It also provides an ordered set of transactions on the

blockchain network and to get updated in the ledgers of the nodes is as shown in Fig. 16.

2. *Channels*: The channels provide privacy and security between different ledgers. These exist in the scope of a channel. It can be shared across an entire network or can be permissioned for a specific number of peers. The smart contract or chaincode is instantiated on a particular channel and installed on the peers of the network to access the world state. The concurrent execution gives performance and scalability is as shown in Fig. 17.
3. *Single-channel network*: In a single-channel network, all peers are connected to the same system channel. All the peers have the same chaincode and maintains the same ledger is as shown in Fig. 18.
4. *Multichannel network*: In a multichannel network, several applications are connected to a specific number of peers. As shown in Fig. 19, two applications are connected to a different number of peers according to their requirements. E1 and E3 are connected to the red channel with chaincode Y and Z whereas E0 and E2 are connected to the red channel with chaincode A and B.

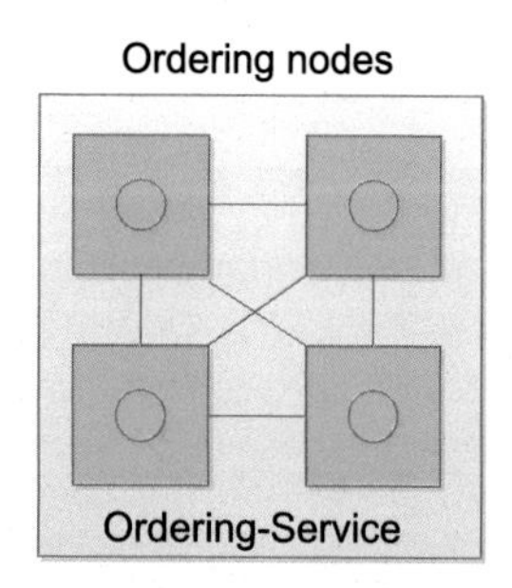

Fig. 16 Ordering-service.

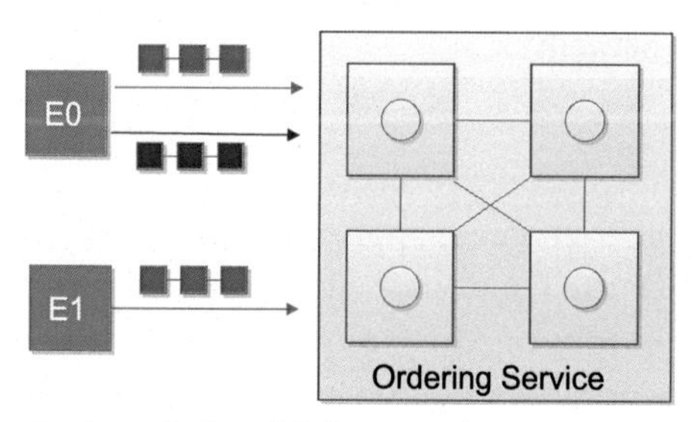

Fig. 17 Channels in the hyperledger fabric network.

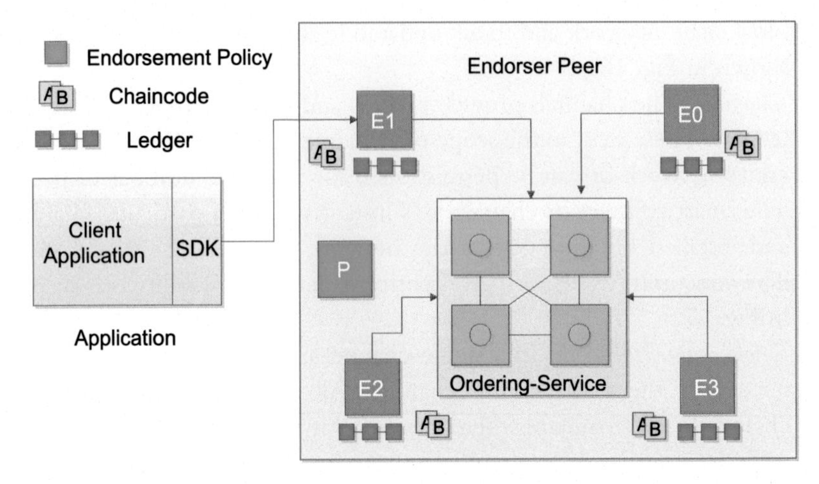

Fig. 18 Single-channel network.

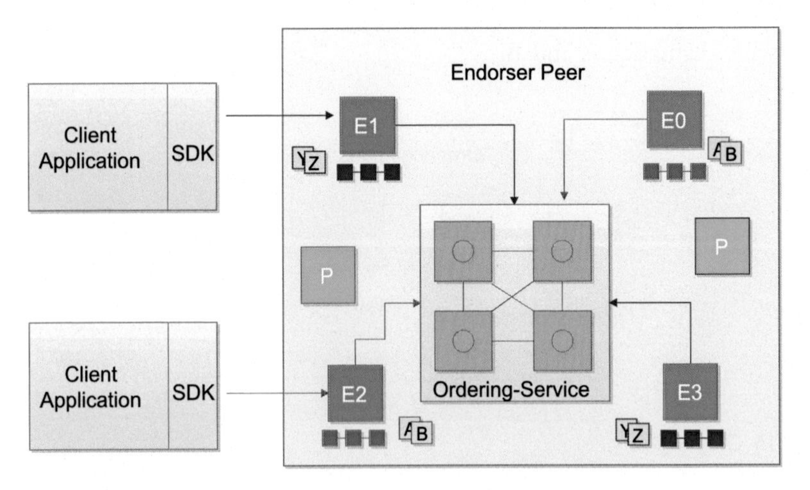

Fig. 19 Multichannel network.

5. *Client application*: Each client application has a fabric software development kit (SDK) for connecting the channels to one or more peers (Fig. 20). It also connects to the orderer nodes through channels and receives events from the peers. A client can be written in different languages like Go, java, python, etc.

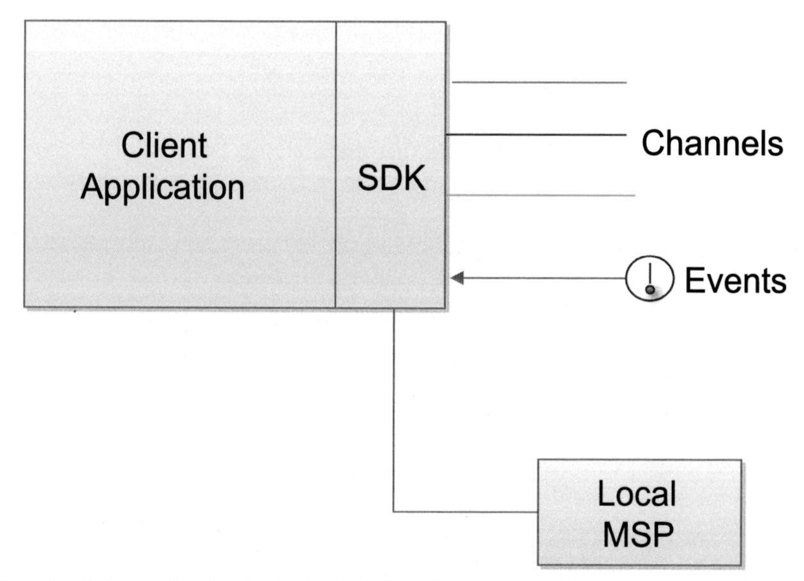

Fig. 20 Client application in the hyperledger fabric network.

2.2 Transaction flow in the hyperledger fabric

The client and the application interact with the blockchain using hyperledger fabric SDK when they want to make a transaction on the network. The authentication and privacy privileges are ensured by the membership services and a certificate authority provides public and private key-pair to all the nodes of the network. The step-wise series of the transaction flow in the hyperledger fabric is as shown in Fig. 21.

1. *Propose transaction*: At the first step of the transaction flow, the client application needs a business agreement so it is going to propose a particular transaction for their automation. Then, the client application sends all the inputs and the transaction to all the endorsing nodes present on the network (E_0, E_1, E_2) is as shown in Fig. 22. There is an endorsement policy in the hyperledger fabric network that is predefined with the smart contract.
2. *Execute proposed transaction*: At the second step of the transaction flow, the broadcasted transaction is executed by all the endorses (E_0, E_1, E_2) present on the network. Each endorser will capture the read and write set of the transaction called RW sets, which will now flow in the hyperledger fabric network. The transactions may be signed and encrypted form. The pictorial representation of execute propose transaction is as shown in Fig. 23.

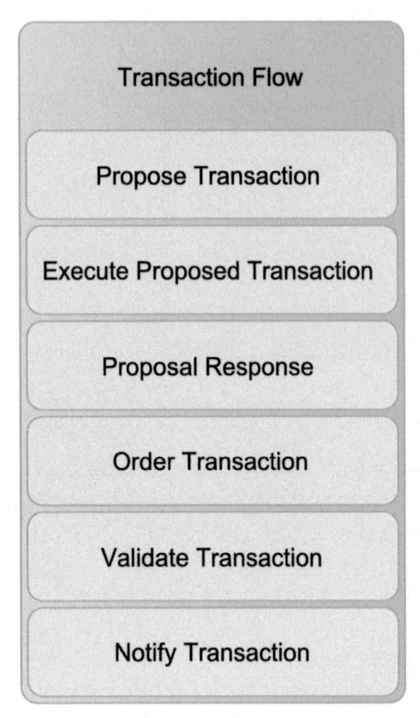

Fig. 21 Transaction flow.

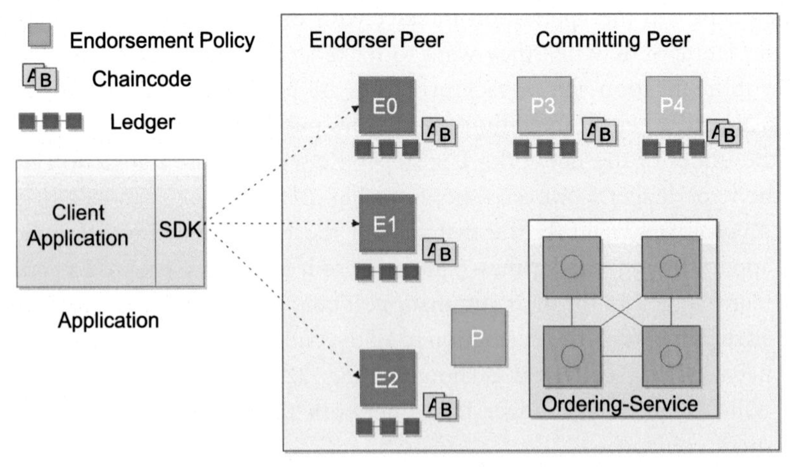

Fig. 22 Propose transaction.

3. *Proposal response*: At the third step of the transaction flow, the endorser sends and returned asynchronously RW sets to the client application with their signatures. But this information is checked later in the consensus process of the blockchain. The pictorial representation of proposal transaction is as shown in Fig. 24.

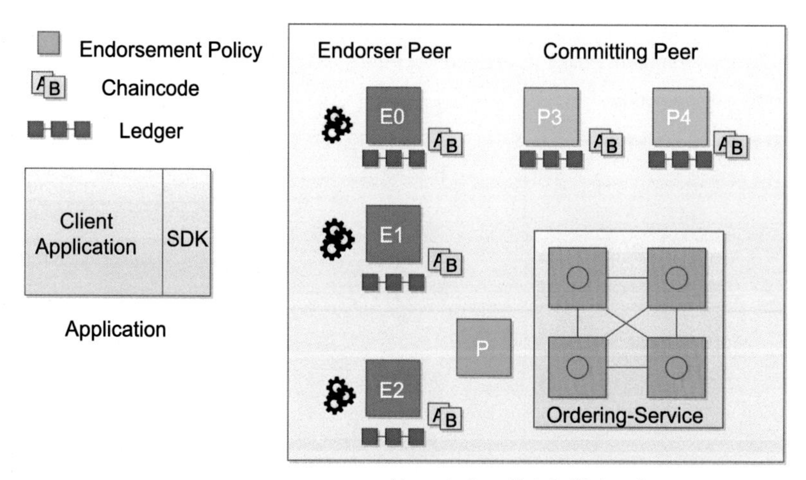

Fig. 23 Execute proposed transaction.

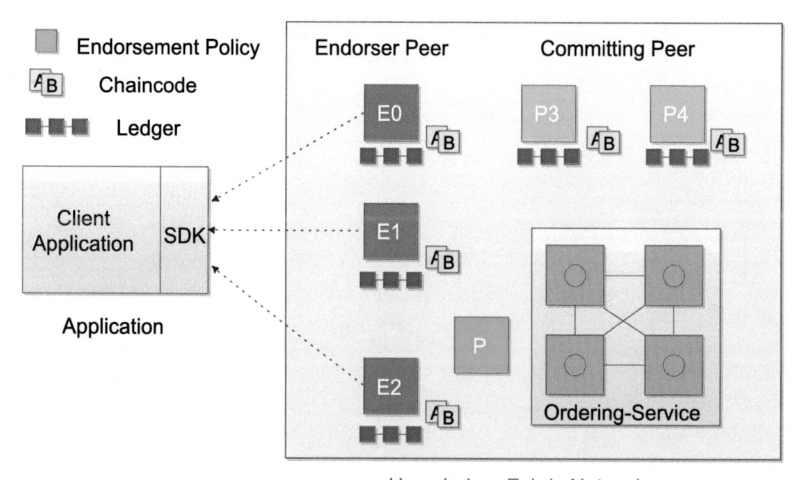

Fig. 24 Proposal transaction.

4. *Order transaction*: At the fourth stage of the transaction flow, once the client application receives the output from all the endorsers with their signatures then, they will submit this response to the ordering-service. At a particular time-period, many such responses (transactions) can be submitted to the ordering-service. Further, ordering-service will determine how to order these transactions and everyone knows the same order of transactions across the network (Fig. 25).

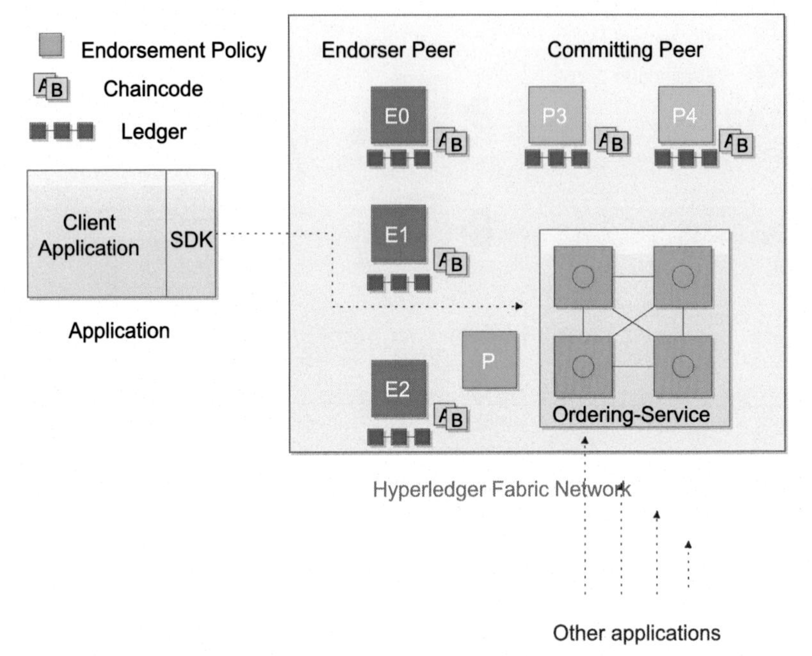

Fig. 25 Order transaction.

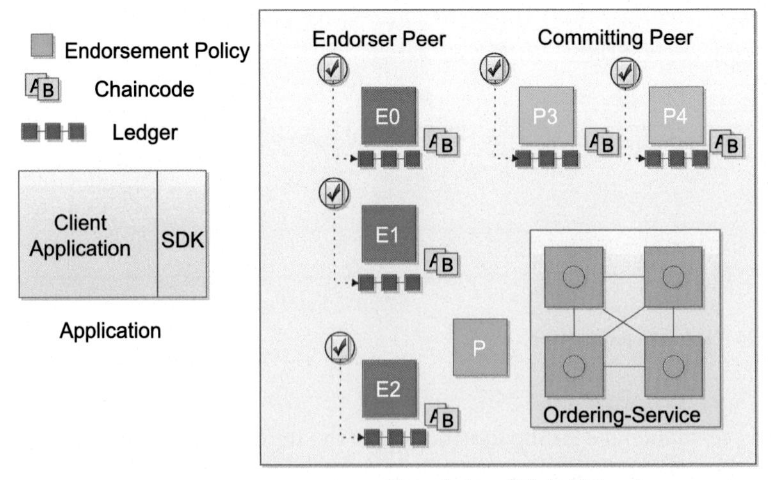

Fig. 26 Validate transaction.

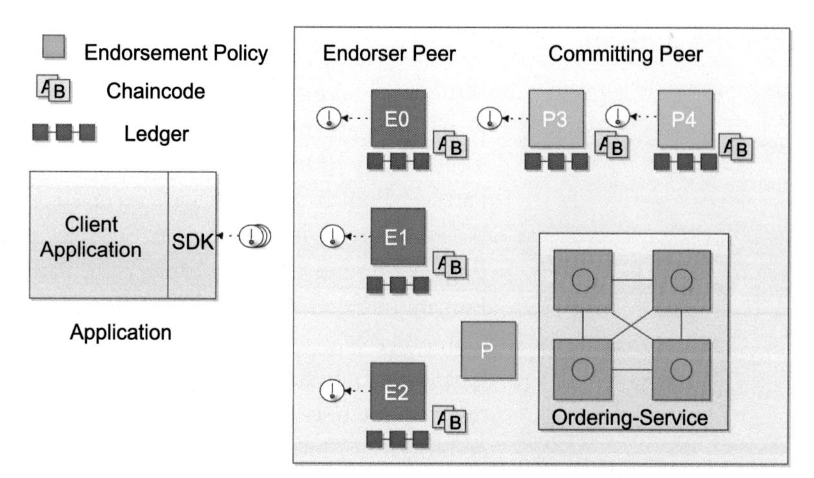

Fig. 27 Notify transaction.

5. *Validate transaction*: Once an ordering-service defines the order then the fifth step of the transaction flow is to validate the transaction starts. At this stage, an ordering-service sends the ordered set of transactions to all the peers of the network. The ordered set of transactions are called blocks. The committing peers validate the blocks against the endorsement policy and also checks that the RW sets are still valid or not for the current world state. The valid transactions are applied to the world state as well as to the ledgers whereas the invalid transactions are kept a hold on the ledger and not updated on the world state (Fig. 26).
6. *Notify transaction*: At the final stage of the transaction flow, all the peers of the network are going to commit to a valid set of a transaction. Then, this set of a transaction is added into the block which is further get added to the blockchain and each peer will transmit the events. The committing peers notify the applications to which they are connected in the hyperledger fabric network (Fig. 27).

About the authors

Shubhani Aggarwal is pursuing PhD from Thapar Institute of Engineering & Technology (Deemed to be University), Patiala, Punjab, India. She received the BTech degree in Computer Science and Engineering from Punjabi University, Patiala, Punjab, India, in 2015, and the ME degree in Computer Science from Panjab University, Chandigarh, India, in 2017. She has many research interests in the area of Blockchain, cryptography, Internet of Drones, and information security.

Neeraj Kumar received his PhD in CSE from Shri Mata Vaishno Devi University, Katra (Jammu and Kashmir), India in 2009, and was a postdoctoral research fellow in Coventry University, Coventry, UK. He is working as a Professor in the Department of Computer Science and Engineering, Thapar Institute of Engineering & Technology (Deemed to be University), Patiala (Punjab), India. He has published more than 400 technical research papers in top-cited journals such as IEEE TKDE, IEEE TIE, IEEE TDSC, IEEE TITS, IEEE TCE, IEEE TII, IEEE TVT, IEEE ITS, IEEE SG, IEEE Netw., IEEE Comm., IEEE WC, IEEE IoTJ, IEEE SJ, Computer Networks, Information sciences, FGCS, JNCA, JPDC, and ComCom. He has guided many research scholars leading to PhD and ME/MTech. His research is supported by funding from UGC, DST, CSIR, and TCS. His research areas are Network Management, IoT, Big Data Analytics, Deep Learning, and Cybersecurity. He is serving as editor of the following journals of repute: ACM Computing Survey, ACM·IEEE Transactions on Sustainable Computing, IEEE·IEEE Systems Journal, IEEE·IEEE Network Magazine, IEE·IEEE Communication Magazine, IEE·Journal of Networks and Computer Applications, Elsevier Computer Communication, Elsevier International Journal of Communication Systems, Wiley. Also, he has been a guest editor of various international journals of repute such as IEEE Access, IEEE ITS, Elsevier CEE, IEEE Communication Magazine, IEEE Network Magazine, Computer Networks, Elsevier, Future Generation Computer Systems, Elsevier, Journal of Medical Systems, Springer, Computer and Electrical Engineering, Elsevier, Mobile Information Systems, International Journal of Ad Hoc and Ubiquitous Computing, Telecommunication Systems, Springer, and Journal of Supercomputing, Springer. He has also edited/authored 10 books with international/national publishers like IET, Springer, Elsevier, CRC: Security and Privacy of Electronic Healthcare Records: Concepts, Paradigms and Solutions (ISBN-13: 978-1-78561-898-7), Machine Learning for Cognitive IoT, CRC Press, Blockchain, Big Data and IoT, Blockchain Technologies Across Industrial Vertical, Elsevier, Multimedia Big Data Computing for IoT Applications: Concepts, Paradigms and Solutions (ISBN: 978-981-13-8759-3), Proceedings of First International Conference on Computing, Communications, and Cyber-Security (IC4S 2019) (ISBN 978-981-15-3369-3). One of the edited text-book entitled, "Multimedia Big Data Computing for IoT Applications: Concepts, Paradigms, and Solutions" published in Springer in 2019 is having 3.5 million downloads till June 6, 2020. It attracts attention of the researchers across the globe (https://www.springer.com/in/book/9789811387586). He has been a workshop chair at IEEE Globecom 2018 and IEEE ICC 2019 and TPC Chair and member for various international conferences such as IEEE MASS 2020 and IEEE MSN 2020. He is a senior member of the IEEE. He has more than 12,321 citations to his credit with current h-index of 60 (September 2020). He has won the best papers award from IEEE Systems Journal and ICC 2018, Kansas City in 2018. He has been listed in the highly cited researcher of 2019 list of Web of Science (WoS). In India, he is listed in top 10 position among highly cited researchers list. He is an adjunct professor at Asia University, Taiwan, King Abdul Aziz University, Jeddah, Saudi Arabia, and Charles Darwin University, Australia.

CHAPTER SEVENTEEN

Blockchain for enterprise☆

Shubhani Aggarwal and Neeraj Kumar
Thapar Institute of Engineering & Technology, Patiala, Punjab, India

Contents

Abstract

Blockchain for enterprise is on its way to pave the way for the future. Many organizations like IBM are trying to invest in an enterprise blockchain platform and enhance their business model. Enterprise blockchains mainly focus on the features of enterprise-grade and solve the issues that the industry faces. All the enterprise blockchain is specially equipped to meet with all organizational demands. Some important features of blockchain that enterprise can utilize are decentralized nature and ensures P2P network, immutability for no corruption, greater transparency increases responsibility, the cheaper cost will save money, the faster network increases efficiency, etc. In this chapter, we have described the need for blockchain technology in the future industry.

Chapter points

- In this chapter, we describe the need for enterprise blockchain in real-world applications.
- Here, we discuss the requirements of a blockchain and high-level overview of blockchain in industry.

The aforementioned chapters describe the concept of blockchain and bitcoin, their working models, requirements, consensus mechanisms, blockchain platforms, cryptographic primitives *etc.* in detail. A big difference between the blockchain and bitcoin is that the bitcoin is just an application of an underlying blockchain technology and blockchain technology is more powerful than the bitcoin. So, this technology

☆ Working model.

Advances in Computers, Volume 121
ISSN 0065-2458
https://doi.org/10.1016/bs.adcom.2020.08.017

Fig. 1 Blockchain definition.

expands to multiple industries for several services like supply chain management, financial services, healthcare services, smart energy management, voting services, transportation services, smart grid services *etc.* The definition of blockchain is as defined in Fig. 1.

1. Track assets in a business network

In the 1950s, when the Internet has emerged, the communication between the organizations is very simple. It is very easy for the people who digitizing and automating the process for communication. Then, move away and the use of physical modes to communicate through digital modes. These modes can use a protocol to send messages from one place to another. So, in today's world, there is a number of TCP/IP protocols to communicate the people with one another. Let us take an example of a payment system in today's scenario. Suppose, *Alice* wants to send some amount of money to *Bob* who is residing in the US. Then, Alice instructs the bank in India to transfer some amount of money to Bob's bank in the US. In this transfer mode of communication, there are at least 4 or 5 banks are involved between them to complete the transaction. And all these banks do point-to-point communication is as shown in Fig. 2.

In this Fig. 2A shows that the bank in India talks to another bank B and then, bank B talks to bank C and so on. There may be 4 or 5 hops between them and each hop permitted the request which is coming from the previous bank and then, finally the money transfers to Bob's account. This process becomes very inefficient, expensive and is also vulnerable to attacks. There could be failures along the way or at any point of communication is as shown in Fig. 2B. Then, these failures have to be manually reconciled correctly which takes a lot of time and cost. So, these all point-to-point

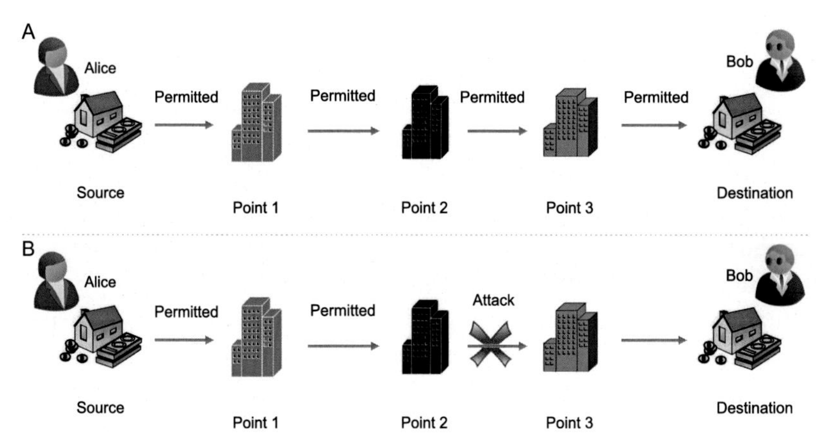

Fig. 2 Point-to-point communication.

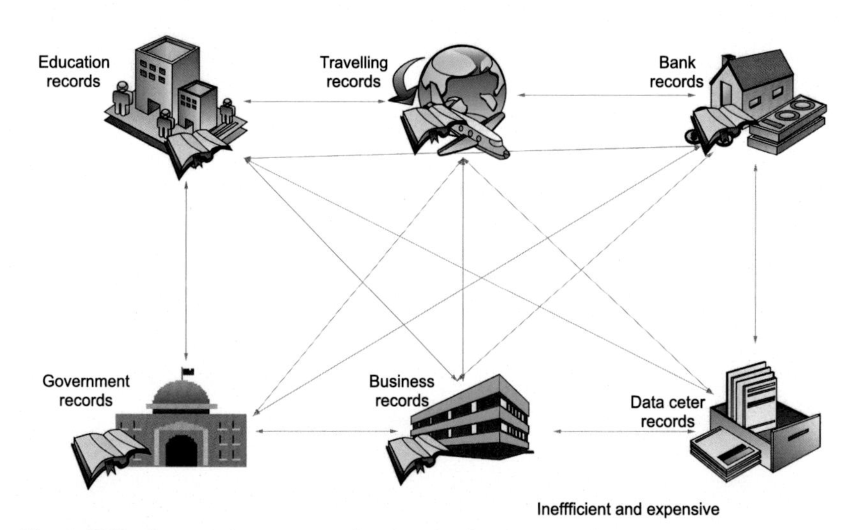

Fig. 3 Difficult to track asset transfers in a business network.

interactions in the current enterprise system becomes very inefficient and expensive. In this way, to track assets in a business network is very difficult to handle and control is as shown in Fig. 3.

The solution for the aforementioned complex and inefficient point-to-point interactions where all the enterprise users are doing business with one another is a common and shared platform. At this platform, each participant or user can share their information and knowledge with the other participants in a secure and right manner. So, this common and shared platform only provides by blockchain technology where all the enterprises or users can come and joined a shared platform, where they can exchange their information at any time with one another to execute their business processes. In this network, everyone is secure and equal, whoever wants to know some piece of information at any period of time, they will immediately get access to that information from the blockchain. The blockchain-based shared, replicated, and permissioned platform in a business network is as shown in Fig. 4.

In this network, every node is going to store a replicated copy of the history of all the transactions done on the network and recorded on the public ledgers of the blockchain. Each node has its own ledger to store the information of all the transactions. This log record is immutable and transparent in nature. So, there is no way to tamper with the previous transactions, the only way is to add the new transactions. And all these transactions get added on the network via consensus mechanism which is an important concept. It is an agreement where all the participants agree on one consensus that this is the legitimate transaction that needs to get added to the ledger and they all simultaneously add the transaction in a consistent manner. Then, this transaction is added to the block and further, the block is added to the blockchain after block verification and validation process. Hence, once the transaction is done on the network, it is there forever right. So, in this overall system, no one can change the data in the future and it reduces the inefficiencies exist in today's system. The pictorial representation of degree of centralization is as shown in Fig. 5.

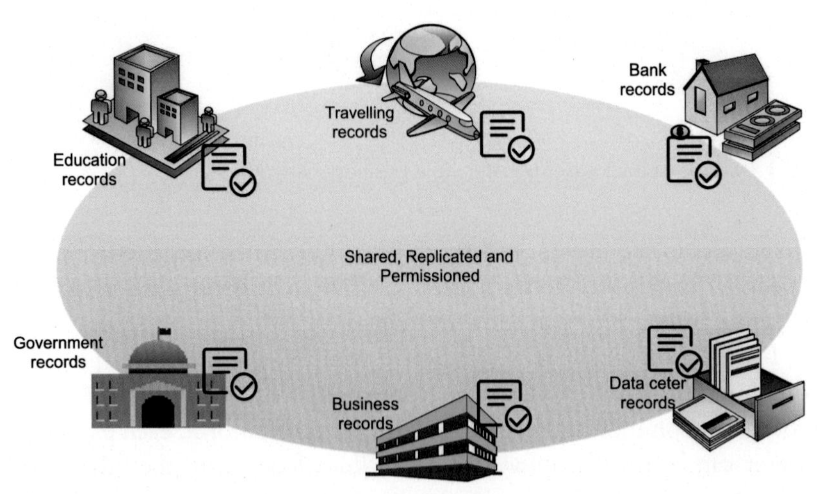

Fig. 4 Shared, replicated, and permissioned network.

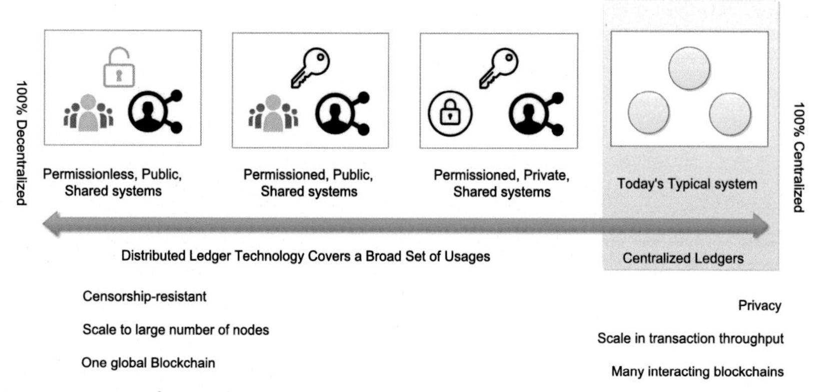

Fig. 5 Degree of centralization.

2. Benefits of blockchain in a business

Now, we discuss the use of blockchain in a business network where all the participants will perform equally. They shared their information with the other participants and also take information from the others whenever they need it to use on their business processes. It is a process of an exchange of messages during the transaction and all the participants agree on that transaction is valid and can be added on to the network. All these transactions are performed by users who are secured and authenticated. So, the exchange of information and execute a business process has been done by using smart contracts. There are two types of blockchain networks used for executing smart contracts, i.e., *Permission-less and Permissioned* as shown in Table 1. In a permission-less system, anyone can join, see, and execute the transactions on the network whereas, in a permissioned system, only a limited number of participants can join, see, and execute the transactions on the network.

For enterprise applications, there is a need for a model that provides some benefits of decentralization with certain security and privacy features on a better scale. So, every transaction on the network has been completely executed and validated by the authorized users. There is no need for a central authority or government agency to validate these transactions. Here, there is a need for a scale to a large number of nodes in one blockchain that validates the transaction rightly. The key concepts and benefits of blockchain for business is as shown in Fig. 6.

Many industries take benefits from the blockchain in terms of greater transparency, enhanced security, improved traceability, increased efficiency and speed, and reduced costs are described as under.

Table 1 Permission-less and permissioned blockchains.

Parameters	Permission-less	Permissioned
Access	Open read/write access to database	Permissioned read/write access to database
Scale	Scale to a large number of nodes but not in transaction throughput	Scale in terms of transaction throughput but not to a large number of nodes
Consensus	PoW/PoS	Closed membership consensus algorithms
Identity	Anonymous/ pseudonymous	Identities of nodes are known, but transaction identities can be private/ anonymous/pseudonymous
Asset	Native asset	Any asset/data/state

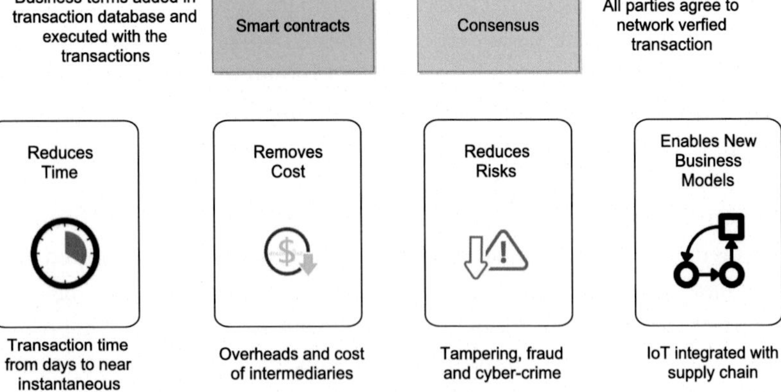

Fig. 6 Key concepts and benefits of blockchain for business.

1. *Greater transparency*: The history of the transaction is becoming more transparent as the use of blockchain technology. The data stored in the public ledgers of the network is more accurate, consistent, and transparent because to change a single transaction record would require the changes in all the corresponding records and the collusion of the entire network.

2. *Enhanced security*: The use of blockchain technology in the industry increases the security and privacy of the data. After a transaction is validated, it is encrypted and linked to the previous transaction that maintains the consistency and security in a business network.
3. *Improved traceability*: To trace the movement of services and goods for supply chain management, blockchain technology is a good solution. When exchanges of goods and services are recorded via blockchain, it ends up with an audit trail that shows the full journey of an asset. It can help to verify the authenticity of an asset and prevent fraud.
4. *Increased efficiency and speed*: Traditional systems use paper-heavy processes that are time-consuming and prone to human error. But, the use of blockchain technology in automating and streamlining the processes gives the transaction faster and more efficient. And when everyone has access to the same information, it becomes easier to trust each other without any involvement of third-party validator.
5. *Reduced costs*: At a business level, reducing cost is an important factor. With the use of blockchain in a business network, there is no need for a third-party validator to make the transaction validates that reduces the cost of the project. Only authenticated and validated user will have permissioned access to a single and immutable version.

2.1 Advantages of blockchain for business

The business could take advantage of this technology by reducing their cost, time and increase accountability. The advantages of blockchain for business are described as under.

- The blockchain technology allows for verification without any involvement of third-parties.
- The blockchain is append-only distributed ledgers. So, the data once be entered in the public ledgers cannot be altered or deleted.
- The blockchain uses hash cryptography to secure the public ledgers. Also, the present ledger is dependent on its adjacent completed block to complete the cryptography process.
- All the transactions and data are added to the block after the verification process. There is a consensus of all the ledger participants on what is to be recorded in the block.
- The transactions are recorded in chronological order. Thus, all the blocks in the blockchain are time-stamped.
- All the transactions in the blockchain are transparent in nature. The individuals who are provided authority can view the transaction.

- The origin of any public ledger present at each node on the network can be tracked along the chain to its point of origin.
- A number of consensus protocols are used to validate the entry of block in the blockchain that removes the risk of duplicate entry or fraud.
- With the smart contracts, the businesses can pre-set conditions on the blockchain. The automatic transactions are triggered only when the conditions are met.

2.2 Businesses take advantage of the blockchain technology

The blockchain technology can be used in any type of industry such as agriculture, banking, healthcare, education, e-governance, property, mining, retail, transport and logistics, media and entertainment, and so on. The type of blockchain architecture will depend on the business type and the area where it is to be implemented. The benefits of blockchain technology from where businesses can take advantage, are described as follows.

1. *Static registry*: In today's world, there is a number of cases of ambiguity in title ownership. With this technology, the records once stored in the public ledgers of the nodes cannot be altered or changed. Any changes in the ledgers are time-stamped. In any case, the title can be tracked through the path of origin. The other places where this technology can be used are in patents and research articles.
2. *Identity*: This is similar to the static registry. But this factor stores a piece of identity information. This factor can be used for identity frauds, voting systems, civil registry, police records, court cases, etc.
3. *Smart contracts*: It eliminates the need for human management that significantly reduces risk. The advancements in technology mean that new types of contracts can be coded and used to communicate during the exchange of important information with each other. It automatically reflects business speed and flexibility. It provides many offers to learn from past contracts and make new concepts and ideas for business processes. It also provides an understanding of the risk-taking capacity of the organization.
4. *Dynamic registry*: In this factor of blockchain, the ledgers keep updating and tracking of goods and services that are exchanged on the blockchain network. This is called supply chain management. It keeps track of all the goods and services from the manufacturer to the distributor and finally to the store and stored it to the public ledgers. The updates provide information on the movement of services that can be used to stop the unusual supply-chain in the market.

5. *Payment registry*: It is a type of dynamic registry that updates the information or cryptocurrency payments that are made on the network. This is advantageous for international payments in busine.

About the authors

Shubhani Aggarwal is pursuing PhD from Thapar Institute of Engineering & Technology (Deemed to be University), Patiala, Punjab, India. She received the BTech degree in Computer Science and Engineering from Punjabi University, Patiala, Punjab, India, in 2015, and the ME degree in Computer Science from Panjab University, Chandigarh, India, in 2017. She has many research interests in the area of Blockchain, cryptography, Internet of Drones, and information security.

Neeraj Kumar received his PhD in CSE from Shri Mata Vaishno Devi University, Katra (Jammu and Kashmir), India in 2009, and was a postdoctoral research fellow in Coventry University, Coventry, UK. He is working as a Professor in the Department of Computer Science and Engineering, Thapar Institute of Engineering & Technology (Deemed to be University), Patiala (Punjab), India. He has published more than 400 technical research papers in top-cited journals such as IEEE TKDE, IEEE TIE, IEEE TDSC, IEEE TITS, IEEE TCE, IEEE TII, IEEE TVT, IEEE ITS, IEEE SG, IEEE Netw., IEEE Comm., IEEE WC, IEEE IoTJ, IEEE SJ, Computer Networks, Information sciences, FGCS, JNCA, JPDC, and ComCom. He has guided many research scholars leading to PhD and ME/MTech. His research is supported by funding from UGC, DST, CSIR, and TCS. His research areas are Network Management, IoT, Big Data Analytics, Deep Learning, and Cybersecurity. He is serving as editor of the following journals of repute: ACM Computing Survey, ACM·IEEE Transactions on Sustainable Computing, IEEE·IEEE Systems Journal, IEEE·IEEE Network Magazine, IEE·IEEE Communication Magazine, IEE·Journal of Networks and Computer Applications, Elsevier Computer Communication, Elsevier International Journal of Communication Systems, Wiley. Also, he has been a guest editor of various international journals of repute such as IEEE Access, IEEE ITS, Elsevier CEE, IEEE Communication Magazine, IEEE Network Magazine, Computer Networks, Elsevier, Future Generation Computer Systems, Elsevier, Journal of Medical Systems, Springer, Computer and Electrical Engineering, Elsevier, Mobile Information Systems, International Journal of Ad Hoc and Ubiquitous Computing, Telecommunication Systems, Springer, and Journal of Supercomputing, Springer. He has also edited/authored 10 books with international/national publishers like IET, Springer, Elsevier, CRC: Security and Privacy of Electronic Healthcare Records: Concepts, Paradigms and Solutions (ISBN-13: 978-1-78561-898-7), Machine Learning for Cognitive IoT, CRC Press, Blockchain, Big Data and IoT, Blockchain Technologies Across Industrial Vertical, Elsevier, Multimedia Big Data Computing for IoT Applications: Concepts, Paradigms and Solutions (ISBN: 978-981-13-8759-3), Proceedings of First International Conference on Computing, Communications, and Cyber-Security (IC4S 2019) (ISBN 978-981-15-3369-3). One of the edited text-book entitled, "Multimedia Big Data Computing for IoT Applications: Concepts, Paradigms, and Solutions" published in Springer in 2019 is having 3.5 million downloads till June 6, 2020. It attracts attention of the researchers across the globe (https://www.springer.com/in/book/9789811387586). He has been a workshop chair at IEEE Globecom 2018 and IEEE ICC 2019 and TPC Chair and member for various international conferences such as IEEE MASS 2020 and IEEE MSN 2020. He is a senior member of the IEEE. He has more than 12,321 citations to his credit with current h-index of 60 (September 2020). He has won the best papers award from IEEE Systems Journal and ICC 2018, Kansas City in 2018. He has been listed in the highly cited researcher of 2019 list of Web of Science (WoS). In India, he is listed in top 10 position among highly cited researchers list. He is an adjunct professor at Asia University, Taiwan, King Abdul Aziz University, Jeddah, Saudi Arabia, and Charles Darwin University, Australia.

Industrial use cases at the cusp of the IoT and blockchain paradigms

Pethuru Raj
Site Reliability Engineering (SRE) Division, Reliance Jio Platforms Ltd. (JPL), Bangalore, India

Contents

Abstract

There are fair amount of reasons and requirements for both the IoT and blockchain concepts to cooperate closely to solve bigger problems at hand. This deadly combination is to result in scores of fresh and fabulous opportunities and possibilities for the total human society. The Internet of Things (IoT) paradigm has made it possible to have digitized and connected things in plenty in our everyday environments. That is, all kinds of physical, mechanical, electrical, and electronics systems are systematically being digitized and connected through proven and potential edge and connectivity technologies. The leading market analysts and researchers have come out with forecasts that there will be billions of connected devices and trillions of digitized entities in the years ahead.

The noteworthy point here is that all these empowered entities, on purposefully collaborating and correlating with one another, can generate massive amounts of multistructured data. The challenge is how to secure IoT devices and data. The arrival of the blockchain technology is being celebrated as the best thing toward convincingly meeting up the IoT security requirements. This chapter is to explore and expound how the cool linkage between IoT and blockchain is to substantially enhance the security and privacy needs of IoT devices and data.

Advances in Computers, Volume 121
ISSN 0065-2458
https://doi.org/10.1016/bs.adcom.2020.08.018

1. Introduction

Devices are not only web-enabled but also are capable of finding any nearby digitized entities in order to avail their unique services. In other words, devices are not only linked up with the Internet but also integrated with all kinds of devices in their vicinity and also with remotely held enterprise, web, mobile and cloud applications, and services. Device to device (D2D) and device to cloud (D2C) integration scenarios are fast evolving. Devices are accordingly instrumented and interconnected to be intelligent in their offerings, outputs and operations. Real-time data capture, ingestion, storage and analytics of devices data are being facilitated through a host of platform solutions and tool chains. That is, devices states and situations emit a lot of useful insights about and for devices to exhibit intelligent behavior. That is, IoT data analytics through batch and real-time data processing comes handy for IoT devices and systems to be cognitive in their deals and deeds. The vast number of IoT devices collect humongous amount of value-adding data. Data analytics methods and artificial intelligence (AI) algorithms are able to make sense out of IoT device data. Wrong data leads to wrong decisions. Hence, data trustworthiness and timeliness are very vital for all the talks around intelligent device services and applications. For arriving at highly accurate decision-making and real-time actuation, we ought to make IoT devices and their data safe and secure.

Increasingly IoT systems are under severe and sustained attacks from external sources. The cyber security attacks are consistently on the rise in the connected world. Therefore, any viable security solution should be truly tamper-proof and has to protect critical IoT data throughout its life cycle. Herein, the blockchain technology pitches in and provides unbreakable and impenetrable IoT security with all the clarity and alacrity.

2. Briefing the Internet of Things Conundrum

With the faster proliferation of edge and digitization technologies, all kinds of common, cheap, and casual things in our midst get systematically digitized. By internally and externally attaching/embedding/embodying diminutive sensors, stickers and specks, invisible chips, codes and controllers, infinitesimal tags, beacons, and LEDs, and disappearing actuators, any concrete item gets digitized easily and quickly. Digitized objects can be minutely and remotely monitored, measured, and managed. Ordinary things get converted into extraordinary artifacts instantaneously. Communication modules/sensors are embedded in devices such as phones, television sets, car's infotainment systems, electrical appliances, traffic lights and cameras, and industrial equipment. These sensors collect data and send them to cloud for further deeper and decisive processing. Feature and cellular phones become smartphones with such remote connectivity getting ingrained. We enjoy multifaceted and state-of-the-art devices due to the continued improvements in cutting-edge technologies such as digitization and digitalization.

The power and value of digitized entities are consistently on the rise with the steady improvements and improvisations in the technology space. In short, the digitization aspect is garnering a lot of attention across the globe.

Especially with the flourish of digital technologies (cloud infrastructures, AI algorithms, blockchain technology, mobility, analytics platforms, digital twins, etc.), digitized objects are going to be the real trendsetter for a host of business verticals in the days to unfold. A bevy of deeper and decisive automation, acceleration, and augmentation can be accomplished so that the traditional business IT is all set to become people IT. The market analysts predict that there will be trillions of digitized entities in the years ahead. That is, all of our physical, mechanical, and electrical systems get digitized methodically to join in the mainstream computing. Second, all kinds of electronics and devices get connected with one another in the vicinity and with cloud-hosted software services and databases. The prediction is that there will be billions of connected devices in the planet earth in the years ahead.

In the beginning, we were enjoying the Internet of computers. Then with service-oriented architecture (SOA) and its direct offshoot "Microservices Architecture (MSA)" flourishing with proper nourishment from a host of worldwide product and tool vendors, we came across the paradigm of the Internet of services (IoS). With the steady growth of the device ecosystem, we started to experience the Internet of devices (IoD). This promising trend has impacted everything else in our daily life. We would have read about the Internet of energy and the Internet of everything (IoE).

Thus, the future generation of the Internet is to comprise digitized entities and connected devices besides having traditional desktops, laptops, smartphones, and server machines. All these converged nicely and neatly and thereby, we are heading toward the era of the Internet of Things (IoT), which is widely touted as the fabulous and futuristic trend.

Precisely speaking, the IoT connects people, places, properties, and products. With this massive-scale Internet, there are fresh openings for value creation and capture. There are significant advancements and accomplishments in the field of data analytics. Therefore, IoT data analytics gets speeded up in order to extract actionable insights in time. That is, the process of transitioning data into information and into knowledge is being optimized and automated. The era of data-driven insights and insights-driven decision-making and action has started with a bang. Newer use cases can be unearthed and realized. The long-term goals of process excellence, infrastructure optimization, operational efficiency, customer delight, etc., can be achieved with ease. Existing challenges and concerns can be adequately addressed. Future expectations can be artistically fulfilled.

Smartphones are the most prominent and dominant IoT device. They are connected with the Internet. They can access web and cloud applications, services, and databases. Real-time operations can be accomplished in phone itself whereas high-end tasks are being delegated to cloud environments, which give an illusion of infinite processing and storage capabilities. There are online repositories for highly handy mobile applications. Because of the connected nature, smartphones are capable of performing a lot of informational and commercial activities. Both resource-constrained as well as intensive devices can be made smarter through web and cloud-enablement.

Types of IoT devices—There are IoT devices for just collecting environmental states and situations and they pass them to nearby or faraway databases, data analytics platforms, and business intelligence (BI) tools to be cleansed and castigated to emit out hidden information out of data heaps. These mostly comprise sensors such as pressure, temperature, gas, humidity, presence, air quality, and light sensors.

There is another category of IoT devices. These can receive information and act upon it. These include devices from smart wearables to smart TVs and 3D printers. Increasingly ground-level IoT devices are empowered to be intelligent in their behavior. There are software libraries, frameworks, and platforms for accomplishing the goal "Intelligence at Edge."

The final category is IoT devices that can do both. For an example, an IoT-based farming system can have a sensor that continuously collects information about soil moisture to clearly decipher and decide how much water the crops need. The knowledge then gets conveyed to the irrigation system to do the needful with all the clarity and confidence. The full lifecycle is pictorially presented in the below diagram. IoT devices stock as well as share a lot of information with integrity and confidentiality in order to fulfill an end-to-end process.

How does the IoT work?—As enunciated above, the IoT refers to a network of digitized objects that are capable of collecting and exchanging data. Also, resource-intensive devices are also hooked into the IoT, and these devices are capable of capturing, stocking and crunching data. Further on, these devices can form ad hoc, dynamic, and small-scale clouds to perform real-time edge device data analytics. There are IoT middleware solutions/hubs/gateways/brokers/queues, etc., for enabling indirect communication. There are IoT-enablement and analytics platforms typically running on cloud environments.

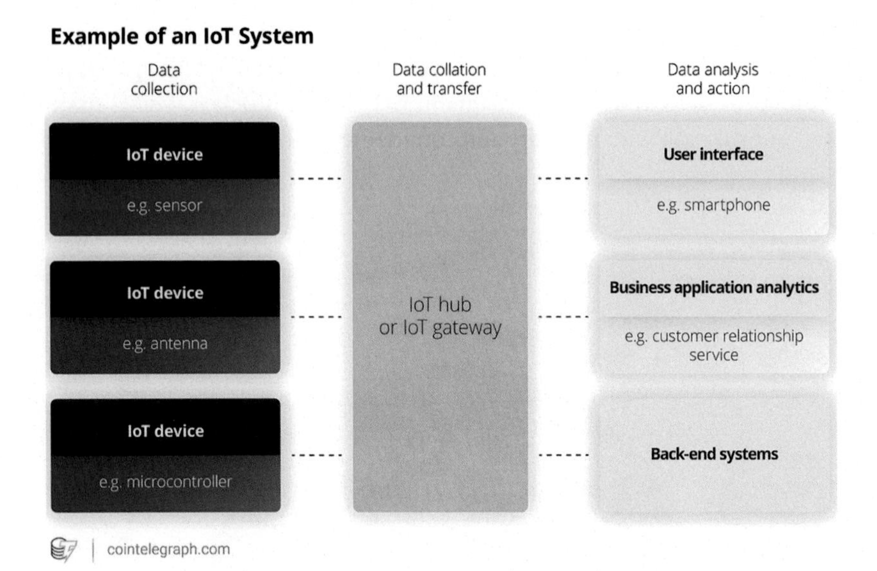

IoT platforms will analyze any captured and cleansed data to extract value-adding information. And the knowledge discovered gets disseminated to actuators and other relevant devices to initiate the best course of actions with all the clarity and confidence. Also, decision-makers, enterprise systems, operators, and stakeholders are being informed accordingly. There are 360-degree knowledge visualization tools and dashboards to supply actionable insights in preferred format in time. The end-result is customer delight, enhanced user experience (UX), deeper automation, and improved productivity.

Almost all the industry verticals are eagerly employing the IoT phenomenon to be customer-centric and competitive. For an instance, in a manufacturing factory, all machineries could be fitted with different sensors in order to capture the latest performance and health state data which can be made known to operators and owners through their mobile applications. Key issues can be proactively and pre-emptively identified and addressed before something serious crops up.

Consumer electronics (connected/cloud-enabled/smart) such as air conditioners can be fitted with a sensor to capture and convey their operational situation so that appropriate remedial action can be comprehensively analyzed and articulated to the repair team in time. The sensor also can continuously pour in data. Both runtime and historical data are blended together and investigated to come out with preventive, predictive, prescriptive and personalized insights. That is, before an issue pops up, the repair team gets all the right and relevant information to prevent any breakdown or slowdown of connected electronics.

As widely reported, security is a big headache for the IoT concept to grow and glow. There is a skepticism and slowdown on large-scale deployment of IoT devices. As everything gets connected with everything else in our places (personal and professional) and with cloud-hosted applications, services, and databases, the security concern is growing up considerably. IoT devices are network-accessible and most often they are resource-constrained. They become an easy target for distributed denial of service (DDoS) attacks. As per news reports, a number of such coordinated attacks have been calculatedly done in order to disrupt service obligations of organizations. IoT devices that are not properly configured and unsecured can be easily exploited by cyber-criminals. An infamous IoT attack is the Mirai Botnet DDoS attack that hugely affected Internet service for the entire East Coast of America. There was also the planned hacking of a Jeep to expose some of the vulnerabilities in IoT devices in cars. Medical and manufacturing IoT devices, when breached, can result in irreparable loss for people and properties.

The IoT paradigm has started in a promising note. Researchers religiously unearth and articulate a dazzling array of use cases across industry verticals. The market for the IoT concept is growing steadily and steadfastly. However, there are some critical challenges and concerns being associated with the spread of IoT devices, networks, systems, and environments. Thus, IoT devices and their data have to be secured while in transit, rest, and utilization. The currently implemented security mechanism is liable to be broken and hence the business world is sincerely and seriously seeking out fool-proof IoT security measures.

3. Entering into the blockchain technology

The blockchain technology is getting big time attention and attraction these days for its uniqueness and usability in fully attending the needs of data security. The blockchain technology is being viewed by many as a disruptive method, which is capable of preparing the Internet for intrinsically fulfilling the fast-evolving needs. The web-scale companies store all their operational, transactional, and analytical data in a centralized location. With the faster acceptance of the decentralization principle of blockchain, there is an enormous scope for things to become truly decentralized. This transition helps our data to be stored and managed across different and distributed systems. As widely known, centralized storage opens up the risks of data getting hacked and unused. Blockchain makes data to be fully secure, authentic, and tamper-proof. Blockchain innately facilitates leveraging decentralized networks to safely and securely record and process transactions.

A blockchain is principally a time-stamped series of immutable records of data. The records of data are batched together to form blocks, and the blocks are then chained together to form a blockchain database, which is, in turn, is being managed by a cluster of computers. The point here is that the blockchain database is not handled by any single entity/authority. Before forming a chain of blocks, each block of data is secured and bound to the other using proven cryptographic principles.

The chain formation steps are as follows. Each blockchain is actually a linked list that contains data (interaction/transaction/operation) and a hash pointer. A hash pointer actually contains the hash value of the data as well as the address of the previous block. This unique process results in a blockchain. When we book railway tickets through any application (mobile or the Web), the credit or debit card company takes its commission for processing the transaction. If we embrace the blockchain technology, then there are only two entities: passengers and the railway department. Each booked ticket is a block that gets continuously added to the ticket blockchain. This guarantees against any tampering and is presented as an independently verifiable record. So, the final blockchain is a complete record containing all the train tickets booked or for a set of trains. Generally, there are one or more intermediaries for establishing and enforcing the much-needed trust between different participants in the transactions. However, with blockchain, the trust is ensured without any intermediary by employing a cluster of compute machines for automating the tasks of any intermediary.

As articulated above, a blockchain system consists of a distributed digital ledger, which is shared between participants in the system. This system resides on the Internet. All kinds of transactions or events are getting validated and recorded in the ledger. It is not possible to bring any amendment or deletion on the already incorporated transactions. Precisely speaking, the system brings a fresh and secure way for information to be recorded and shared by a community of users. Within this community, selected members could maintain their copy of the ledger and must validate and approve if there is any new transaction coming in. A variety of consensus mechanisms are emerging in order to accept and add new transaction.

Blockchain, a promising and potential distributed ledger technology (DLT), has the inherent wherewithal to address the serious and severe IoT security and scalability challenges. An enterprise-grade IoT network is supposed to create and capture massive amount of multistructured data. Further on, it has to have the facility to process millions of transactions per day in order to emit out outliers and any abnormal condition. The operational and management complexities are to be on the higher side. The scalability and security challenges of IoT device networks are therefore a bit tedious, tough, and time-consuming to detect and repair before something unthinkable happens.

Blockchain technology is now getting a lot of attention. It can bring in a series of innovations and disruptions for the future Internet. Besides creating new and sustainable value, it creates a solid decentralized system by removing the indulgence of central servers. This often-repeated decentralization means the facilitation of peer-to-peer (P2P) interaction. Also, it creates a fully transparent distributed database, which is open to all. This groundbreaking and sophisticated technology is being inherently blessed with the following important and interesting elements.

1. *Consensus*—This is for providing the proof of work (PoW) and verifying the action in the networks.
2. *Ledger*—This provides the complete details of all the transactions conducted within networks.
3. *Cryptography*—This is for ensuring the encryption of data in ledger (at rest) and networks (at transit). Also, only authorized users can decrypt the information.
4. *Smart contract*—This is used to verify and validate the participants of the network.

All the participating nodes in the system are actually connected through the Internet, which is the most open, all-encompassing, publicly and globally available and affordable communication infrastructure for data sharing and storage. These nodes maintain all the transactions made on a blockchain network collaboratively. The authenticity of a transaction is checked by a protocol. When a new transaction comes up, its records are added to the ledger of past transaction. The other nodes verify the PoW. The database is composed of blocks of transactions and is copied to every node of the system. The ledger is made publicly available and is incorruptible which is updated every time a transaction is made.

A new digital transaction is first cryptographically protected. Then, miners compete with one another to validate the transaction and once it is validated, it gets time stamped and is added to the blockchain, which is chronologically ordered. The acceptance of block by nodes is communicated when a new block is created in the chain, using the hash of the earlier accepted block.

- *Public blockchain*—Each and every node in the Internet can send or read transaction without requiring any permission. Consensus is also open to the public.
- *Consortium blockchain*—This is for a group of authorized nodes only.
- *Private blockchain*—This is for organizations that own the network of blockchain.

Thus, the blockchain technology is literally mesmerizing and meticulously used for producing scores of decentralized applications and smart contracts. In a system wherein if there are heterogeneous participants, then the contributions of this technology are enormous.

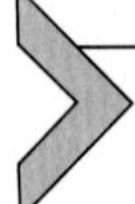

4. The convergence of the IoT and the blockchain paradigms

The typical architecture of IoT systems today is primarily client-server. IoT sensors and devices communicate with cloud systems through an intermediary (say, IoT middleware (gateway, broker, queue, processor, etc.)). This centralized and brokered communication model is problematic in the long run. In this traditional setup, information and service access and data sharing happens via the public and open Internet. This is risky not only for IoT devices but also for IoT data. Further on, it is quite easy and fast to collectively attack centralized systems such as clouds. With zillions of IoT devices are to participate in the future Industrial Internet, cloud centers are definitely going to be under immense pressure. Besides cloud infrastructure, the IoT platforms, which have to manage such a large number of IoT devices, can face the heat. Blockchain comes handy in alleviating the IoT security and scalability concerns.

- The distributed ledger in a blockchain system is famous for granting the tamper-proof capability and this facility guarantees the much-needed trust among the involved parties. Conventionally the trust is being provided by an intermediary, which does the centralized monitoring and management. Here, there is no central authority to have the tighter grip over the vast amount of IoT devices.
- IoT data records are more secure with blockchain, which provides a robust level of encryption.

- Blockchain provides transparency in the sense that it allows anyone, who is authorized to access the network, to track the transactions that happened in the past. This offers a dependable way to pinpoint a specific source of any data leakage and to take an appropriate remedial action quickly.
- As the number of instrumented and interconnected devices is huge, blockchain is able to speed up transaction processing by enabling device coordination.
- There will be a sharp reduction in processing overheads and communication costs as IoT gateways are not needed anymore to guarantee the trust. That is, without IoT gateways, blockchain enables proper coordination among millions of IoT devices.
- Smart contracts are a formal agreement established between two and more parties, and this is being stored in blockchain. When the agreed criteria getting fulfilled, the contractual agreements get executed automatically without any interpretation and involvement of humans.

5. Prominent blockchain–IoT platforms and products

The faster proliferation of IoT devices with the voluminous production of tiny and implantable chips, controllers, and sensors has brought in a paradigm shift. IoT devices are increasingly interacting with blockchain ledgers. Already we are being bombarded with a growing array of IoT device services. Now with the strategically sound combination with blockchain, we can visualize hitherto unheard device services. A few blockchain and IoT players are briefed here.

- *Chain of Things (CoT)*—This is a consortium deeply investigating how the combination of IoT and blockchain can lead to a new set of business, operational, transactional, and analytical applications. CoT has built Maru, an integrated blockchain and IoT hardware solution to solve issues with identity, security, and interoperability. More decisive details can be found in this site (https://www.chainofthings.com/).
- *Riddle & code* (https://www.riddleandcode.com/) provides cryptographic tagging solutions for blockchains in smart logistics and supply chain management. Riddle & Code offers a combined hardware and software solution that enables secure and trusted interaction with IoT devices by giving the devices a "trusted digital identity."
- *The Hyundai Digital Asset Company (HDAC)* (https://www.hdactech.com/en/index.do) is applying the blockchain technology to quickly

and effectively communicate, handle identity verification, authentication, and data storage between IoT devices. The system incorporates a double-chain system (public and private) to increase transaction rate and volume. This feature makes it ideal for IoT devices. The technology is increasingly applied to different verticals such as smart factories, smart homes, and smart buildings for machine-to-machine transactions and operation between IoT devices.

- *VeChain* (https://www.vechain.org/) is an enterprise-level public blockchain platform. This blockchain platform is specially for IoT devices interaction. Cold-chain logistics companies primarily use proprietary IoT devices to track key metrics such as temperature throughout the entire journey. In addition, the platform can hold automobile passports by creating digital records of cars. The records can hold the car repair history, insurance, registration and even driver behavior throughout its lifecycle.

Fresh medical and healthcare applications can be realized by using end-to-end tracking of production processes of medical devices. Patients can securely share their biometric data with their doctors to enable real-time monitoring. VeChain also uses IoT technology for luxury goods by embedding smart chips within the luxury products. The product company can minutely monitor their sales details in real-time. The remote monitoring, measurement, and management of goods can be made possible.

- *Waltonchain* (https://www.waltonchain.org/en/) is created through a combination of RFID and blockchain technologies for effective IoT integration. They primarily focus on tracking processes and products in the supply chain. This can be applied to high-end clothing identification, food and drug traceability, and logistics tracking. All that are getting identified and tracked are implanted with RFID tags, RFID reader and writer module. The details captured can then be sent for data analytics.
- *Streamr* (https://streamr.network/) is an open-source blockchain infrastructure to power the world's data economy and to give people back control of their own information. This can be implanted into everyday objects such as cars to record data including traffic, potholes, and local fuel prices. The user can then sell this data to fellow car users or highway agencies or buy information from other users that will help them make real-time and accurate decisions in a connected city. Information gets transmitted through the decentralized P2P network to get posted on the network nodes and is powered by the network's native cryptocurrency (DATACOIN).

- *IOTA* is a promising protocol for faster transaction settlement and data integrity. Tangle ledger, which is the distributed ledger that supports IOTA, eliminates the need for expensive mining. As indicated above, mining is all about validation of transactions. IOTA is touted as a promising infrastructure for IoT devices to process large amounts of small and fast data. The prominent features of the Tangle ledger are M2M communication, fee-less micropayments, and quantum resistant data. More information is found in this site (https://www.iota.org/).

  IOTA was designed specifically for the IoT and the Tangle platform is being widely described as the one that goes beyond blockchain. It is being termed as a blockless, cryptographic, and decentralized network. In this network, instead of outsourcing network verification, users verify the transactions of other users. The advantages are truly winsome. It facilitates greater scalability and there is no need to pay anything to the so-called miners. These benefits are being greatly appreciated because, in a standard IoT network, there can be millions of devices participating and billions of microtransactions getting accomplished between devices a day. IOTA has already entered into several important partnerships including.

1. The Bosch XDK (Cross Domain Development Kit) (https://xdk.bosch-connectivity.com/) is a programmable sensor device and IoT prototyping platform used to collect specific and real-time data which can be sold out via the IOTA Data Marketplace.
2. Fujitsu Japan is using the IOTA protocol in a proof-of-concept and immutable data storage medium for audit trails across industrial production environments and supply chains.
3. Volkswagen is working with IOTA on a project called "Digital CarPass." This is a report card for cards stored on a distributed ledger that ensures critical factors such as mileage are reliable and up-to-the-date.

There are other noteworthy projects including.

- *Ambrosus* (https://ambrosus.com/) is a blockchain-powered IoT network for food and pharmaceutical enterprises. This is mainly for enabling secure and frictionless dialogue between sensors, distributed ledgers, and databases to optimize supply chain visibility and quality assurance
- *IoT Chain* (https://iotchain.io/) is a secure IoT light operating system driven by blockchain technology.
- *Atonomi* (https://atonomi.io/) provides IoT developers and manufacturers with an embedded solution to secure devices with blockchain-based immutable identity and reputation tracking. They also help provide the identity and trust required for our increasingly connected world

- *IoTeX* (https://www.iotex.io/)—We are anticipating that a machine-driven, automated, and connected future. As we all know, IoT sensors and devices are being fit into every tangible thing in our everyday environments. For example, IoT sensors are being attached onto vehicles to power them up to be fully autonomous and onto robots to empower them to be collaborative and cognitive. AI chips are being embedded into humans to assist in their everyday decision-making, deals, and deeds. As IoT idea is greatly penetrative and pervasive, there is a pertinent and paramount question. *Can we fully trust the rapidly evolving IoTs?* IoTeX is for building the Internet of Trusted Things. This is an open network where all physical and virtual things (humans, machines, and businesses) can interact with full trust and privacy. IoTeX brings in the missing trust by combining blockchain, secure hardware, and decentralized identity (DID) innovations.
- *OriginTrail* (https://origintrail.io/) is an ecosystem dedicated to making global supply chains work together by enabling a universal, collaborative, and trusted data exchange.
- *Slock.it* (https://slock.it/)—This connects *IoT devices to the blockchain with secure interoperability*.
- *BlockMesh* (https://www.blockmesh.io/)—The BlockMesh's vision is to create the world's first decentralized and cost-free communications network. The BlockMesh platform supports mesh-based devices focusing on communication and the loT. This platform will provide the user the ability to communicate and digitally transact in the P2P network, crushing data costs, and banking fees. The aim is to create a self-sustainable financial ecosystem for communication and loT. Users and loT devices can earn "Mesh Token" for supporting the network and pay in "Mesh Token" for access to the conventional Internet at a fraction of the cost. Users with unlimited Wi-Fi can offer their community or clients Internet access with no passwords needed. With "mesh," the internet connection can then be passed beyond the hardware range.
- *Helium* (https://www.helium.com/)—With a Helium Hotspot, anyone can easily earn cryptocurrency by building a wireless network in their city. This is leading to a more connected world. This provides a wireless coverage for low-power IoT devices and helps to earn a new cryptocurrency, Helium, from our living room.
- *FOAM* (https://foam.space/)—The stock market, supply chains, the airline industry, the electrical grid, communication networks, and others rely on GPS, which is vulnerable to hacking and spoofing. This is one

of the critical infrastructures liable to be damaged. Therefore, secure location verification is widely demanded across an array of industries. FOAM solves the problem of location verification with an alternative and fault-tolerant system provided through an open network of terrestrial radios that anyone can operate.

There are other blockchain-driven IoT-enablement products and platforms being envisioned and implemented. There are a number of articles and blogs published in this web site (https://theblockchainland.com/) explaining how the combination of IoT and blockchain technologies is expected to bring forth scores of innovations and disruptions to lay a stimulating foundation for real digital transformation of our enterprises, organizations, institutions, governments, and other establishments.

6. The challenges of blockchain–IoT combination

As explained previously, one of the fundamental and foundational problems with the current IoT systems is their security architecture. The architecture is a centralized client-server model. There is a central authority managing everything, and this leads to the widely quoted issue (single point of failure, SPoF). This can be overcome by integrating with the hugely popular blockchain paradigm. The core and central aspect of the blockchain paradigm is the outright decentralization of the decision-making (which has been centralized for long) to a consensus-based shared network of devices. However, there are a few hitches/hurdles here when we try to integrate the blockchain technology with the IoT world.

1. *Scalability*—As inscribed above, there will be millions of IoT devices and the amount of data generated, collected, and processed is simply humungous. These can increase transaction processing time. But the need is to have less latency.
2. *Network privacy and transaction confidentiality*—The IoT network has to be secured for ensuring data privacy. Also, the confidentiality of any transaction that spans across multiple and distributed IoT devices and sensors has to be maintained.
3. *Sensors*—The reliability of IoT sensors and their data is another worrisome factor. Due to battery leakage, the data output of any IoT sensor may go wrong or incomplete. The physical security of sensors is another challenge and concern expressed across.

Decentralization of IoT Networks is the way forward—Employing the P2P communication model is being widely recognized as the robust and resilient response for all the notified ills affecting the conventional client/server architecture. A P2P solution has to support P2P messaging, distributed file sharing, and autonomous device coordination as vividly illustrated in the below diagram.

The P2P approach is slated to considerably lessen the financial implications on the centralized and consolidated cloud model. In the blockchain era, processing and storage are being distributed.

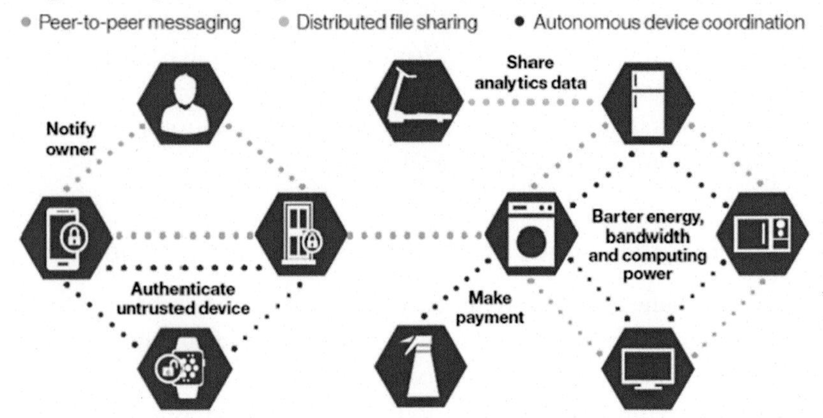

Even in P2P communication, security and privacy are quoted as serious issues. It has to be sufficiently strengthened to be tamper-proof. As the blockchain ledger is decentralized and tamper-proof, it looks very promising for ensuring tighter security for IoT devices and data.

The decentralization of the blockchain ledger means that data processing and storage happen across millions of resource-intensive IoT devices and not on one cloud server. Therefore, any server crash does not bring down the entire IoT network. As quoted above, there will be billions of connected devices. There are platform solutions for enabling smart coordination among these devices so that there is no need for every communication to be routed via a central server. The leverage of devices as a means for computation and storage remarkably reduces the costs of server ownership and operations. With this decentralization, accommodating big number of IoT devices and networks is a simple and sustainable affair.

For guaranteeing IoT data security and confidentiality, blockchain technology depends on cryptographic algorithms. As tampering is not possible due to the decentralization mechanism, IoT device networks can escape from any cyber-attack. Similarly, man-in-the-middle attacks are also not possible because of the data encryption while in transit.

The blockchain ledger will keep a tamper-proof record of IoT devices and their historical interactions. Devices in the IoT network can embark on secure and trustless messaging with one another. There is no intermediary needed to be trusted by participating parties (herein, it is IoT devices) in a blockchain environment. Still, the utmost security is being guaranteed in a trustless environment through the blockchain's ingrained decentralization feature. For instance, smart devices in an oil platform can exchange data in order to adjust their working based on weather conditions detected by sensors. Thus, IoT devices are made smart in their operations, outputs, and offerings through blockchain-enabled networking. Resource-constrained IoT devices such as sensors can transmit their data to resourceful IoT devices, which can store and process the received data to discover and disseminate knowledge in time. Without any central authority, data transmission happens securely thereby right decision can be taken to activate and accelerate right action.

The blockchain functions as a distributed transaction ledger for various IoT transactions

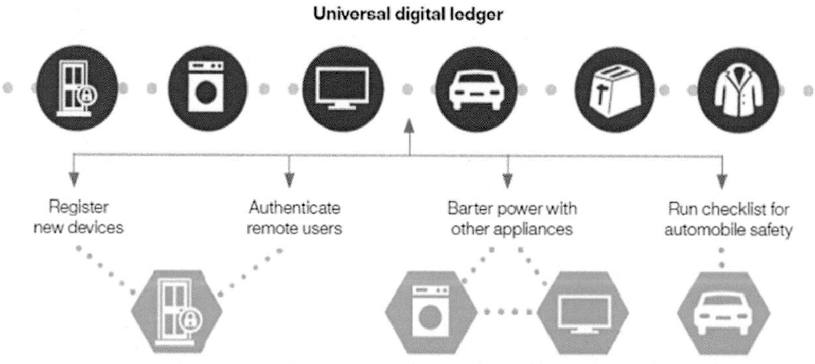

The innovative, disruptive, and transformative blockchain technology is being keenly explored, experimented, and expounded by different business verticals. Blockchain technology, a kind of DLT, is about to bring forth tectonic shifts in digital business. Any digital data and assets can be secured from hackers and evil doers. As it turns out, the future Internet is to be composed of trillions of digital entities, billions of connected devices, and millions of software services. The IoTs is to benefit immensely through the smart application of blockchain phenomenon.

Blockchain technology has led to the realization of decentralized applications (DApps) and smart contracts. Smart contracts mean some agreed operations are automatically performed when a specific condition is met. That is, when some environmental or goods' conditions are within the decided threshold, then the associated actions are getting initiated and implemented in an automated manner without any instruction and involvement of humans. IBM gives an example for this peculiar aspect of smart contracts. Smart contracts can be programmatically attached in complex trade lanes and logistics. Everything that has happened to individual items and packages is being registered. Then the immediate benefits out of the incorporation of smart contracts include audit trails, accountability, new forms of contracts and speed.

To enable exchanging messages, blockchain empowers IoT devices to have the wherewithal to leverage smart contracts fluently, blockchain enables P2P contractual behavior without any third party to certify the IoT transaction. The blockchain-inspired autonomous interaction among IoT devices without the need for centralized authority is the real differentiator. This can be further extended and expanded to enable human-to-human. From there, human-to-object interaction can get also automated in due course of time.

Leveraging the blockchain features for the ensuing IoT era brings in a number of distinct benefits. First, it offers newer ways for automating business processes. That is, business process consolidation toward optimization is being simplified through blockchain, which functions beautifully without the need for highly expensive, complex, and centralized IT infrastructure. Not only server machines but also IoT devices can participate in. The three key benefits of blockchain for IoT are building trust among devices, cost reduction, and the acceleration of transactions. It can answer the perpetual challenges of massive scalability, SPoF, time stamping, record, privacy, and trust in a very consistent way.

With the faster proliferation of IoT devices, enterprise-scale, people-centric, and mission-critical applications are being realized through the leverage of IoT devices, which can collectively and insightfully participate in implementing complex industrial processes. In order to safely and sagaciously experience all the originally expressed benefits of the IoT paradigm, there is a need for seamless and spontaneous synchronization with other two prominent technological paradigms such as artificial intelligence (AI) and blockchain. AI is able to incorporate the much-needed intelligence aspect into IoT applications and services whereas blockchain is capable of bringing in the mandated security to IoT devices, applications, and data. Additional advanced features can be accrued out of IoT through the linkage with AI and blockchain. Not only the trio but also other breakthrough technologies also fuse well to facilitate newer possibilities and opportunities. The list include digital twins, edge/fog computing for real-time analytics and applications, advanced processors such as graphics processing unit (GPU), tensor processing unit (TPU), vision processing unit (VPU), etc., in-memory database, etc.

IoT compliance and cost-efficiency can be guaranteed through the smart application of the blockchain technology. Several established product vendors have therefore plunged into this phenomenon. For instance, IBM blockchain allows to extend private blockchain into cognitive IoT. Blockchain platforms and applications are being meticulously made to run on digital infrastructures in order to prepare worldwide enterprises to be digitally transformed in their service engineering and delivery aspects.

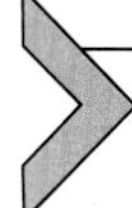

7. The IoT and blockchain convergence: Next-generation applications

With IoT and blockchain together, we are destined to achieve pretty unlimited possibilities. Several business verticals are keenly involving blockchain technology. Manufacturing, supply chain, retail, healthcare, financial services, cybersecurity, etc., are exuberantly doing blockchain pilots in order to gain deeper confidence on emerging technologies.

Intelligent ERP application—With actionable insights being emitted by analytics platforms, autonomous decisions are being taken consistently in order to have automated interactions among participants (IoT devices, cloud-hosted applications, etc.). Cameras, robots, drones, gateways, consumer electronics, medical instruments, defense equipment, appliances, communication gateways, machineries, utensils and wares, etc., are the prominent IoT elements. Software systems are empowered with the incorporation of IoT devices to bring additional people-centric features. According to an IDC report, the combination of IoT, AI, and blockchain comes handy in transitioning conventional and classic software solutions into intelligent software packages. For an example, any traditional ERP solution (say, SAP) in the market can become intelligent ERP through this technological combination.

Smart buildings and grids—Advancements in building automation and building energy management are gaining importance these days. The technological convergence facilitates leveraging energy coming from different sources such as local solar or wind systems and microgrids. This sharply enhances energy efficiency and leads to the much-expected facility optimization. The recording of autonomous and machine-to-machine transactions on electricity use is another blockchain application. This can bring in fresh business models for tracing transactions on the smart grid. As energy becomes an increasingly digital, decentralized, and IoT-enabled, excess renewable energy could be traded via blockchain.

Blockchain for the pharmaceutical industry—We all know that fake and expired medicines are being transmitted across. This results in sharp deterioration of medicine quality and there is no transparency at all in the medicine distribution. The budding potential of blockchain along with IoT is being projected as trend-setting and transformative. It can track the shipment of goods efficiently and maintain the legal ownership of permissible products through smart contracts. A popular example is *Mediledger*, which is designed to track the legal change of ownership of prescription medicines.

Another interesting example is as follows. Modum sensors (https://modum.io/) can record environmental conditions that sensitive goods are subjected to while in transit. When the goods change ownership, the sensor data gets verified against predetermined conditions in a smart contract. The contract validates that the conditions meet all the requirements agreed between seller/sender and buyers. Once validated automatically, a series of actions are being triggered. That is, appropriate notifications are being sent to the right and relevant stakeholders in this supply chain. A kind of proof has to be sent to a concerned regulator vouching that the shipped medicines were not exposed to a hot temperature during the transit. Herein, Modum typically combines IoT sensors and the smart contract capability of the blockchain technology to produce and offer this proof.

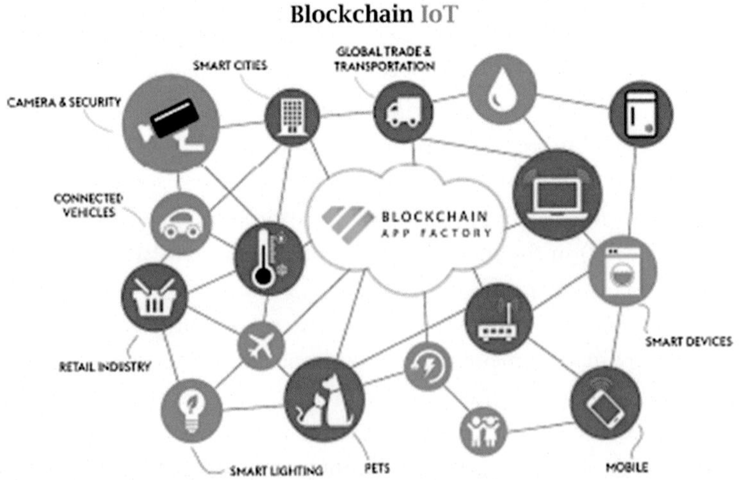

Blockchain IoT

Smart cities—Through IoT, it is possible to enhance the security of our homes, buildings, hotels, hospitals, campuses and cities by having centralized platforms in our smartphones. But there is a flaw in centralized monitoring and management solutions. Therefore, the world is seriously tending toward decentralization in order to ensure the unbreakable and impenetrable security for our cities and their prime components. Blockchain is the prominent method in leveraging the decentralization concept toward the fool-proof security for our properties. Blockchain is the techno mix of blockchain and IoT can help in resolving such issues. Primarily, the perpetual authentication and security issues could be resolved by smartly applying the distinct principles of cryptography.

Sharing economy—This sharing economy is flourishing steadily with the faster maturity and stability of corresponding technologies and tools. Especially the arrival ofblockchain has totally changed the situation. This facilitates P2P transactions over decentralized and distributed systems with all the trust. We can share goods and shares as physical entities over a network.

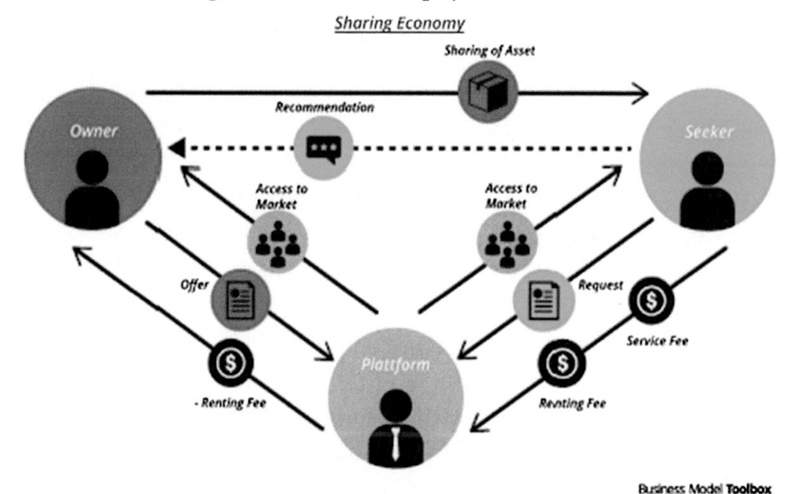

Blockchain could help create decentralized, shared economy applications to earn considerable revenue by sharing the goods seamlessly. Can you imagine an Airbnb apartment which leases itself? Slock.it is using blockchain technology for sharing of IoT-enabled objects or devices. They have planned to develop a Universal Sharing Network (USN) to create a secure online market of connected things. With USN, any object can be rented, sold, or shared securely without requiring intermediaries. It could be possible for third parties like manufacturers to onboard any object to the USN without seeking permission. Smart contracts ensure data privacy and transparency by controlling access to information.

Food industry—There has been a steady increase in the world population. This has put the extreme pressure on agriculture research scholars and scientists to unearth viable and venerable methods to substantially increase the food production volume with less resource. The mantra of more with less is definitely appropriate for the agriculture domain. This agenda has mandated nations, department officials, and agriculture experts for investing their talents, treasures, and time for producing and sustaining competent technologies-centric solutions. Blockchain and IoT are intrinsically capable of fusing well to create a deeper and decisive impact on the agriculture industry while remarkably enhancing the food chain supply management.

A global supply chain network generally involves many stakeholders such as raw material providers, product producers, middle men, etc. This multiplicity definitely is a barrier for an end-to-end visibility. Also, the supply chain can extend over months of time and consist of a multitude of payments and invoices. There is a possibility for uncertainty and delivery delays. Therefore, business houses are working on using IoT-enabled vehicles and trucks to precisely and concisely monitor the movement throughout shipment process.

However, due to the lack of transparency and trust along with increasing complications in the way the current supply chain and logistics activities are being performed, there is a need for better approach. The blockchain technology, when it gets combined with IoT, can substantially enhance the reliability and traceability of the network.

As we know, multifaceted IoT sensors are being attached not only on passing vehicles but also the on their path toward the destination. That is, connected vehicles are getting further empowered through multiple sensors such as motion sensors, GPS, and temperature sensors. All kinds of right and relevant data about any shipment can be procured and subjected to a variety of investigations. Sensor information is then stored in a blockchain database. Once the data gets saved on the blockchain, stakeholders enlisted in the smart contracts can avail correct details in time to initiate the subsequent actions. Supply chain participants can accordingly prepare for transhipment and perform cross-border transactions with all the clarity and confidence.

Golden State Foods (GSF) is a diversified supplier and widely known for manufacturing and distribution of food products. With the aim of producing and delivering high-quality products, GSF has tied up with IBM to optimize the business processes using blockchain and IoT. Sensors data are collected and stocked in blockchain database in order to ensure the issues are properly addressed and reported automatically before they create serious problems. GSF could create tamper-proof, immutable, and transparent ledger.

Automotive industry—Advanced cars are fit with a number of multipurpose sensors to automate, accelerate, and augment a number of tasks. The connectivity, cognition, and comfort levels have gone up significantly with the leverage of digitization and digitalization technologies and tools. Better choice and convenience are being provided to car owners, drivers, caregivers, and occupants through the consistently growing innovations and disruptions in the field of automotive electronics. Without an iota of doubt, the industry is moving toward smart and self-driving products. Connecting IoT-enabled vehicles with the decentralized blockchain network for data storage enables multiple users and stakeholders to exchange crucial information with ease and elegance. The other use cases include automated fuel payment, autonomous cars, smart parking, and automated traffic control.

For an example, NetObjex (https://www.netobjex.com/) has demonstrated the smart parking solution using blockchain and IoT. The strategically sound integration eases and quickens the task of finding a vacant space in the parking place. Further on, through smart contracts, it is possible to automate the payments using crypto wallets. The company also has collaborated with a parking sensor company "PNI" for real-time vehicle detection and for finding the availability of the parking area. IoT sensor applications calculate the parking charges for the parking duration, and the billing takes place directly through the crypto wallet.

Smart homes—Modern homes are boasting of several simple and complex devices ranging from microcontrollers, actuators, humanoid robots, communication gateways, smartphones, computers, displays, kitchen utilities and wares, medical instruments, consumer electronics, home security solutions, to a variety of sensors including autonomous cameras. We read that these digitized and connected artifacts are being further empowered through AI-inspired solutions such as Amazon Echo & Alexa Devices. Thus, natural interfaces such as voice, gesture, etc., are being fit with our everyday devices to be self-contained and serving. Thus, unmistakably smart IoT-enabled devices play a crucial role in our day-to-day lives. The combination of IoT and blockchain solves a number of weaknesses and brings forth additional capabilities. It is possible to secure access, assess, and manage home appliances, equipment, and machines remotely. This combination gains prominence as the traditional centralized approach to exchange information generated by IoT devices lacks the security standards and ownership of information. Blockchain could elevate the smart home to the next level by solving security issues and removing centralized infrastructure.

Telstra, the Australian telecommunication company, provides smart home solutions. The company has implemented blockchain and biometric security to ensure no one can manipulate the data captured from smart devices at any cost. Sensitive user data such as biometrics, voice recognition, and facial recognition are stored on the blockchain for improved security. Once the data is saved on the blockchain, it cannot be modified, and the access is only provided only to authenticated and authorized people.

Pharmacy industry—The issue of the counterfeit medicines in the pharmaceutical sector is increasing with every passing day. The pharmacy industry is responsible for developing, manufacturing, and distributing drugs; therefore, tracking the complete journey of drugs is difficult. The transparent and traceable nature of the blockchain technology can help to monitor the shipment of drugs from its origin to the destination of the supply chain.

Mediledger is one of the blockchain–IoT use cases, designed to track the legal change of ownership of prescription medicines. Transparency and traceability are essential when it comes to monitoring sensitive healthcare products. The data stored on the distributed ledger is immutable, timestamped, and accessible to manufacturers, wholesalers, dispensers, and end-customers involved in the supply chain. Mediledger is the blockchain-based platform, offering simplified payment process, controlled users' access and stop counterfeiting drugs from invading the supply chain.

Agriculture—Growing more food for the increased population while minimizing environmental footprints and ensuring transparency across the supply chain is essential for maximum customer satisfaction. Blockchain coupled with IoT has the potential to reshape the food production industry—from farm to grocery to home. By installing IoT sensors in the farms and sending its data directly to the blockchain can help enhance the food supply chain to a greater extent.

Offering a new and smart farming approach to farmers, Pavo is one of the blockchain–IoT use cases that brings unparalleled transparency. The information collected from Pavo's IoT hardware device installed in farms gets saved on the blockchain. It enables farmers to enhance farming techniques by looking at the captured data, while distributors, retailers, and consumers to make informed decisions about buying a specific crop or food item. Also, Pavo marketplace allows farmers to presell crops through smart forward contracts that can prevent farmers from having to wait for payment after harvest.

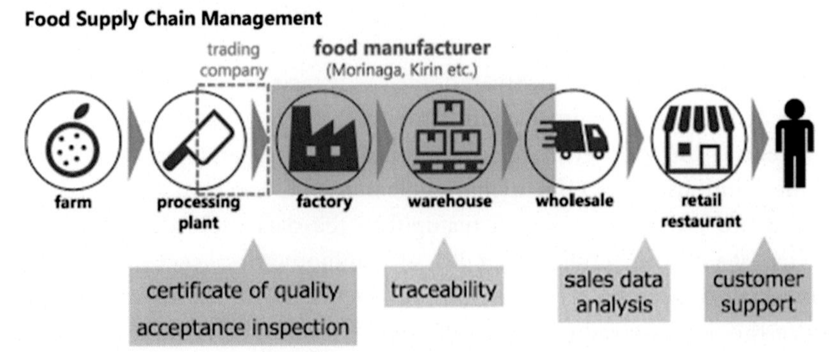

Several industries have begun to explore the potential applications of IoT and blockchain to improve efficiency and bring automation.

As industries are strategizing to become digitally transformed, the adoption of IoT, AI, and blockchain technologies is being fast-tracked. A right mix of digitization and digitalization technologies would guarantee improved transparency, security, and traceability. Some of the emerging industry use cases are pictorially represented in the diagram below.

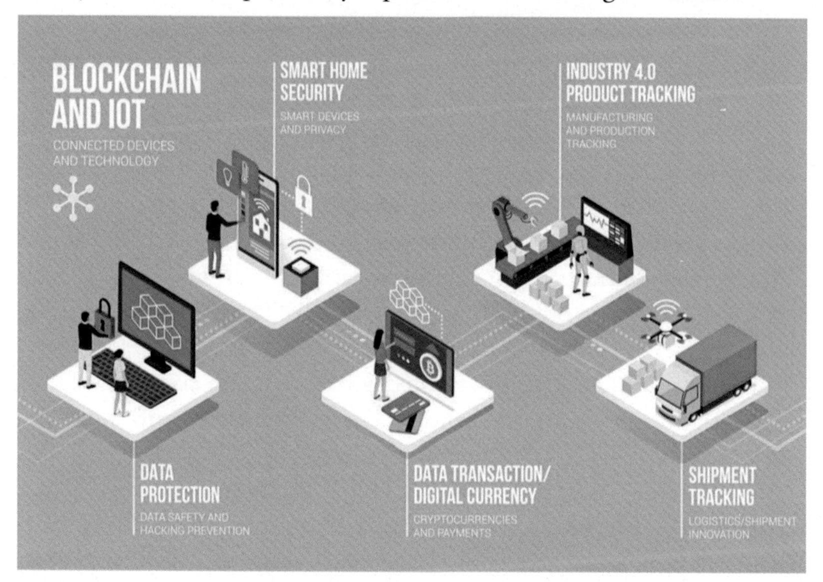

Supply chain—The blockchain technology is increasingly integrated with the IoT paradigm in order to surmount the existing IoT problems and to visualize next-generation use cases at the cusp of these two pioneering phenomena. For the supply chain domain, the integration of blockchain with IoT sharply improves the traceability of goods. IoT sensors like temperature, positioning, and motion sensors embedded on the transport vehicles provide the exact and timely details about the shipment status. Data fetched from the employed sensors gets stored in the blockchain database. This fulfills the need for trust, traceability, transparency, and auditability.

By merging IoT and blockchain technologies, the capabilities such as controlling temperature and container weight enable shippers, shipping pathways, freight forwarders, port, and terminal operators, inland transportation and customs authorities to interact intelligently through real-time access to shipping data and shipping documents, which are able to give a single shared view of a transaction with all the right and relevant details. Further on, all the parties involved in any international trade can easily collaborate, correlate, and corroborate in cross-organizational business processes and information exchanges. The real beauty is that all these complex things get accomplished in a secure and safe manner.

Logistics—Trust in the veracity of information is being touted as a huge problem in the logistics industry. The implementation of a private blockchain gives each party involved in a transaction a distinct and decentralized access to vital information in the transport process. With a high degree of accuracy and uncompromising tamper protection, these blocks in the digital ledger can give controlled access. As each party has its own copy of data, which cannot be changed or accessed by outsiders, the history of transactions remains intact and transparent.

Smart contracts are gaining a lot of attention these days as it is being projected as a secure and guaranteed way to automate legal agreements in the logistics process. Hyperledger Fabric and Sawtooth are the top platforms enabling smart contracts to be monitored at each step in the process and check for the rules laid out in the code. This is to ensure that each contract has been done fully. Using the digital ledger system of blockchain, assets getting transmitted can be elegantly tracked and paired with claimants without any confusion on the ownership status. This permanent record solves asset verification.

The scalability aspect of blockchain comes handy in ensuring that even the largest delivery tasks can be performed without any hitch. Even same-day delivery services can be accomplished nicely and neatly with blockchain.

Blockchain, IoT, and mobile technologies can be integrated into a single powerful solution for real-time delivery tracking. A Raspberry Pi is connected to blockchain in order to retrieve GPS data from an AI-Thinker A7 GPRS/GSM module and send it to a Node.js API through a Caddy server. This project involves Python SDK and CLI tools, while the Sawtooth REST API was used to update delivery status with new geolocations.

With the assistance of the blockchain technology, the IoT paradigm can automate a number of human-centric activities across industry verticals. For example, with several specialized IoT sensors being verified against predetermined conditions, medicines can be given to on specific conditions.

By applying the latest technologies into the logistics and supply chain industry, we can have far-reaching ramifications. By reducing costs across the board substantially and allowing entities in the logistics process to act with more individual agency, logistics in its entirety will see sweeping improvements in the days ahead.

Insurance—In the recent past, the complex tasks such as claims and fraud management, healthcare insurance, and property and casualty insurance have gone through a bevy of optimizations, rationalization, consolidation, and automation. The smart contracts combined with IoT data acquired from multiple machineries, appliances, equipment, instruments, wares, tools, etc., can elegantly automate many things across business domains. Today we are being bombarded with a family of sensitive and responsive (S & R) systems, scores of wearables, handheld and implantable devices, a growing array of sensors-embedded objects such as vehicles, shipping containers, etc. Further on, location-identifying sensors are sprinkled on various environments such as factories, warehouses, hospitals, hotels, and homes. There are edge IoT devices such as alarms, cameras, industrial control systems, etc., fitted with the capability of extracting their current location details. Precisely speaking, the blockchain's smart contract feature in synchronization with IoT devices and their context data comes handy in envisioning and experiencing next-generation applications.

Advertisement industry—This is a money-spinning vertical with the flourish of electronic and social media. For brands and advertising agencies, blockchain offers an independently verifiable and decentralized way to verify spend throughout the supply chain. Blockchain is being positioned as a system that runs in parallel and complements existing ad delivery, data targeting and yield-management functions. All participants in the end-to-end system gain full transparency into the supply chain and a clear understanding of where the advertiser's money is going.

As data production from popular social media platforms skyrocketed, there is a greater possibility for data breaches. We have read about the case involving Cambridge Analytica and Facebook. In this context, blockchain-enabled social media are emerging to guarantee heightened security. As most of the traditional social media platforms are centralized, this new phenomenon of decentralized social media platforms brightens social media users.

Here come the specific upsides of using blockchain in advertising and marketing.

1. Eliminates Intermediaries
2. Enhances the Consumer Trust
3. Ensures the Consumer Data Privacy

Blockchain for the Energy Industry—This DLT has gained a lot of interest from institutions, individuals, and innovators. The energy sector is keen in embracing this paradigm to be modern in its operations, offerings and outputs as there is a greater understanding that the coming together of blockchain and energy is indisputably categorized as one of the prime technologies in the recent past.

Blockchain immensely contributes in trading and crediting for the energy industry interested energy suppliers easily leverage the widely reported distinctions of the blockchain technology to facilitate energy transactions on a partitioned or wholesale level. This is capable of creating and sustaining virtual grids. Consumers can trade among their own devices and resources. Also, they can do the same with their neighbors and with the grid. This end-to-end process could be automated by applying the smart contract feature.

A "Health Coin" provides a way to use Smart Contracts to motivate patients to perform certain activities. Since it depends on the patient to meet health goals, an additional incentive can be planned to help patients achieve the goals. The Health Coin can be used for the purpose. Once earned, the health coins can be exchanged to avail various health services.

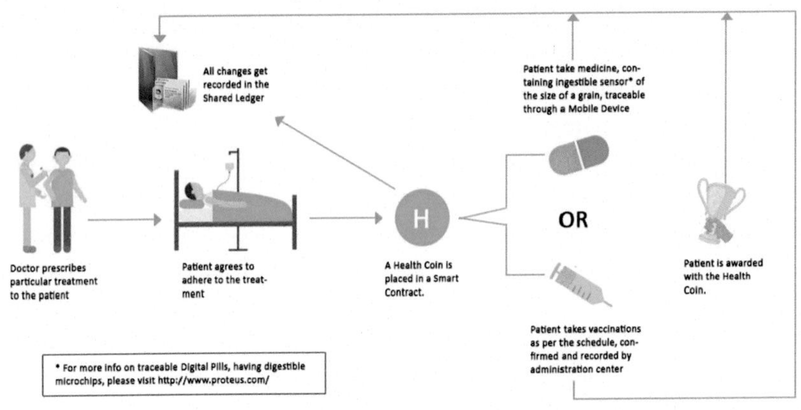

Here are the five ways in which blockchain is transforming the manufacturing sector.

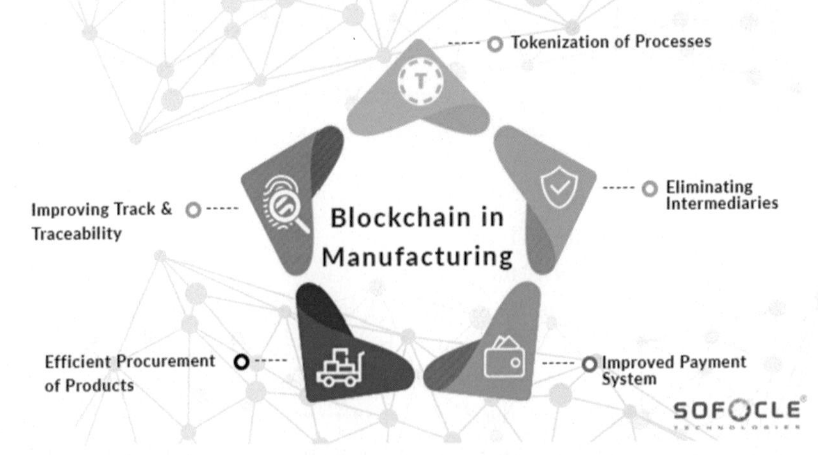

Blockchain-enabled IoT systems are capable of simplifying business processes, improving transparency, and enhancing productivity. Blockchain enables secure and trustless messaging between devices in an IoT network. This P2P coordination among digitized devices has laid down a scintillating and sparkling foundation for designing, developing, and deploying people-centric applications.

8. How blockchain and IoT can work together

There are several areas wherein this combination is to work wonders.

Security—Blockchain technology has proven its mettle and might in providing direct payment services without the involvement of any centralized third-party financial organization through cryptocurrencies. So, this trust-enabling and automatic solution can orchestrate highly distributed and decentralized IoT devices and their security.

Data encryption—Blockchain uses encryption to ensure tighter security for data and leverages distributed storage to securely record data in IoT devices. This ultimately automates transactions among the participating IoT devices without any intervention from human beings.

Communication—Blockchain can allow IoT devices to securely communicate and transact with each other. It also guarantees that the transaction will be accomplished as per the predefined rules of engagement. As indicated above, blockchain's smart contract facility helps to keep the terms of engagement defined and saved. For an example, an IoT device that continuously monitors solar power generator could automatically place a sale order of extra power generated back to the power grid if the buying price of power exceeds a certain threshold.

Cost reduction—Blockchain eliminates middlemen yet does business transactions securely and sagaciously. This eventually leads to the cost reduction for enterprising businesses. By leveraging smart contract capability, blockchain empower IoT devices to perform transactions in an automated manner.

Tracking—Blockchain can keep immutable records of the history of an IoT device. In an IoT network, this capability empowers smart devices self-function without any need for a centralized authority. The faster proliferation of self-contained/autonomous IoT devices can open up fresh possibilities and opportunities.

9. Companies using IoT: Blockchain combination

1. ArcTouch (https://arctouch.com/) builds decentralized applications for blockchain-related solutions that can connect with IoT devices (like Alexa and Facebook Messenger).
2. Blockset (https://blockset.com/) helps building enterprise-class blockchain solutions. They have developed a few ready-to-use blockchain tools to simplify and streamline blockchain application development.
3. Chronicled (https://www.chronicled.com/) has implemented the use of blockchain in its IoT devices to notify all parties involved in its food supply or medicine shipping process about the chain of custody.
4. GridPlus (https://gridplus.io/) relies on the Ethereum blockchain so consumers can access energy-saving IoT devices. The company is creating the world's first blockchain-based energy retailer.

10. Conclusion

Blockchain and IoT are the two technologies capable of bringing in a series of innovations, disruptions, and real transformations. The combination is bound to work wonders in eliminating a few hard-to-crack problems and in opening fresh possibilities and opportunities. In this chapter, we have discussed about the enormous potential of the IoT idea across industry verticals. However, there are a few crucial issues for the IoT domain in the increasingly connected world. Now with the arrival of the blockchain technology, the widely quoted IoT challenges get overcome easily. Further on, there will be hitherto unheard and unknown use cases emerging and evolving toward the ensuring era of smart things, smarter devices, and the smartest humans.

About the author

Pethuru Raj is working as the chief architect and vice-president in the Site Reliability Engineering (SRE) division of Reliance Jio Platforms Ltd., Bangalore. His previous stints are in IBM global Cloud center of Excellence (CoE), Wipro consulting services (WCS), and Robert Bosch Corporate Research (CR). In total, he has gained more than 19 years of IT industry experience and 8 years of research experience.

Finished the CSIR-sponsored Ph.D. degree at Anna University, Chennai and continued with the UGC-sponsored postdoctoral research in the Department of Computer Science and Automation, Indian Institute of Science (IISc), Bangalore. Thereafter, he was granted a couple of international research fellowships (JSPS and JST) to work as a research scientist for 3.5 years in two leading Japanese universities. Published more than 30 research papers in peer-reviewed journals such as IEEE, ACM, Springer-Verlag, Interscience, etc. Has authored and edited 20 books thus far and contributed 35 book chapters thus far for various technology books edited by highly acclaimed and accomplished professors and professionals.

Focuses on some of the emerging technologies such as the Internet of Things (IoT), Artificial Intelligence (AI), Big and Fast Data Analytics, Blockchain, Digital Twins, Cloud-native computing, Edge/Fog Clouds, Reliability Engineering, Microservices architecture (MSA), Event-driven Architecture (EDA), etc. His personal web site is at www.tinyurl.com/peterindia.

Blockchain components and concepts☆

Shubhani Aggarwal and Neeraj Kumar
Thapar Institute of Engineering & Technology, Patiala, Punjab, India

Contents

Abstract

The blockchain architecture consists of various components like peer network, smart contract, membership, events, ledger, system integration, wallet, and system management. Several actors such as developer, user, architect, regulator, operator, membership services, etc. are present in a blockchain that provides a solution to a blockchain-based business network. Each actor and component present in a blockchain plays a major role to make the network more secure and reliable. In this chapter, we have described the functionality of blockchain components and concepts.

Chapter points

- In this chapter, we discuss some basic components and concepts used in the blockchain network.
- Here, we discuss the interaction and coordination of components and concepts with each other in a blockchain.

☆ Working model.

Advances in Computers, Volume 121
ISSN 0065-2458
https://doi.org/10.1016/bs.adcom.2020.08.019

1. Actors in a blockchain

Several actors present in a blockchain are represented as shown in Fig. 1 and their functionality is as shown in Fig. 2. These actors are used to provide architecture and solution to a blockchain-based business network and their description is described as under.

- *Blockchain architect*: This actor is used to responsible for the architecture and design of the blockchain solution.
- *Blockchain developer*: This actor of the blockchain is used to develop various applications and smart contracts that interacts with the blockchain and are used by blockchain users.
- *Blockchain user*: This actor is used to operate in a business network. This can interact with the blockchain using an application.
- *Blockchain regulator*: This actor of the blockchain is used for overall authority in a business network. These may require broad access to the ledger's data present on the blockchain.

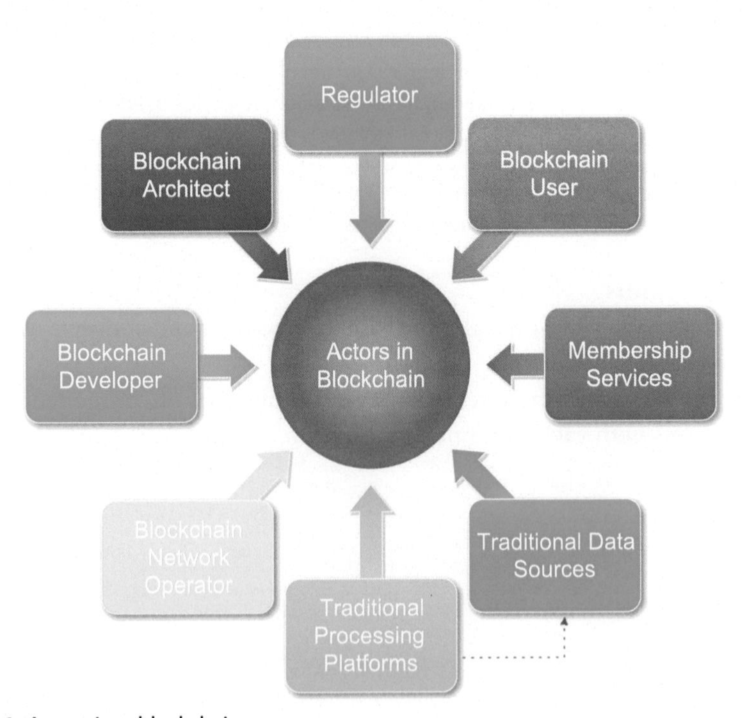

Fig. 1 Actors in a blockchain.

Fig. 2 Functionality of actors in a blockchain.

- *Blockchain operator*: This actor is used to manage, control, and monitor the blockchain network. Each business network has its own blockchain network operator.
- *Membership services*: Manages the different types of certificates required to run a permissioned blockchain.
- *Traditional processing platform*: Traditional systems may be used blockchain to automate their system processing. These systems may require a program that initiate requests into the blockchain.
- *Traditional data sources*: Traditional systems may provide data to the blockchain to control the actions of smart contracts.

2. Components used in a blockchain

Several components are present in a blockchain that helps to make the transaction successful are defined in Fig. 3. Their functionality and description are described as under.

- *Ledger*: A ledger is a current and previous state data storage maintained by each node on the network is as shown in Fig. 4.

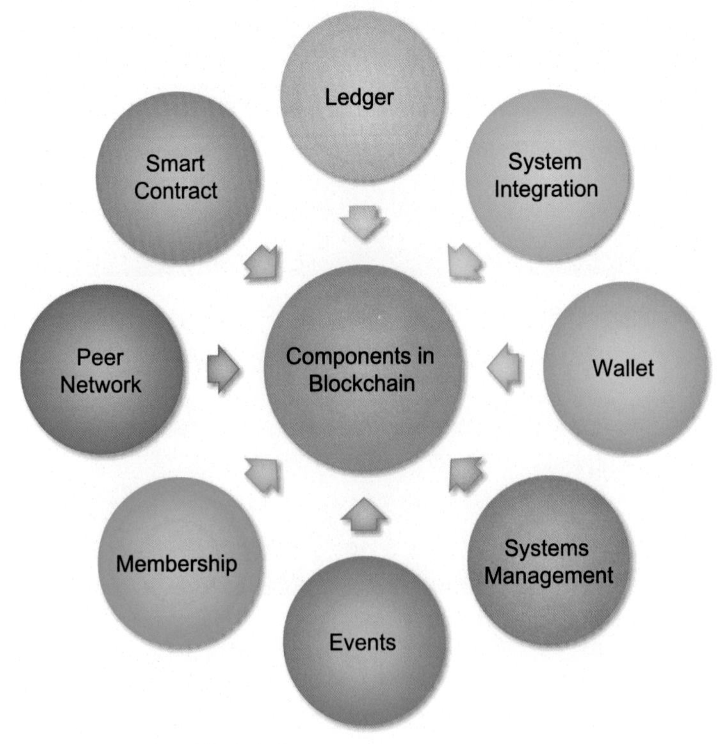

Fig. 3 Components in a blockchain.

Fig. 4 Ledger used in a blockchain.

Fig. 5 Smart contract used in a blockchain.

- *Smart contract*: Smart contracts are lines of code that are stored on the ledgers on the blockchain and automatically execute when predetermined terms and conditions are met. These are the programs to run the application designed by the people on a blockchain network is as shown in Fig. 5.
- *Peer network*: A P2P network is used in a blockchain network where each node is connected to every other node and share resources with each other without any involvement of third-party authenticator or server. All the nodes are equally privileged in the application of the blockchain network is as shown in Fig. 6.
- *Membership*: This component of the blockchain is used for permissioned blockchain that authenticates, authorizes, and manages identities on a blockchain network is as shown in Fig. 7.
- *Events*: This component is used to create notifications on a blockchain network. The notification consists of smart contracts, transaction execution, payment activation, or any significant operations operated on the

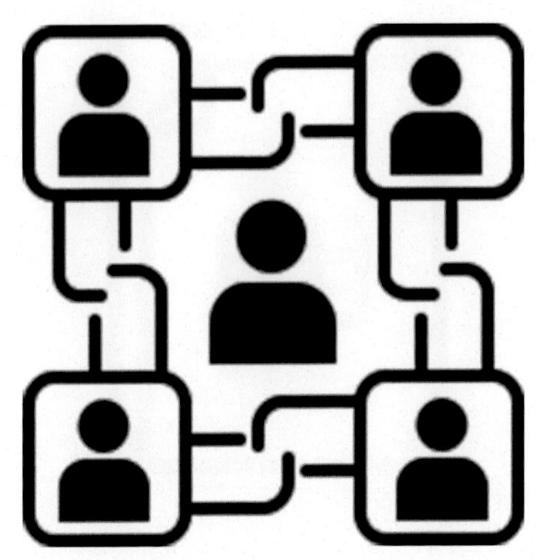

Fig. 6 Peer network used in a blockchain.

Fig. 7 Membership services used in a blockchain.

Fig. 8 Events used in a blockchain.

blockchain network is as shown in Fig. 8. The execution of an event depends on the internal processing of the organization.

- *System management*: This component of the blockchain has the capability to create, change, manage, and control the blockchain components is as shown in Fig. 9.

Fig. 9 System management in a blockchain.

Fig. 10 Wallet used in a blockchain.

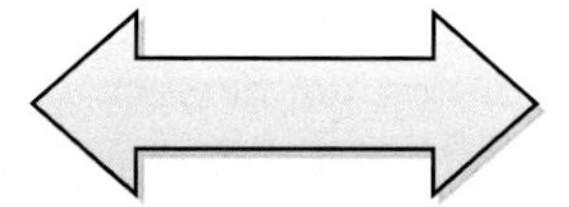

Fig. 11 System integration used in a blockchain.

- *Wallet*: This component is used to store the user's credentials secure and confidential is as shown in Fig. 10.
- *System integration*: This component is not a part of blockchain but used in a blockchain. It is used to take responsibility to integrate blockchain with the outside applications bidirectionally (Fig. 11).

3. Applications interact with the blockchain ledger

A ledger consists of two data structures, i.e., (i) blockchain and (ii) world state is as shown in Fig. 12. A *blockchain* data structure contains a linked list of hashed blocks that store a set of transactions, which are immutable and transparent in nature whereas *world state* data structure stores the recent history of smart contracts that may apply delete, add, modify operations, and stored it in a traditional database.

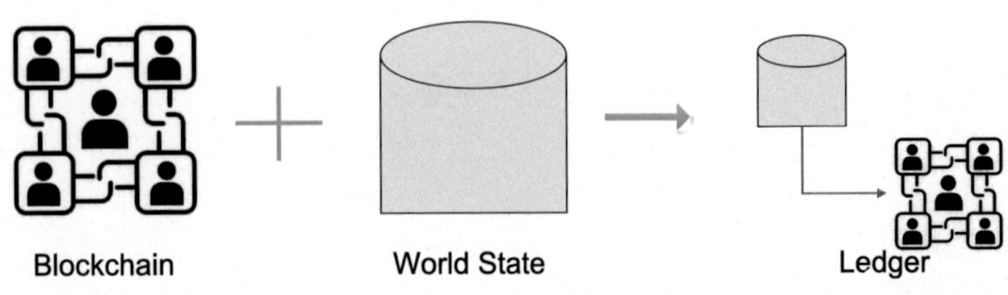

Fig. 12 Blockchain ledger.

A developer wants to develop client applications and smart contracts. Further, they will do the following considerations for developing client application.

- Implore the functions and write it in the smart contracts.
- Use software development kit to enter the transactions on the blockchain.
- To find the committed transactions and history of the transactions by using accessing the blockchain.

Then, they will do the following consideration for developing smart contracts.

- Record all the transactions on the blockchain.
- Record the modified data elements on the world state for every operation such as put, get, or delete.

Whenever a developer develops client application then, a client application submits data to the smart contracts that are executed on all nodes of the network and updates the world state simultaneously. Then, they both will agree on the updates and concludes the output is valid and consistent. After this, the updated data is added to the block, which further added it to the blockchain network (Fig. 13).

4. Blockchain events

Once the transaction has been committed on all the ledgers of the blockchain then, an event can be generated. Then, this event can be used anywhere for additional processing in the internal of the organization where an organization wants to use and perform it is as shown in Fig. 14.

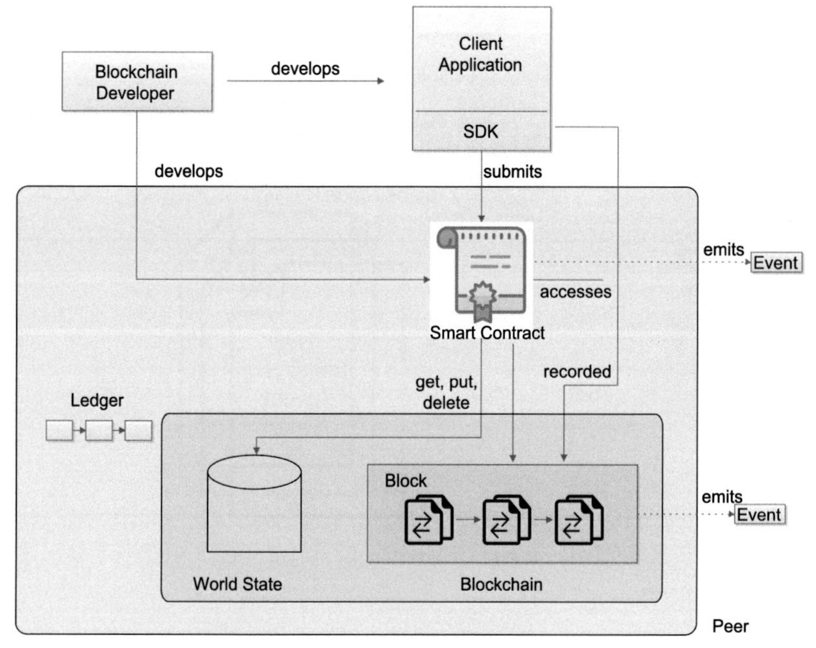

Fig. 13 Applications interact with the blockchain ledger.

An event is an occurrence that can trigger handlers when the disk is full, message received, temperature too cold, a transaction failed, etc. These events are important in asynchronous processing systems like blockchain. So, this blockchain can execute events anytime that are helpful for application programmers in an organization. Even, the external events may also change or trigger the blockchain activities.

4.1 Integrating with the existing systems

In an organization model, the blockchain can be integrated with the existing systems so that the information can be sent to the blockchain and can also get from the blockchain called a two-way exchange. In this, events of the blockchain network can create actions in the existing systems. These actions (cumulative) create interaction between the existing systems and the blockchain network is as shown in Fig. 15.

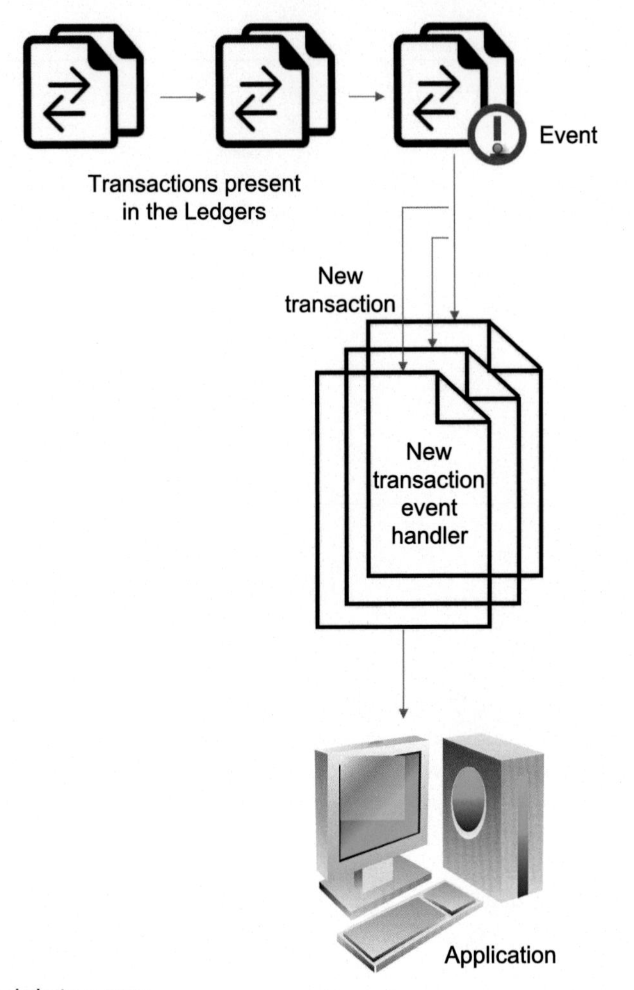

Fig. 14 Blockchain events.

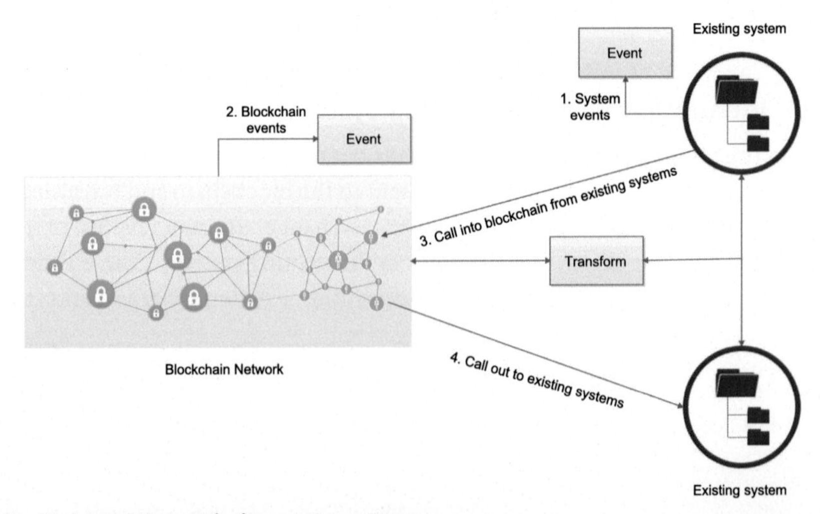

Fig. 15 Integrating with the existing systems.

About the authors

Shubhani Aggarwal is pursuing a PhD from Thapar Institute of Engineering and Technology (Deemed to be University), Patiala, Punjab, India. She received the BTech degree in Computer Science and Engineering from Punjabi University, Patiala, Punjab, India, in 2015, and the ME degree in Computer Science from Panjab University, Chandigarh, India, in 2017. She has many research interests in the area of Blockchain, cryptography, Internet of Drones, and information security.

Neeraj Kumar received his PhD in CSE from Shri Mata Vaishno Devi University, Katra (Jammu and Kashmir), India in 2009, and was a postdoctoral research fellow in Coventry University, Coventry, UK. He is working as a professor in the Department of Computer Science and Engineering, Thapar Institute of Engineering and Technology (Deemed to be University), Patiala (Pb.), India. He has published more than 400 technical research papers in top-cited journals such as IEEE TKDE, IEEE TIE, IEEE TDSC, IEEE TITS, IEEE TCE, IEEE TII, IEEE TVT, IEEE ITS, IEEE SG, IEEE Netw., IEEE Comm., IEEE WC, IEEE IoTJ, IEEE SJ, Computer Networks, Information sciences, FGCS, JNCA, JPDC, and ComCom. He has guided many research scholars leading to PhD and ME/MTech. His research is supported by funding from UGC, DST, CSIR, and TCS. His research areas are Network Management, IoT, Big Data Analytics, Deep Learning, and Cybersecurity. He is serving as editors of the following journals of repute: ACM Computing Survey, ACM·IEEE Transactions on Sustainable Computing, IEEE·IEEE Systems Journal, IEEE·IEEE Network Magazine, IEE·IEEE Communication Magazine,

IEE·Journal of Networks and Computer Applications, Elsevier·Computer Communication, Elsevier·International Journal of Communication Systems, Wiley. Also, he has been a guest editor of various International Journals of repute such as—IEEE Access, IEEE ITS, Elsevier CEE, IEEE Communication Magazine, IEEE Network Magazine, Computer Networks, Elsevier, Future Generation Computer Systems, Elsevier, Journal of Medical Systems, Springer, Computer and Electrical Engineering, Elsevier, Mobile Information Systems, International Journal of Ad hoc and Ubiquitous Computing, Telecommunication Systems, Springer and Journal of Supercomputing, Springer. He has also edited/authored 10 books with International/National Publishers like IET, Springer, Elsevier, CRC. Security and Privacy of Electronic Healthcare Records: Concepts, paradigms and solutions (ISBN-13: 978-1-78561-898-7), Machine Learning for cognitive IoT, CRC Press, Blockchain, Big Data and IoT, Blockchain Technologies across industrial vertical, Elsevier; Multimedia Big Data Computing for IoT Applications: Concepts, Paradigms and Solutions (ISBN: 978-981-13-8759-3), Proceedings of First International Conference on Computing, Communications, and Cyber-Security (IC4S 2019) (ISBN 978-981-15-3369-3). One of the edited text-book entitled, "Multimedia Big Data Computing for IoT Applications: Concepts, Paradigms, and Solutions" published in Springer in 2019 is having 3.5 million downloads till 6 June, 2020. It attracts attention of the researchers across the globe (https://www.springer.com/in/book/9789811387586). He has been a workshop chair at IEEE Globecom 2018 and IEEE ICC 2019 and TPC Chair and member for various International conferences such as IEEE MASS 2020, IEEE MSN2020. He is senior member of the IEEE. He has more than 12,321 citations to his credit with current h-index of 60 (September 2020). He has won the best papers award from IEEE Systems Journal and ICC 2018, Kansas-city in 2018. He has been listed in the highly cited researcher of 2019 list of web of science (WoS). In India, he is listed in top 10 position among highly cited researchers list. He is adjunct professor at Asia University, Taiwan, King Abdul Aziz University, Jeddah, Saudi Arabia and Charles Darwin University, Australia.

Attacks on blockchain☆

Shubhani Aggarwal and Neeraj Kumar
Thapar Institute of Engineering & Technology, Patiala, Punjab, India

Contents

Abstract

The blockchain network is generally to be considered as secure and scalable but their security level is directly proportional to the amount of hash computing power that supports the blockchain. As the miners increase in the mining process the more difficult for an attacker to attack the blockchain. Several ways a blockchain can be attacked. Performing these attacks such as Finney attack, race attack, 51% attack, eclipse attack, Sybil attack, DDoS, routing attack, DAO attack, parity multisig parity attack on a blockchain becomes more difficult as more computing power is added to the network. The attacks described in this chapter are not endemic to blockchains except 51% attack and there are many measures and solutions to mitigate the risk of these attacks.

☆ Working model.

Advances in Computers, Volume 121
ISSN 0065-2458
https://doi.org/10.1016/bs.adcom.2020.08.020

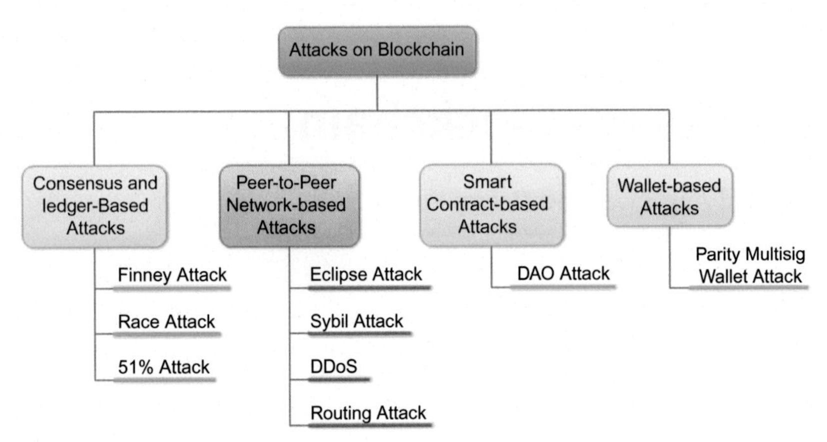

Fig. 1 Attacks on blockchain.

Chapter points

- In this chapter, we discuss the various type of network security attacks, i.e., consensus and ledger-based, P2P network-based, smart contract-based, wallet-based attacks on the blockchain network.
- Here, we discuss the mitigation process of these attacks.

The blockchain technology is generally used in various types of applications like healthcare system, agriculture, smart grid, smart city, financial system, etc. for its security and privacy features. It is considered to be very secure, but the level of security provided by blockchain is directly proportional to the amount of hash power that supports the network. The more miners there are, the more powerful mining is there, which produces difficulty for an attacker to attack the network. But some security attacks are emerging, which are very sophisticated and can cause huge irreparable damages. So, anyone needs to understand these security attacks who are developing and deploying blockchain solutions. Fig. 1 shows the attacks on blockchain that are categorized into four different groups, which are described as follows.

1. Consensus and ledger-based attacks

Double-spending is a problem, which means spending the same cryptocurrency twice as shown in Fig. 2. In order to prevent the system from this problem, the network must stay decentralized so that one party cannot take control of all the transactions in the network. Bitcoin uses a consensus mechanism to avoid central authority that verifies the transaction are not double-spends. In this, decentralized group of individuals known as bitcoin miners has performed the verification process. The transactions are recorded

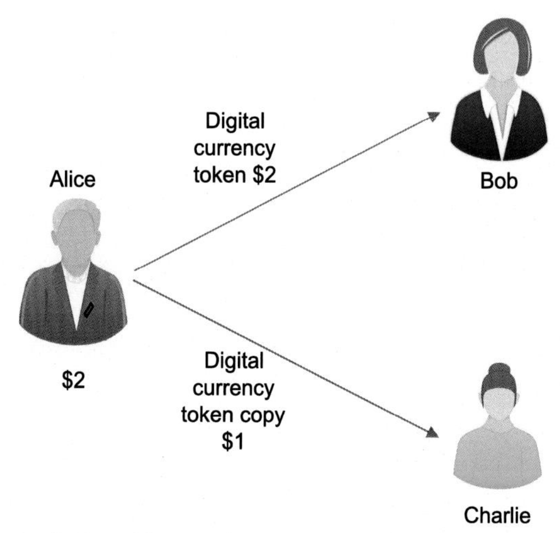

Fig. 2 Double-spending problem.

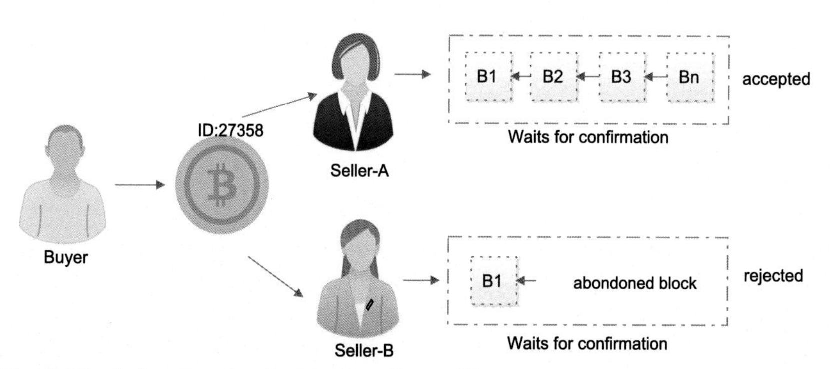

Fig. 3 Bitcoin handles the double-spending problem.

as valid once it has been added into the blocks of the blockchain. As the more blocks are added into the blockchain, it becomes difficult to go back and to double-spends the transaction. It means a very high computational power is required as same amount of computing power that has been used during mining. All the transactions are shared in the public ledgers ensures that any node wishing to spend bitcoin really is in custody of that bitcoin. Fig. 3 shows that bitcoin handles the double-spending problem.

The confidence that the transaction cannot be double-spend is directly proportional to the number of block confirmations that the transaction has received. The increased number of confirmations will increase the chance of transaction free from double-spending. The bitcoin network has approximately spent 1 h to protect one transaction against double-spends.

If the transaction has six confirmations than an attacker would require a high amount of hashing power to come back six blocks and double-spends the transaction. However, it is possible in certain cases for double-spends the transaction to occur. For example, a Finney attack, race attack, and 51% attack. The description of each attack is as follows.

1.1 Finney attack

This attack occurs when payments are accepted at zero confirmations. The execution of this attack requires a miner that has already mined a block but not yet broadcast to the rest of the network. In this, the miner could include a transaction of payment from address *A* to address *B* into the mined block. Then, again the miner could buy a service from the wholesaler by making the payments from address *A* to address *C*. The wholesaler could sell the service to a miner in the expectation of bitcoin payments. The miner could defraud the wholesaler's service by broadcasting previously mined block to the network that includes the transaction from *A* to *B* in case of the transaction from *A* to *C*. In this way, this attack happens that includes double-spending on the network.

1.2 Race attack

This attack also happens to the wholesalers and the other particulars who accept the payment transaction with zero confirmations. It could happen by sending two conflicting transactions in a fast sequence on the network. For example, the malicious actor could send a bitcoin transaction for a service to wholesaler. At the same time, he sends a conflicting transaction to the network spending same bitcoin to himself. In this case, the second conflicting transaction is mined into the block and showed as a genuine transaction by the network nodes. This would be the damage of the wholesaler that service to the actor in the expectation of the bitcoin payments. The pictorial representation of race attack is as shown in Fig. 4.

1.3 51% attack

It is also known as majority attack. An attacker can make double-spends if he has control more than half of the network. With this attack, he can generate blocks faster than the other nodes on the network. He could spend funds on the network that is being built by the honest miners but then, not included in the private blockchain. The attacker then broadcast the private blockchain and be able to make the transaction with their funds again. The pictorial representation if 51% attack is as shown in Fig. 5.

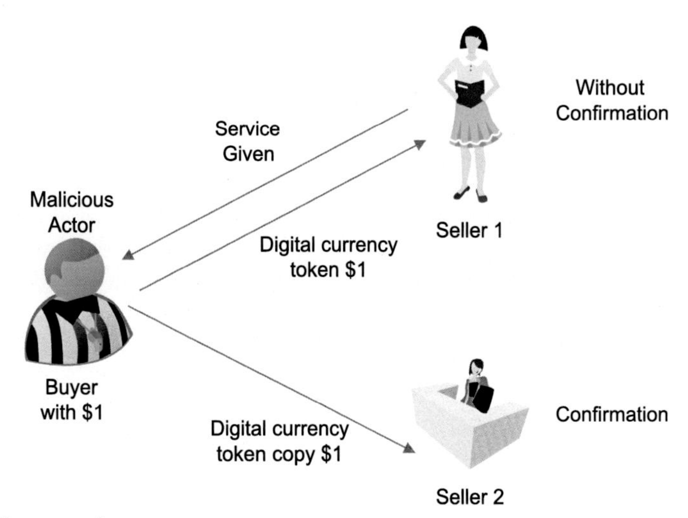

Fig. 4 Race attack.

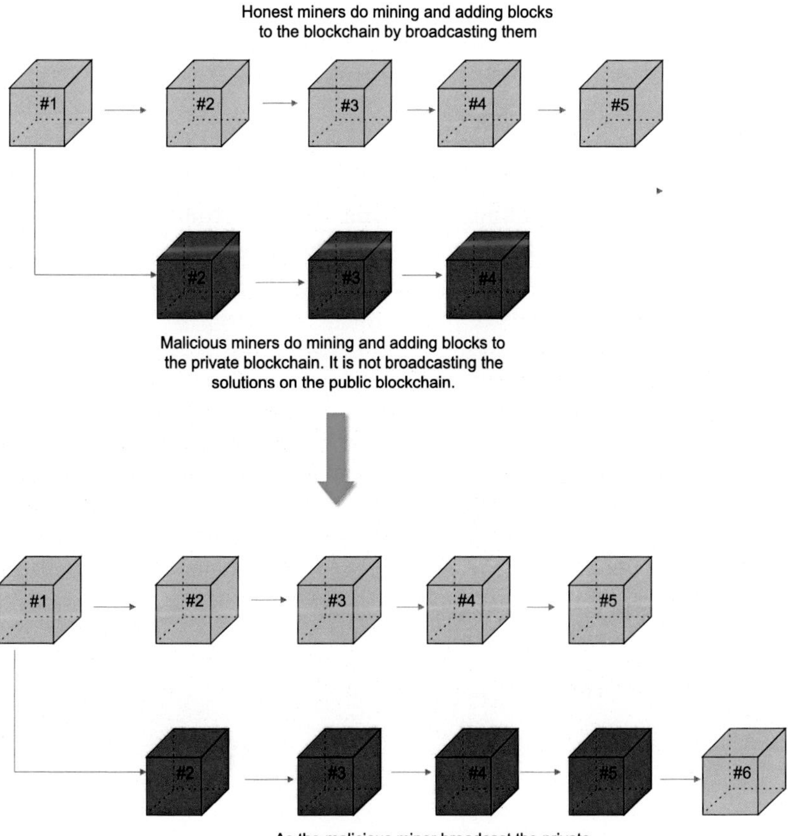

Fig. 5 51% attack.

2. Peer-to-peer network attacks

The important feature of blockchain technology is the enhancement of security and privacy without the involvement of a third-party authenticator. A P2P network of computers that all run the protocol and hold an identical copy of the ledger of transactions, enabling P2P value transactions through a consensus mechanism. Here, we describe network security in blockchain by introducing four major attacks: Distributed Denial-of-Service, Sybil, Eclipse, and Routing attacks.

2.1 Sybil attack

A Sybil attack is one where an attacker pretends to be so many people at the same time. It is one of the biggest issues when connecting to a P2P network. It manipulates the network and controls the whole network by creating multiple fake identities. To a single view, these different identities look like regular users, but behind the scenes, a single entity is called an unknown attacker who controls all these fake entities at once. While the Eclipse attack is about eclipsing certain nodes whereas the Sybil attack targets the whole network.

Fig 6 shows a large number of Sybil nodes surround one node and prevent that node from connecting to the honest nodes on the network.

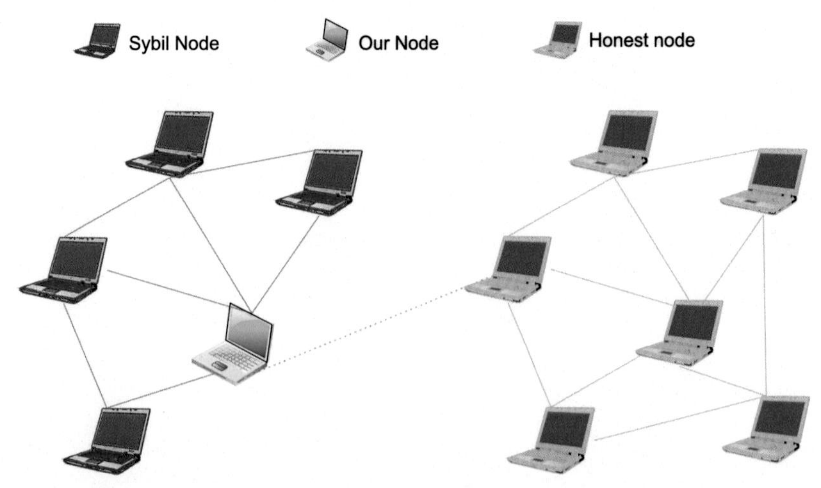

Fig. 6 Sybil attack.

In this way, one could try to prevent either sending or receiving information to the network. The only way to mitigate Sybil attacks is to raise the cost to create an identity. This cost should be balanced so that new participants are not restricted from joining the network and create legitimate identities. It should also be so high enough that creating a large number of identities in a short time-period becomes expensive. For example, In PoW blockchains, the miner nodes do the validation and verification process that needs a large amount of computational power. So, the associated cost of computational power makes it hard for the Sybil attack.

2.2 Eclipse attack

Eclipse attacks are a type of network attack that aims at eclipsing certain nodes from the entire P2P network. It is a type of attack, which manage the node's connection in such a way that only attacking nodes receive the information than other nodes. It is mainly focused on attacking single node rather than the entire network at once. In this context, an attacking node could easily perform a double-spending attack. It can be done by sending the victim node a transaction showing proof of payment and eclipsing it from the network. Finally, sending another transaction to the entire network that spent the same tokens again. Fig 7 shows the network in an Eclipse attack. The attacking nodes in red that isolate one of the normal nodes in blue from the entire network by monopolizing and controlling its connections.

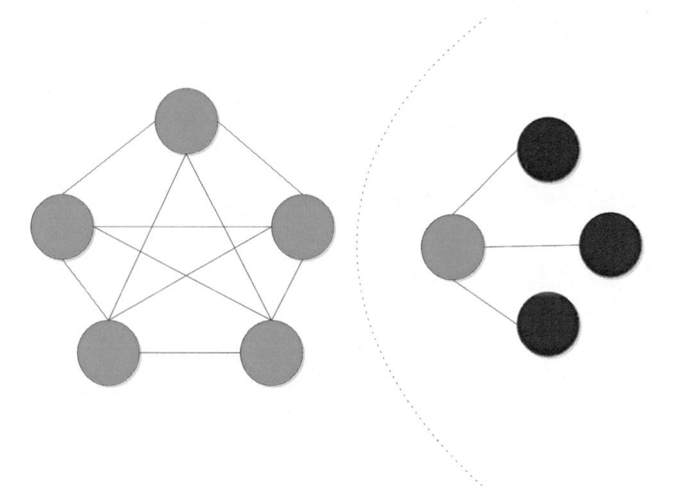

Fig. 7 Eclipse attack.

2.3 Distributed denial-of-service

DDoS attack is an attack, where an attacker overloads the network by flooding with a large number of requests in one attempt and makes the network resources unavailable to its users. In a DoS attack, all these requests are generated from a single source. But in the case of a DDoS attack, all these requests are generated from a large number of different sources. In this context, it is very difficult to tackle these types of attacks because to do so we first need to differentiate between legitimate and malicious requests. The mitigation of this attack is to introduce a transaction fee that automatically decreases the illegal transaction requests [1, 2]. The pictorial representation of the DDoS attack is as shown in Fig. 8.

2.4 Routing attack

Routing attacks rely on blocking the messages propagating through the network and make changes with them before transferring them to their peers. The only way to detect such types of attacks is when a receiver

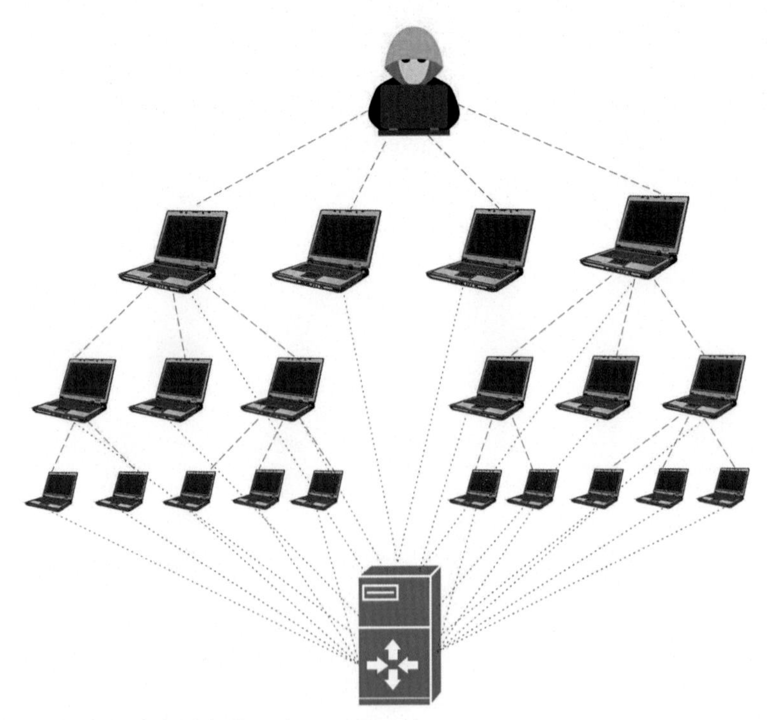

Fig. 8 Distributed denial-of-service attack.

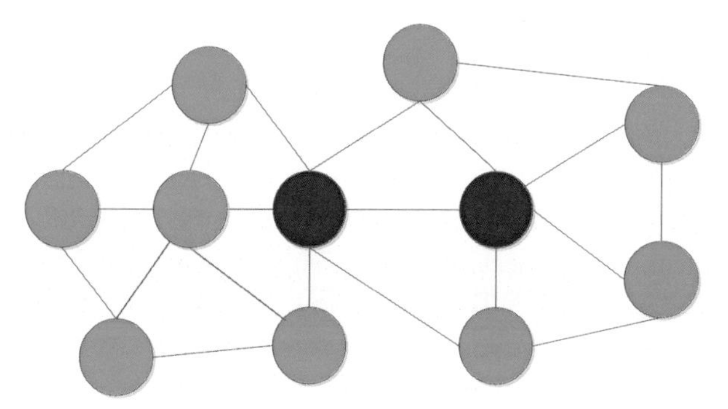

Fig. 9 Routing attack.

receives a different copy of it from another node. In other words, an attacker node divides the network into two or more partitions that cannot communicate with each other. Routing attacks are categorized into two smaller attacks.

- *Partitioning attack*: An attacker tries to divide the network into two or more disjoint groups. This can be done by attacking certain points within the network, which acts as the linking point between the two groups.
- *Delay attack*: An attacker picks up the propagating messages, make changes with them, and finally transfers them to the side of the network, which has not seen it before.

The mitigation method from this attack is by continuously diversifying the network connections. This will make it difficult for an attacker to find link points to hijack and split the network into two or more disjoint groups. The pictorial representation of the routing attack is as shown in Fig. 9.

3. Smart contract-based attacks

Smart contracts are automated programming lines used in blockchain technology to run decentralized applications. They have executed the transaction between the participants according to an agreement agreed, with inputs from the real-world and without the involvement of middleman. So, once the smart contract started, it cannot be stopped. The transaction once is completed on the blockchain network, it becomes immutable. So, if smart contracts have faults, millions of currency are in stake and no one can be changed. Therefore, look at such type of attack that affects smart contracts, which is described as follows.

3.1 DAO attack

A DAO is a decentralized autonomous organization. It aims to write the code of lines of a company to eliminate the need for documents, and people in guiding controlling the decentralized structures. It was an ambitious feature of an Ethereum. An organization called *Slock* started crowdfunding for a "*DAO*" project. This project collects 12.7 million Ether ($150 million) with a great number of responses. But, an unknown attacker finds a vulnerability in a code. So, the attacker starts attack by contributing a small amount and requesting withdrawal $70 million with a recursive withdraw function. The Ethereum foundation warned an attacker to freeze an account but an attacker's response that he took part in the contract and was playing as per an agreement through a soft or hard fork. Later, he stopped the attack. But from this attack, a DAO project raised many concerns about the autonomous of smart contracts for the future.

4. Wallet-based attacks

Wallet contracts are logic than can be built on user wallets for regular automated payments. Each node on the network has its wallet to make transaction payments. An attacker may attack the node's wallet on the blockchain for doing a malicious activity on the network. The wallet-based attacks are described as follows.

4.1 Parity multisig wallet attack

The parity multisig wallet attack was the case of blame in which the wallet of parity client had been hacked by an attacker amounts 500,000 Ether ($77 million). The functionality of the party multisig wallet is o reduce the transaction fees having a centralized library contract. But they left some loopholes in this function, which was taken advantage by an attacker. An attacker added his account as owner in the library contract and became a joint owner of all the wallets. Then, an attacker changes the library function that froze the currencies in the wallet. He locked $155 million from cryptographically inaccessible wallets.

References

[1] A. Bose, G.S. Aujla, M. Singh, N. Kumar, H. Cao, Blockchain as a service for software defined networks: a denial of service attack perspective, in: 2019 IEEE Intl. Conf. on Dependable, Autonomic and Secure Computing, Intl. Conf. on Pervasive Intelligence and Computing, Intl. Conf. on Cloud and Big Data Computing, Intl. Conf. on

Cyber Science and Technology Congress (DASC/PiCom/CBDCom/CyberSciTech) IEEE, 2019, pp. 901–906.
[2] R. Singh, S. Tanwar, T.P. Sharma, Utilization of blockchain for mitigating the distributed denial of service attacks, Secur. Priv. 3 (3) (2020) e96.

About the authors

Shubhani Aggarwal is pursuing PhD from Thapar Institute of Engineering & Technology (Deemed to be University), Patiala, Punjab, India. She received the BTech degree in Computer Science and Engineering from Punjabi University, Patiala, Punjab, India, in 2015, and the ME degree in Computer Science from Panjab University, Chandigarh, India, in 2017. She has many research interests in the area of Blockchain, cryptography, Internet of Drones, and information security.

Neeraj Kumar received his PhD in CSE from Shri Mata Vaishno Devi University, Katra (Jammu and Kashmir), India in 2009, and was a postdoctoral research fellow in Coventry University, Coventry, UK. He is working as a Professor in the Department of Computer Science and Engineering, Thapar Institute of Engineering & Technology (Deemed to be University), Patiala (Punjab), India. He has published more than 400 technical research papers in top-cited journals such as IEEE TKDE, IEEE TIE, IEEE TDSC, IEEE TITS, IEEE TCE, IEEE TII, IEEE TVT, IEEE ITS, IEEE SG, IEEE Netw., IEEE Comm., IEEE WC, IEEE IoTJ, IEEE SJ, Computer Networks, Information sciences, FGCS, JNCA, JPDC, and ComCom. He has guided many research scholars leading to PhD and ME/MTech. His research is supported by funding from UGC, DST, CSIR, and TCS. His research areas are Network Management, IoT, Big Data Analytics, Deep Learning, and Cybersecurity. He is serving as editor of the following journals of repute: ACM Computing Survey, ACM·IEEE Transactions on Sustainable Computing, IEEE·IEEE Systems Journal, IEEE·IEEE Network Magazine, IEE·IEEE Communication Magazine, IEE·Journal of Networks and Computer Applications, Elsevier Computer Communication, Elsevier International Journal of Communication Systems, Wiley. Also, he has been a guest editor of various international journals of repute such as IEEE Access, IEEE ITS, Elsevier CEE, IEEE Communication Magazine, IEEE Network Magazine, Computer Networks, Elsevier, Future Generation Computer Systems, Elsevier, Journal of Medical Systems, Springer, Computer and Electrical Engineering, Elsevier, Mobile Information Systems, International Journal of Ad Hoc and Ubiquitous Computing, Telecommunication Systems, Springer, and Journal of Supercomputing, Springer. He has also edited/authored 10 books with international/national publishers like IET, Springer, Elsevier, CRC: Security and Privacy of Electronic Healthcare Records: Concepts, Paradigms and Solutions (ISBN-13: 978-1-78561-898-7), Machine Learning for Cognitive IoT, CRC Press, Blockchain, Big Data and IoT, Blockchain Technologies Across Industrial Vertical, Elsevier, Multimedia Big Data Computing for IoT Applications: Concepts, Paradigms and Solutions (ISBN: 978-981-13-8759-3), Proceedings of First International Conference on Computing, Communications, and Cyber-Security (IC4S 2019) (ISBN 978-981-15-3369-3). One of the edited text-book entitled, "Multimedia Big Data Computing for IoT Applications: Concepts, Paradigms, and Solutions" published in Springer in 2019 is having 3.5 million downloads till June 6, 2020. It attracts attention of the researchers across the globe (https://www.springer.com/in/book/9789811387586). He has been a workshop chair at IEEE Globecom 2018 and IEEE ICC 2019 and TPC Chair and member for various international conferences such as IEEE MASS 2020 and IEEE MSN 2020. He is a senior member of the IEEE. He has more than 12,321 citations to his credit with current h-index of 60 (September 2020). He has won the best papers award from IEEE Systems Journal and ICC 2018, Kansas City in 2018. He has been listed in the highly cited researcher of 2019 list of Web of Science (WoS). In India, he is listed in top 10 position among highly cited researchers list. He is an adjunct professor at Asia University, Taiwan, King Abdul Aziz University, Jeddah, Saudi Arabia, and Charles Darwin University, Australia.

Financial system ☆

Shubhani Aggarwal and Neeraj Kumar
Thapar Institute of Engineering & Technology, Patiala, Punjab, India

Contents

Abstract

The financial system is one of the most useful applications of blockchain technology in the banking sector. All the involved parties such as complex transactions can be on-boarded on a blockchain and the information can be shared by exporters, importers, and banks on one common distributed ledger. This ledger can store facts like, who owns a particular piece of land or any information transfer on the network. This technology offers transparency and verifiable financial transaction with ease. With blockchain in the financial industry, individuals, and banks can access their transactions and it helps to improve the system's efficiency. In this chapter, we have described the transaction management in the banking sector and the role of blockchain in the financial system.

Chapter points

- In this chapter, we discuss the blockchain that can be used in various applications like the banking system.
- Here, we discuss the blockchain-based transaction management in a financial system and the role of blockchain in crowdfunding.

☆ Applications.

Advances in Computers, Volume 121
ISSN 0065-2458
https://doi.org/10.1016/bs.adcom.2020.08.021

Blockchain technology is transforming everything from the transaction of payments to how money is elevated to a higher level in the private sector. This technology provides a way of untrusted parties to come to one consensus without any involvement of third-party. It provides a ledger to each node of the network that nobody can change or alter the data. It can provide specific financial services like payments or securitization like the banking system without the use of a middleman. In the following bank services where blockchain could disintermediate the key services.

- *Payments*: Blockchain provides a decentralized environment for making a payment transaction. It provides a distributed ledger for payment (e.g., bitcoin). It could facilitate faster payments, high security, and low cost of sending payments than banking. It cuts down the verification process from an intermediator and decreases the processing time of the transaction than the traditional bank transfers. As from the observation, 90% of the members of the European council believed that the blockchain technology will fundamentally change the industry by 2025.
- *Clearance and settlement systems*: Distributed ledgers provided by blockchain technology can reduce operational costs and bring us close to the real-world transactions. It can allow a transaction to be settled directly on the network without the use of a third-party and can keep track of all the information used in the transaction.

  As shown in Fig. 1, the fact in the banking system is that the existing bank transfers system can take an average of 3 days to settle and make a complete transaction between the clients whereas the blockchain technology has to bypass the complicated intermediaries and to make

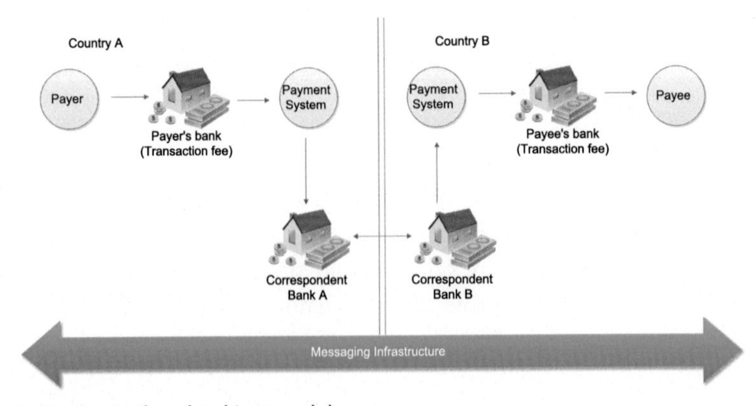

Fig. 1 Correspondent banking model.

a complete transaction from one account to another account within a fraction of seconds.

- *Fund-raising*: Blockchain provides privacy and security to the network. It enables fast and transparent transactions at a low cost that has paved the way for fund-raising through initial coin offerings (ICO). At the same time, ICO represents a paradigm shift in how private and public companies develop the finance system using blockchain technology.
- *Securities*: Blockchain provides security and privacy to the network nodes by removing the use of a middleman in assets right transfers. It lowers the assets exchange fee and giving access to wider global markets. It could more efficient and interoperable transaction as a comparison to the existing banking system. By the use of blockchain in finance system could save from $17 billion to $24 billion per year in global trade processing costs.

  As shown in Fig. 2A, the central or traditional bank tracks payments between the clients whereas the blockchain banking in Fig. 2B, transactions are recorded on the multiple computers and settled by many participants of the network.
- *Loans and credit*: The blockchain-enabled network provides a more secure way to offer and borrow money for personal loans to a large number of customers and make the loan process cheaper and more efficient than traditional methods.
- *Trade and finance*: The blockchain technology can support cross-border trade transaction by giving the properties of distributed ledger and security to the network. It would reduce the processing time and delivery time of the product and also reduces the use of paper. It enables the transaction details, keeps track of product from delivery station to the arriving station, digitally prove country of origin, etc. securely and transparently. The absence of a third-party reduces the risk of integrity and confidentiality on the flow of goods and documentation.

  As shown in Fig. 3, the adoption of blockchain technology in cross-border could mean greater trust between the trade parties that increase the global business. The payments between the buyer and the seller are through blockchain in a tokenized form after the delivery of goods. The use of smart contracts in the network would ensure the automatic payments according to the predefined set of rules in the contract. It cuts off the possibility of missed, repeatedly debt, and lapsed shipments. Here, the following are the blockchain use cases in the financial system as described in Table 1.

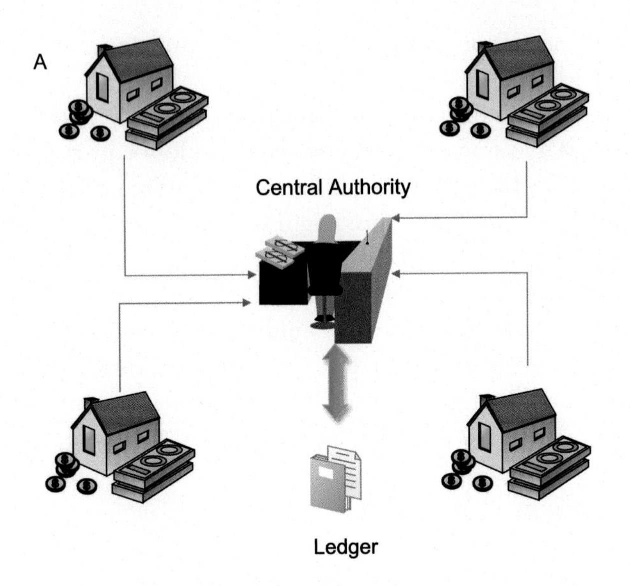

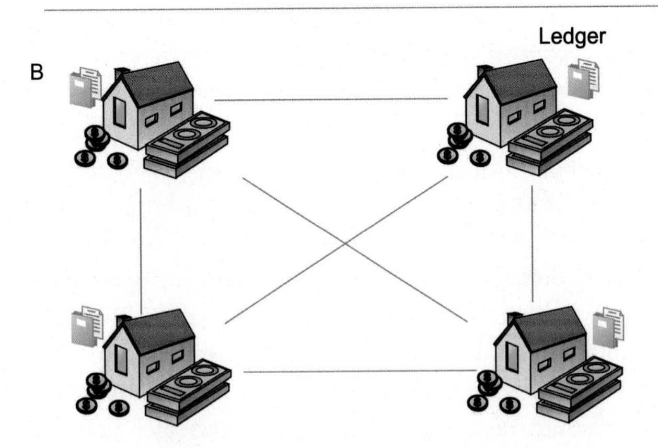

Fig. 2 (A) Centralized system; (B) Decentralized system.

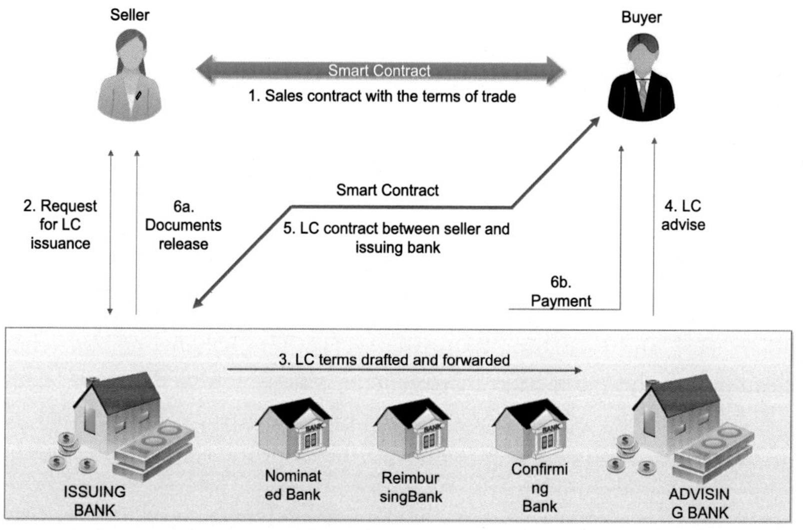

Fig. 3 Blockchain-based cross-border trade.

Table 1 Blockchain use cases in financial services.

Blockchain use case	Financial services
Capital markets	• Issuance • Sales and trading • Clearing and settlement • Posttrade services and infrastructure • Asset servicing • Custody
Capital markets	• Issuance • Sales and trading • Clearing and settlement • Posttrade services and infrastructure • Asset servicing • Custody
Investment management	• Fund launch • Cap table management • Transfer agency in asset management • Fund administration
Payments and remittances	• Domestic retail payments • Domestic wholesale and securities settlement • Cross-border payments • Tokenized fiat, stablecoins, and cryptocurrency
Banking and lending	• Credit prediction and credit scoring • Loan syndication, underwriting, and disbursement • Asset collateralization
Trade finance	• Letters of credit and bill of lading • Financing structures
Insurance	• Claims processing and disbursement • Parametrized contracts • Reinsurance markets

1. Transaction management in financial system

Blockchain is a distributed and shared database of transactions that facilitates the exchange of value. It is like a giant global spreadsheet which eliminates the need for third parties to validate transactions, reducing intermediary fees, and increasing transparency [1]. The benefits of blockchain technology in the transaction management is as shown in Fig. 4.

Blockchain can solve a lot of problems like privacy and security for transaction management. It possesses all the important characteristics that are needed to make a payment transaction. It is safe, private, secure, transparent, and decentralized in nature. It provides a high-level of security when it comes to exchanging data, information, and money during a transaction. It allows users to take advantage of the transparent network infrastructure along with low operational costs without the use of a third-party validator. These characteristics of blockchain would help to make the network reliable and appropriate for the bank transaction system. In blockchain banking, it performs an important function that makes and stores the payment transaction secure and transparent. It aims to make a secure transaction so that the overall experience of the customer more satisfactory and consuming less money during transaction processing. It offers many rewards to the customers and cutting out inefficient banking intermediaries could save billions for consumers and the financial services industry. The pictorial representation of blockchain banking is as shown in Fig. 5 and the transaction done on the blockchain network is as shown in Fig. 6.

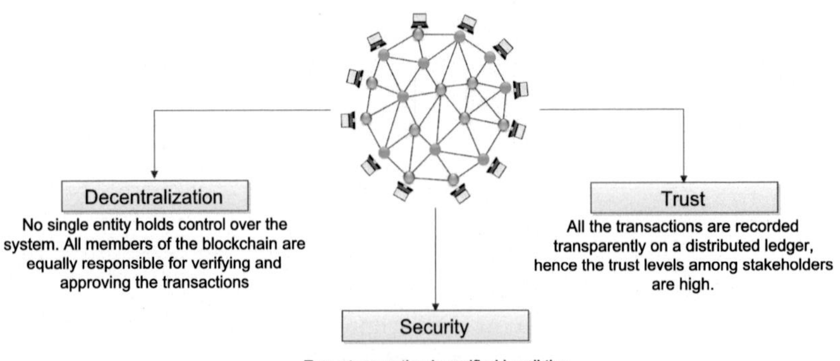

Fig. 4 Key benefits.

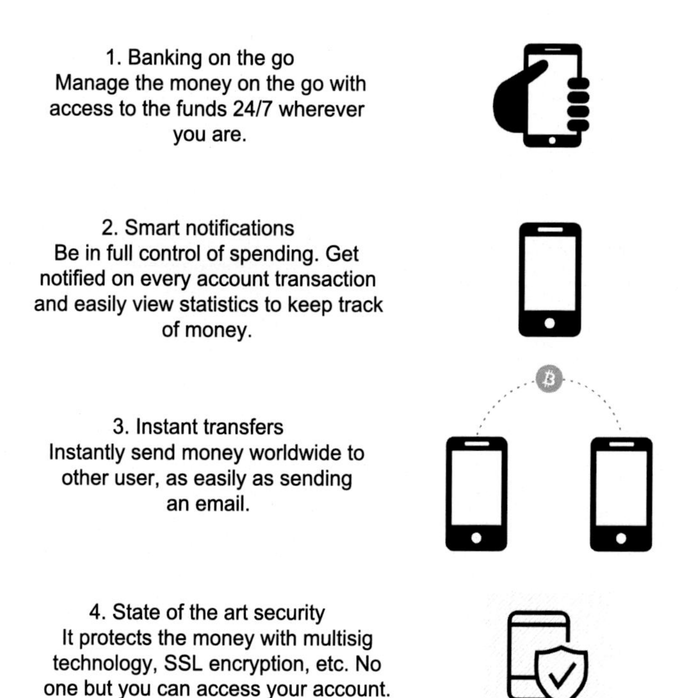

Fig. 5 Blockchain for banking industry.

2. Benefits of blockchain in the banking system

The blockchain platform enables secure business networks, more efficient and effective processes, open and secure transactions, reduces costs, new services, and products in the banking system. It provides digital services within a short period at low costs and a great level of customization. From the past few years, this technology has grown for business-grade use determining the following benefits is as shown in Fig. 7.

1. *Security*: The consensus-based distributed architecture provides security from a single point of failure in transferring payments, messaging systems, inefficient monopolistic utilities, etc.
2. *Transparency*: It provides mutualized standards, protocols, and shared processes, that act as a single shared source of trust platform for all the participants of the network.
3. *High-performance*: The private and public access to networks in the blockchain is managed to approve hundreds of transaction per second and periodically grows in the network activity.

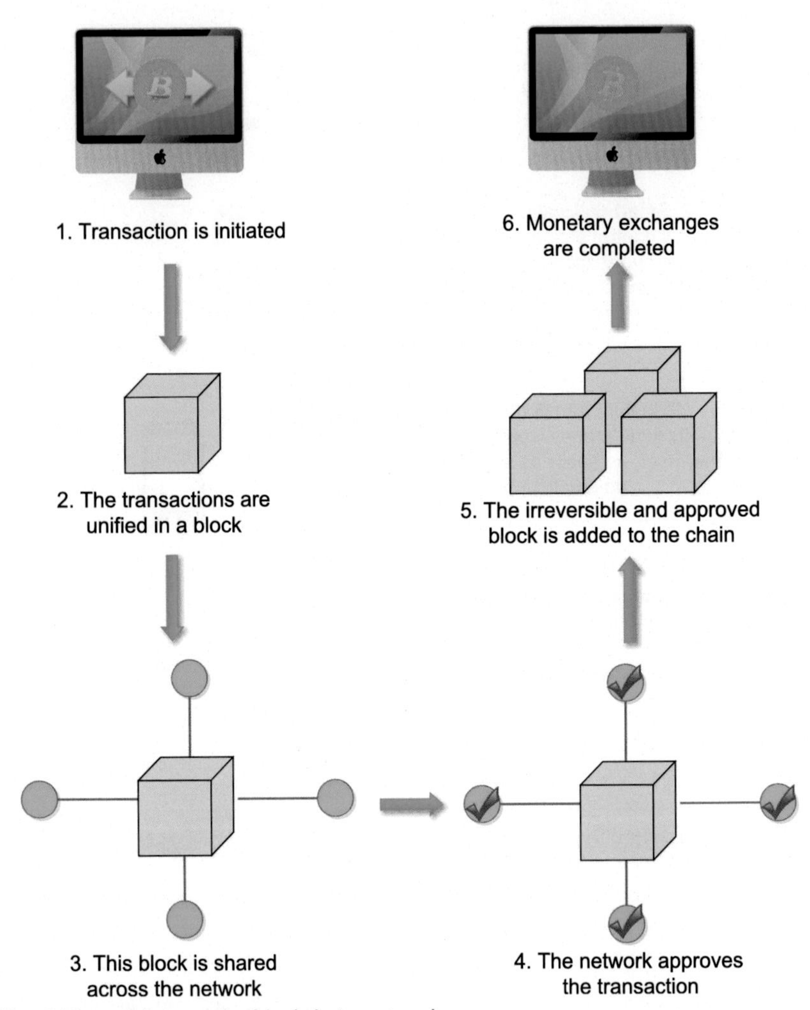

Fig. 6 Transaction on the blockchain network.

4. *Trust*: The transparent and immutable ledger of the blockchain makes it easy for different types of parties to collaborate, manage, and reach agreements in an untrusty business network.
5. *Programmability*: Blockchain technology supports the creation and execution of smart contracts that enable tamper-proof and automates the business logic in a trust and efficient way.
6. *Privacy*: Blockchain provides various types of tools for granular data privacy across every layer of the software stack that allows the sharing of data in a business network. This improves transparency, trust, and efficiency during the privacy, integrity, and confidentiality of the data.

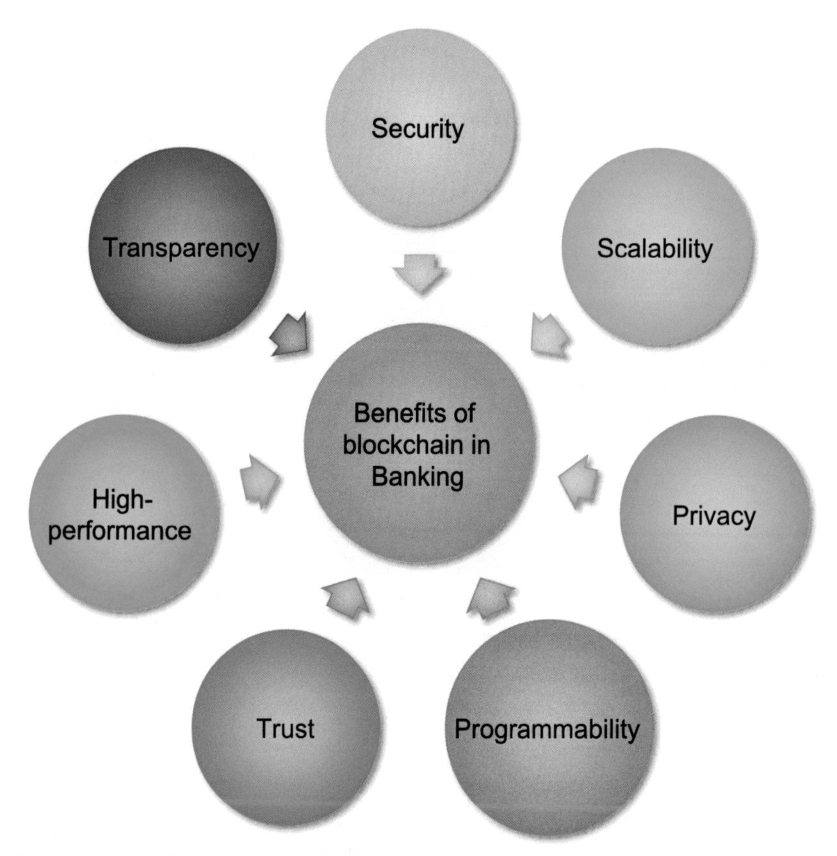

Fig. 7 Benefits of blockchain in the banking system.

7. *Scalability*: Blockchain provides interoperability between the private and public networks that offer business solutions for global reach and high-integrity when cryptocurrency transactions are being broadcasted, verified, and recorded.

3. Digitization impact the banking system

The digitization in the banking system including digital assets, smart contracts, and digital verification takes the benefits of blockchain (Fig. 8). This technology provides a new paradigm and will redefine the processes of business and financial markets. The digitization of the blockchain impacts the banking system by the following benefits.

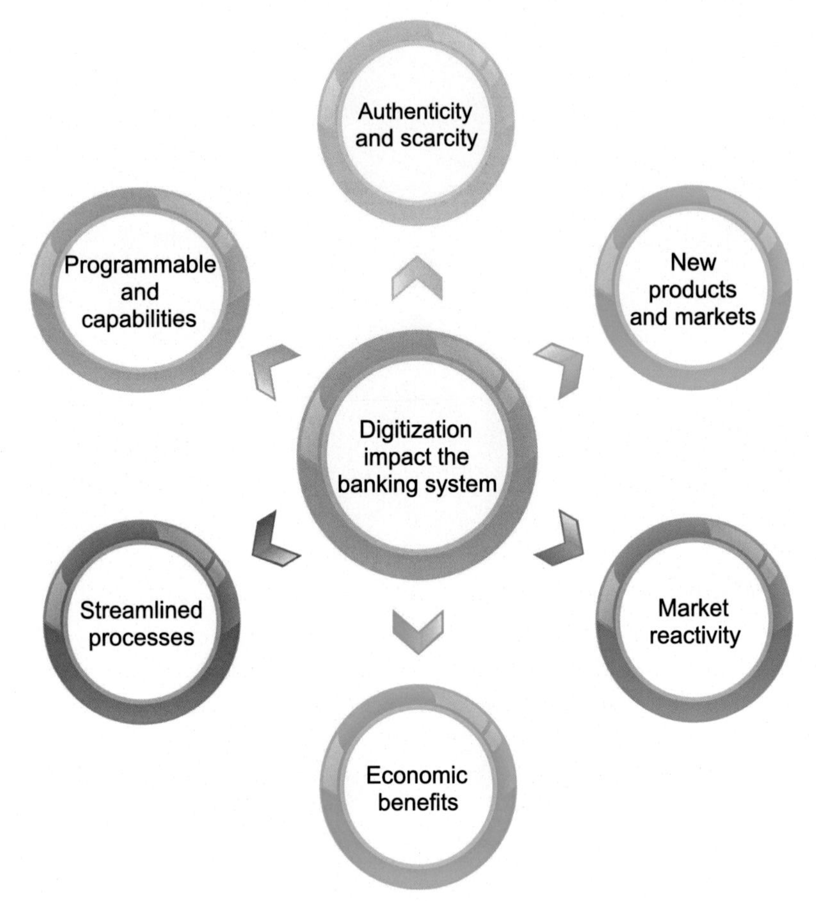

Fig. 8 Digitization impact the banking system.

1. *Authenticity and scarcity*: The digitization in the banking system provides a single shared trust platform that provides data integrity and enables asset provenance.
2. *Programmable capabilities*: The blockchain has a code capability that addresses governance, data privacy, identity management, security, and business processes automation.
3. *Streamlined processes*: High-level of automation increases overall operational efficiency. It enables real-time settlement, audit, and reporting. It reduces processing times, the potential for error and delay, and the number of steps and intermediaries required to achieve the same levels of confidence in traditional processes.
4. *Economic benefits*: The automated and efficient processes by blockchain triggers reduced infrastructure costs, operation costs, and transaction costs.

5. *Market reactivity*: Digital securities provide better customization than standardized securities. It can be issued within short time-periods.
6. *New products and markets*: Secure, scalable, and rapid asset transfers provide the ownership of real-world assets, and more.

   Combine, all these benefits result in the more transparent governance system, more efficient business processes, lower costs of capital, reduced single point of failures, greater privacy, and security, reduced counterparty risk, etc. to all digital financial systems.

4. Crowdfunding

Crowdfunding is the practice of funding a project or enterprise from a large number of participants who raise small amounts of capital through the Internet. It makes easy access to vast networks of people via social media. Its websites bring investors and entrepreneurs to work together. It is a form of crowdsourcing and alternative finance. The viewpoint of the world's crowdfunding could be changed when blockchain technology will be used for creating, distributing, and exchanging currency. Over the past few decades, crowdfunding has risen in popularity. At present, there are more than 375 platforms used for crowdfunding. Using blockchain, crowdfunding can become an even more legitimate means of funding a vast spectrum of projects and purposes. There are eight types of use cases defined by blockchain for crowdfunding is as shown in Fig. 9.

4.1 Limitations of the crowdfunding

Crowdfunding offers major advantages to the inventors and the entrepreneurs. It adjusts the process of investment even if it is a business, an invention, or work. It gives life to new products but there are some faults with this method that are described as follows.

- *The platform cash-grab*: Every crowdfunding platform is dependent on one or more devoted crowdfunding platforms. These platforms do not provide free opportunities for people to join the projects. They take enough fees from people for joining the projects under the crowdfunding. However, the inventors who have a critical level of funding of projects can be a major point on this side.
- *Rules and regulations*: Most of the crowdfunding platforms serve as gatekeepers for the projects they fund. These projects must follow some rules and regulations. For example, on *Kickstarter*, every project must

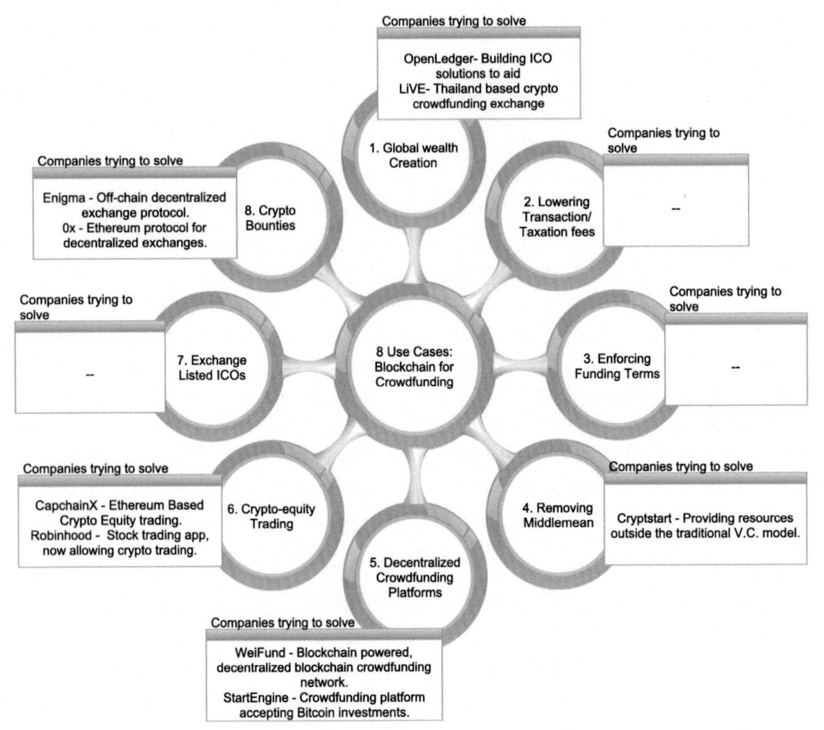

Fig. 9 Eight use cases: blockchain for crowdfunding.

create something that shares with others. It must be honest and clearly understood. It cannot fundraise for charity, offer equity, and involve prohibited items.

- *The burden of marketing and advertising*: Making good ideas for the project under the crowdfunding platform does not provide a guarantee of success. To attract new people to the project, you should know how to make the crowdfunding page more visible. For advertising, marketing, branding, and strategies like search engine optimization for the projects is a practical requirement. However, it takes time from the actual invention or business but it is helpful to make projects more popular.
- *Empty promises*: Most of the crowdfunding agencies contribute money in exchange for a promise. Unfortunately, those promises can be broken. People never get the products they paid for years ago. Often, they forget about their contributions, so they never end up taking legal recourse.
- *Copyright vulnerability*: Some inventors and creators have seen their entire project under the crowdfunding before they got a chance to start production. When their ideas got popular on the circuit, the entrepreneurs

got inspired and rushed to bet them in the market. Even with copyright, they may only be giving more ideas and more inspiration to our top competition.

4.2 Blockchain changes the crowdfunding

The advantages of blockchain technology can support and improve crowdfunding in various ways are described as follows.

- *Decentralization*: Blockchain is a fully decentralized system that does not rely on trusted third-party to raise funds for enabling creators. It positively affects the crowdfunding community. For starters using blockchain, creators will no longer follow the rules and regulations of the crowdfunding platform on the internet. Using blockchain on the platform makes the crowdfunding less expensive for creators and investors.
- *Accessible equity*: The blockchain-based crowdfunding could rely on asset tokenization instead to enable preorders of upcoming tangible products. It provides investors and creators with equity of ownership. In this way, investors will feel success proportional to the conditional success of the company. This could potentially open whole new worlds of investment opportunity.
- *Universal availability*: In a blockchain-based crowdfunding model, any person can connect and contribute to any project through an Internet connection. These projects can potentially get funded from blockchain-based models. There may be some demand for discovery and visibility platforms also.
- *Immediate provision*: In a blockchain-based crowdfunding model, there is no worry about the empty promises. Instead of contributing money and waiting weeks to receive the promised product, contributors will immediately receive fractional enterprise or product ownership.
- *Peer-to-peer exchanges*: In a blockchain-based crowdfunding model, there is impossible to change or alter the product stakes. It is a peer-to-peer network that generates more interest in the overall crowdfunding projects because of transparency and security. This change could finally accelerate the production of a new type of marketplace.

5. Scope of blockchain in financial system

The blockchain technology could save billions in cash by reducing processing costs and enhancing trust. It holds great promise to give the impact for the financial system is as shown in Fig. 10.

Fig. 10 Scope of blockchain in financial system.

6. Decentralized finance on Ethereum

Let us take an example of a Coinbase wallet, which is a mobile crypto wallet that supports multi-coin assets as well as ERC-20 tokens and ERC-721 collectibles. ERC-20 tokens are designed and deployed only on the Ethereum platform, which is used to share, exchange for other tokens, or transfer tokens to the wallet. For this, ERC-20 follows a list of standards in which three optional rules, i.e., token name, symbol, decimal and six mandatory rules, i.e., totalSupply, balanceof, transfer, transfer from, approve, allowance are present [2]. ERC-721 is a free, open standard that describes the unique tokens on the Ethereum platform. The abstract used to generate the unique tokens is defined in Listing 1 [3].

LISTING 1 Abstract of ERC-721.

```
1  pragma solidity ^0.4.20;
2  interface ERC721
3  {
4  event Transfer(address indexed _from, address indexed _to,
       uint256 indexed _tokenId);
5  event Approval(address indexed _owner, address indexed
       _approved, uint256 indexed _tokenId);
6  event ApprovalForAll(address indexed _owner, address
       indexed _operator, bool _approved);
7  function balanceOf(address _owner) external view returns (
       uint256);
8  function ownerOf(uint256 _tokenId) external view returns (
       address);
9  function safeTransferFrom(address _from, address _to,
       uint256 _tokenId, bytes data) external payable;
10 function safeTransferFrom(address _from, address _to,
       uint256 _tokenId) external payable;
11 function transferFrom(address _from, address _to, uint256
       _tokenId) external payable;
12 function approve(address _approved, uint256 _tokenId)
       external payable;
13 function setApprovalForAll(address _operator, bool
       _approved) external;
14 function getApproved(uint256 _tokenId) external view
       returns (address);
15 function isApprovedForAll(address _owner, address
       _operator) external view returns (bool);
16 }
```

6.1 Token generation

Ethereum platform is used to generate tokens, which are used in a Coinbase wallet. It is a decentralized group of computers having two basic functions, i.e., (i) blockchain that record transactions, (ii) virtual machine that generate smart contracts. Due to these two functions, this platform can support decentralized applications. They charged the developers for the computation work in their network. They have used Ether currency. So, ERC-20 and ERC-721 tokens to function as a currency or share in an organization or company. These tokens are generated with the use of

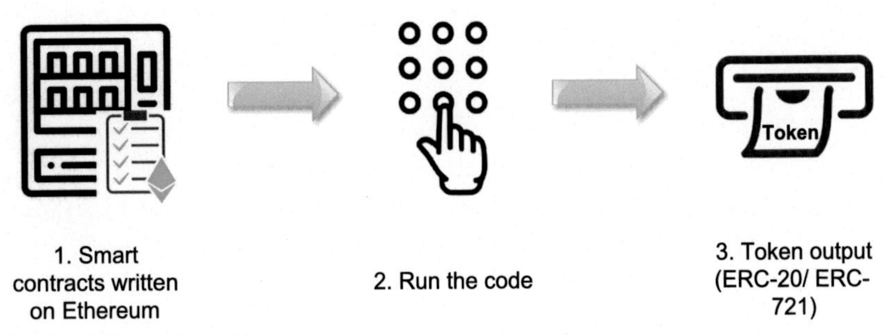

Fig. 11 Token generation.

smart contracts. These smart contracts are used to help transactions of token and record the balances of a token in the user's account as shown in Fig. 11. These are written in the Solidity programming language. After the token has been generated, it can be traded, spent, or given to someone else.

6.2 Use case–DApp: Coinbase wallet

The Coinbase Wallet DApp is used to make transactions secure and transparent. This DApp was developed by the Ethereum DeFi Ecosystem used to transform old financial transactions into a trustless, transparent and secure transactions without the use of third-party validator. The steps to download and use the Coinbase Wallet DApp are described in Fig. 12, which shows download the application on our personal computer or mobile device to make all the transaction payments decentralized [4]. All the digital transactions in one place where we take full control of our tokens and collectibles and storing them on our devices. This DApp has several benefits, which are described as follows.

- *Trade trustlessly*: This Coinbase Wallet gives access to leading and deploying decentralized exchanges or traded, where we can buy or sell the tokens.
- *Collect unique digital items*: We use and trade our favorite collectibles like cats, stickers, robots, fine art, etc. in market places and games.
- *Earn cryptocurrency*: We can earn cryptocurrency using this DApp by doing tasks, answer questions, solve puzzles, participate in communities, and more.
- *Easy send and receive cryptocurrency*: We can pay anywhere and anyone in the world with the Coinbase wallet usernames. It is the easiest way to send and receive cryptocurrency without the involvement of a third-party validator.

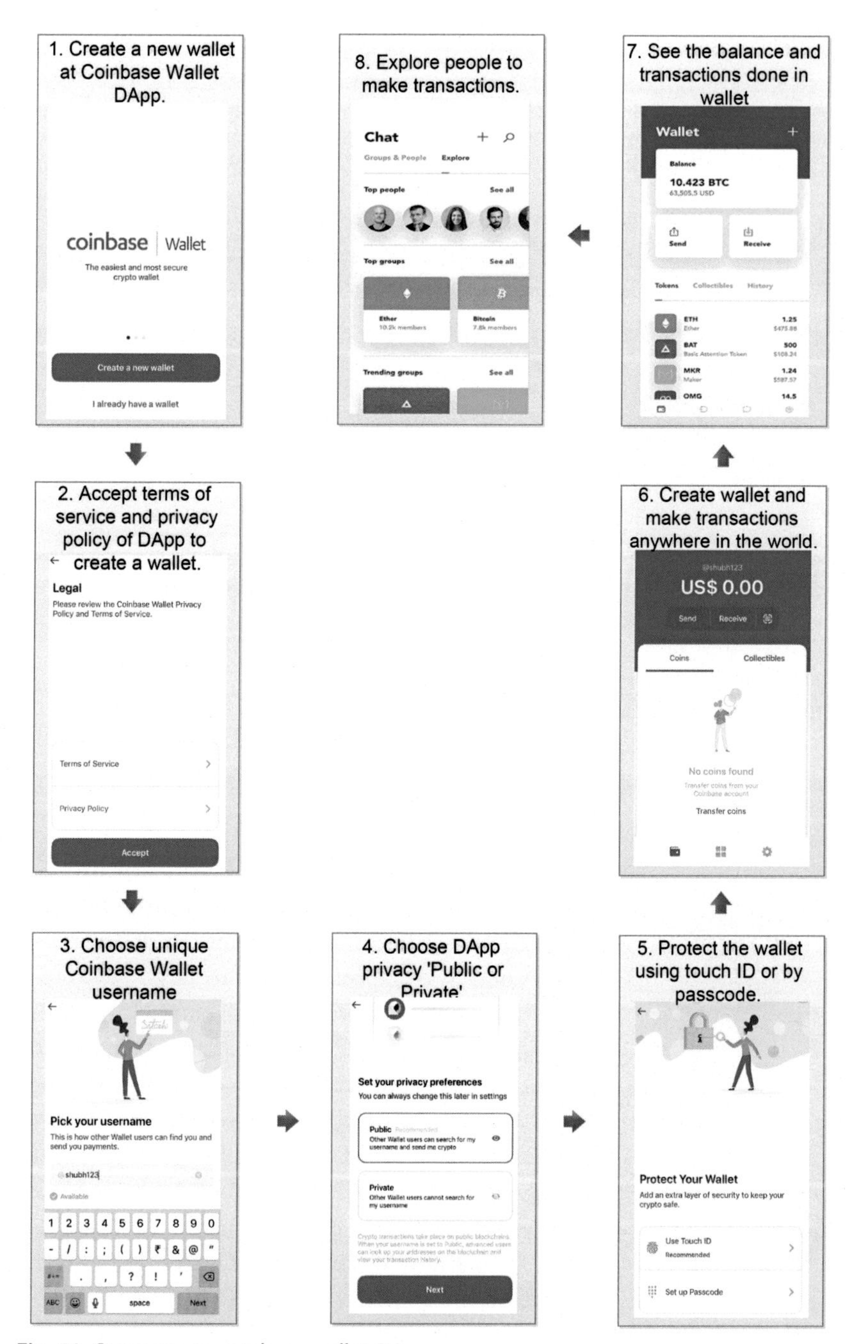

Fig. 12 Steps to use coinbase wallet DApp.

References

[1] N. Kabra, P. Bhattacharya, S. Tanwar, S. Tyagi, Mudrachain: Blockchain-based framework for automated cheque clearance in financial institutions, Futur. Gener. Comput. Syst. 102 (2020) 574–587.
[2] ERC-20, Available: https://cointelegraph.com/explained/erc-20-tokens-explained (accessed 4 April 2020).
[3] ERC-721, Available: http://erc721.org/ (accessed 4 April 2020).
[4] Coinbase Wallet, Available: https://wallet.coinbase.com/ (accessed 10 April 2020).

About the authors

Shubhani Aggarwal is pursuing PhD from Thapar Institute of Engineering & Technology (Deemed to be University), Patiala, Punjab, India. She received the BTech degree in Computer Science and Engineering from Punjabi University, Patiala, Punjab, India, in 2015, and the ME degree in Computer Science from Panjab University, Chandigarh, India, in 2017. She has many research interests in the area of Blockchain, cryptography, Internet of Drones, and information security.

Neeraj Kumar received his PhD in CSE from Shri Mata Vaishno Devi University, Katra (Jammu and Kashmir), India in 2009, and was a postdoctoral research fellow in Coventry University, Coventry, UK. He is working as a Professor in the Department of Computer Science and Engineering, Thapar Institute of Engineering & Technology (Deemed to be University), Patiala (Punjab), India. He has published more than 400 technical research papers in top-cited journals such as IEEE TKDE, IEEE TIE, IEEE TDSC, IEEE TITS, IEEE TCE, IEEE TII, IEEE TVT, IEEE ITS, IEEE SG, IEEE Netw., IEEE Comm., IEEE WC, IEEE IoTJ, IEEE SJ, Computer Networks, Information sciences, FGCS, JNCA, JPDC, and ComCom. He has guided many research scholars leading to PhD and ME/MTech. His research is supported by funding from UGC, DST,

CSIR, and TCS. His research areas are Network Management, IoT, Big Data Analytics, Deep Learning, and Cybersecurity. He is serving as editor of the following journals of repute: ACM Computing Survey, ACM·IEEE Transactions on Sustainable Computing, IEEE·IEEE Systems Journal, IEEE·IEEE Network Magazine, IEE·IEEE Communication Magazine, IEE·Journal of Networks and Computer Applications, Elsevier Computer Communication, Elsevier International Journal of Communication Systems, Wiley. Also, he has been a guest editor of various international journals of repute such as IEEE Access, IEEE ITS, Elsevier CEE, IEEE Communication Magazine, IEEE Network Magazine, Computer Networks, Elsevier, Future Generation Computer Systems, Elsevier, Journal of Medical Systems, Springer, Computer and Electrical Engineering, Elsevier, Mobile Information Systems, International Journal of Ad Hoc and Ubiquitous Computing, Telecommunication Systems, Springer, and Journal of Supercomputing, Springer. He has also edited/authored 10 books with international/national publishers like IET, Springer, Elsevier, CRC: Security and Privacy of Electronic Healthcare Records: Concepts, Paradigms and Solutions (ISBN-13: 978-1-78561-898-7), Machine Learning for Cognitive IoT, CRC Press, Blockchain, Big Data and IoT, Blockchain Technologies Across Industrial Vertical, Elsevier, Multimedia Big Data Computing for IoT Applications: Concepts, Paradigms and Solutions (ISBN: 978-981-13-8759-3), Proceedings of First International Conference on Computing, Communications, and Cyber-Security (IC4S 2019) (ISBN 978-981-15-3369-3). One of the edited text-book entitled, "Multimedia Big Data Computing for IoT Applications: Concepts, Paradigms, and Solutions" published in Springer in 2019 is having 3.5 million downloads till June 6, 2020. It attracts attention of the researchers across the globe (https://www.springer.com/in/book/9789811387586). He has been a workshop chair at IEEE Globecom 2018 and IEEE ICC 2019 and TPC Chair and member for various international conferences such as IEEE MASS 2020 and IEEE MSN 2020. He is a senior member of the IEEE. He has more than 12,321 citations to his credit with current h-index of 60 (September 2020). He has won the best papers award from IEEE Systems Journal and ICC 2018, Kansas City in 2018. He has been listed in the highly cited researcher of 2019 list of Web of Science (WoS). In India, he is listed in top 10 position among highly cited researchers list. He is an adjunct professor at Asia University, Taiwan, King Abdul Aziz University, Jeddah, Saudi Arabia, and Charles Darwin University, Australia.

Transportation system☆

Shubhani Aggarwal and Neeraj Kumar
Thapar Institute of Engineering & Technology, Patiala, Punjab, India

Contents

Abstract

Blockchain technology is used adeptly to simplify the complex and fragmented processes present in the transportation system like mobility-as-a-service and supply chain management. Companies across the trucking industry have integrated blockchain in transportation and mobility service. A blockchain is a distributed, digital ledger that records transactions in a series of blocks. It exists in multiple copies spread over multiple computers, and track all the records of the system. It records transactions, tracks assets, and creates a transparent and efficient system to manage all documents involved in the transportation system. In this chapter, we have described the role of blockchain in the transportation system in terms of mobility-as-a-service and supply chain management.

Chapter points

- In this chapter, we discuss the role of blockchain in the transportation system.
- Here, we discuss the role of blockchain in the mobility and supply chain management of the transportation system.

The blockchain technology is one of the most promising technologies in the modern world. The reason behind the success of blockchain technology is

☆ Applications.

Advances in Computers, Volume 121
ISSN 0065-2458
https://doi.org/10.1016/bs.adcom.2020.08.022

the benefits that have to offer and applications grow multiple folds. This technology was originated from the cryptography exchange, i.e., digital transactions and now it has been used in several applications like healthcare, financial system, smart grid, energy management, transportation system [1, 2].

In this chapter, we discuss the role of blockchain technology in the transportation system and make it intelligent and smart, which is an integral part of mobility service and supply chain management. Before going to know how blockchain technology can revolutionize the transportation system, it is important to understand the challenges faced by the transportation system [3].

1. Challenges faced by the transportation system

A list of challenges faced by the transportation system and the blockchain has the solution to those challenges, which are described as follows.

- *Tracking*: The main challenges faced by the transportation system is of payment and dispute resolution to track the order. Due to this, every day around the world approx. $140 billion in money expenses in dispute payment settlement, which involves a lot of time and intermediators. But with the use of blockchain technology to track the order of the transportation systems, there is no need of intermediators. Also, the tracking system of blockchain technology can easily track the vehicles on the road and update the status on the network, which is immutable and tamper-proof.
- *Transportation of temperature-controlled products*: The next challenge of the transportation system is the transportation of temperature-controlled goods and services. As per the research, around 8% of the sensitive pharmaceutical products experience a difference in temperature, which adversely affects the shipment of past custom products and leads to the wastage of transportation cost. But with the help of blockchain technology, one company can collaborate with the other companies to launch temperature-controlled products on time with the required efficiency, authenticity, and transparency in the transportation system.
- *Smart contracts*: Smart contracts are one of the best applications of blockchain technology. It is a preprogrammed code that has some predefined rules and conditions. When these conditions are met, the payment transaction is executed. With the help of smart contracts of blockchain in the transportation system, the payment method is initiated,

processed, and executed automatically after the delivery of a product to the destination point. This payment transaction has no data tampering and data manipulation.

- *Easier carrier on-board*: The blockchain technology can be helpful in carrier on-board by validating the records of every single entity. It is beneficial for a load broker to reach the capacity of a load at a location. If a broker identifies a new carrier, first it validates a carrier using blockchain then, only assign a load to that carrier. So, with the help of blockchain technology, all the records of the carriers can be kept on the public ledgers across the industry having a decentralized platform.
- *Makes load board more reliable*: The other major challenge in the transportation system faces is the load board. By the use of traditional systems in supply chain management, the data can be altered or changed by an adversary. But with the use of blockchain technology, the load data information is stored on public ledgers with time-stamping cryptography, which is tamper-proof.
- *Vehicle communication*: In today's world, vehicle-to-vehicle communication is implemented for easy transportation of goods and services, which ensure energy management in the transportation system. When this system uses blockchain technology, the whole data is stored on public ledgers that can help transportation companies to distribute and assign their operations accordingly.

2. Blockchain relevant to the transportation system

The goods and services when moved from its origin to its destination, takes a product, a carrier, and a middle-man who might be irrelevant and disappear. So, blockchain provides a relevant feature to the transportation system without any involvement of a third-party.

Blockchain technology was developed out of a branch of mathematics called cryptography that produces a complex hash value, which is difficult to calculate for an adversary. It is a replicated, secure, and immutable ledger in which to modify the data in one block is impossible without modifying the data of the entire chain. In this way, it is difficult to perform malicious and falsifying activities on the blockchain and it can potentially be applied in the transportation system. The traditional transportation system and future with blockchain-based transportation systems are described as follows.

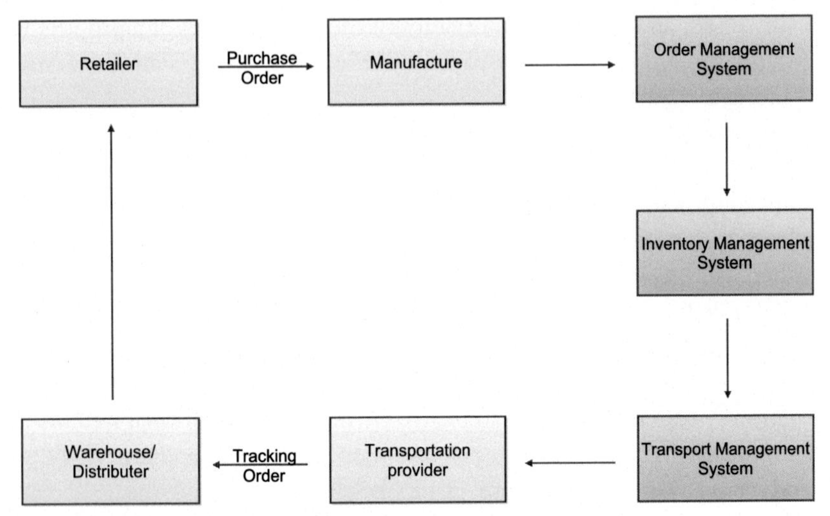

Fig. 1 Traditional transportation system.

2.1 Traditional/current transportation system

The traditional transportation system used electronic data interchange and application program interface to provide and record the transportation data [4]. This data can be altered, changed, or manipulated by a third-party authenticator, which can have critical consequences on the global transportation system. The pictorial representation of traditional transportation system is as shown in Fig. 1.

2.2 Future transportation system with blockchain

With the help of blockchain technology, there is no involvement of third-party authenticator. So, there is no single point of failure in the system. This technology is used for data authentication where the whole network can contribute and validate data, which makes the system tamper-proof and transparent.

Fig. 2 shows if blockchain is implemented and shared by all entities then, it eliminates redundancy by having a single source of truth. In this context, a new platform called blockchain in transport alliance (BiTA) is working for applying blockchain technology to solve the problems of an the transportation system. This platform has been used for creating a decentralized framework to activate the development of blockchain applications for logistics management, asset tracking, transaction processing, and more.

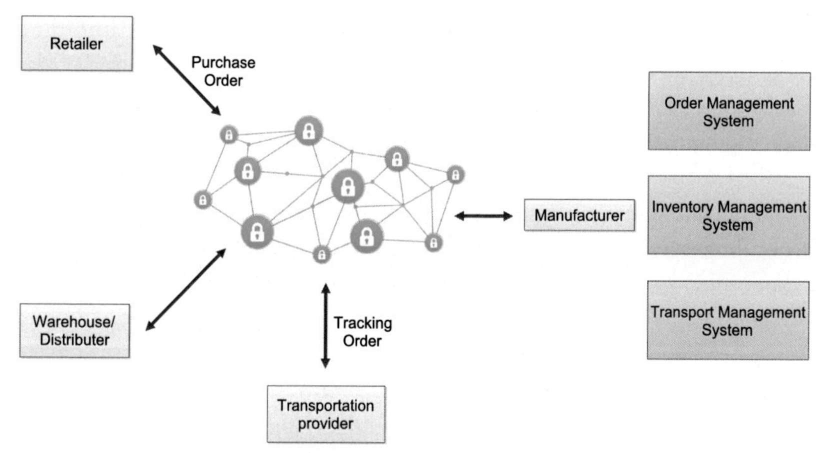

Fig. 2 Future transportation system with blockchain.

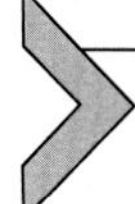

3. Reasons to implement blockchain in the transportation system

The benefits of implementing blockchain technology in the transportation system can be summarized as under.

- *Break down silos*: The largest benefit of blockchain technology in the transportation system is transparency. A distributed ledger helps to provide efficient and effective communication through a carrier and maintains redundancy between different entities in the transportation system.
- *Better traceability*: The blockchain technology records and traces all the information on goods and services during the shipment process. It also provides rich information about the product in the transportation system at every step that includes manufacturing, supplying, logistics, distribution, and other details.
- *Faster payments, easier audits*: In traditional systems, intermediators are involved in the payment transaction system and it takes a lot of time in transaction processing. It may also have some errors and flaws that can be resolved and diagnosed easily by the traditional systems. So, blockchain technology is a viable solution that can easily diagnose the errors in payments processing. It keeps track of all the information on the payment transactions in the distributed ledgers on the network. With this, smart contracts can also reduce delays and errors during the processing of payment transactions.

- *Easier identification of attempted frauds*: At every second, a large amount of data is created. So, it is very difficult to handle a large amount of data from an adversary who makes changes or frauds on the data. Hence, blockchain technology makes it impossible to make changes or alterations in the past transaction data stored in the distributed ledgers on the network. Even, slight changes in the data will change the signature completely. Therefore, the attempt of fraud on the blockchain network becomes very difficult but identification of frauds is very easy.
- *Greater consumer trust*: The blockchain technology provides a decentralized platform where untrusted parties collaborate and coordinate that makes the platform more efficient. So, this technology creates greater trust between the producer and the consumer by providing realistic delivery expectations in the transportation system.
- *Real-time feedback from consumers*: The benefits of blockchain technology also involve the real-time consumer's feedback to the producer's product. This will directly be connected to the information of the supplier and the producer, which will further help to create more accurate forecasts in a real-time transportation system.
- *Better scalability*: Blockchain technology can be used to identify potential trends in the transportation system and help the industry to grow a business to respond to a possible increase in demand.

3.1 Blockchain improves transportation management

Blockchain technology has broad significance to improve transportation management in the transportation system and how the public ledger leads to benefit for this management is described as follows.

- *Better accountability*: Blockchain technology provides better accountability with transportation management providers, systems, and contractors. On average, depending on the volume of payload and carriers, every day hundreds of thousands of dollars in erroneous charges and lost opportunities.
- *Need for more back-office*: Blockchain technology eliminates the need for more back-office staff that supports transportation management. It also reduces internal costs where appropriate, and outsourcing payload considering verification, validation, and payment transaction. The blockchain technology with smart contracts can automate and execute the entire process, which encourages trust between supply chain partners and networks.

- *Access more information about the transportation system*: Payload and carriers have access to more information for transaction management and making forecasting more accurate, which results in better load planning, effective use of payload opportunities, and analysis of the most cost-effective means of transport. The blockchain technology can be identified and rerouted to meet sudden demand changes, essential in cross-channel supply chains.
- *ELD implementation and fleet maintenance*: Electronic logging device (ELD) implementation and agile maintenance are ready for the use of blockchain technology. So, better tracking of shipments and transport goods in the supply chain improves maintenance planning, reducing breakdowns, and increasing return on investment.

The integration of blockchain and transportation management technologies in the transportation system will lead to greater adoption of the technology, alignment of business goals with actions, improved shipper-carrier relationships, as well as business-to-business sales, and increase trust between partners. The attributes of blockchain technology used in the transportation system are described as follows (see Fig. 3).

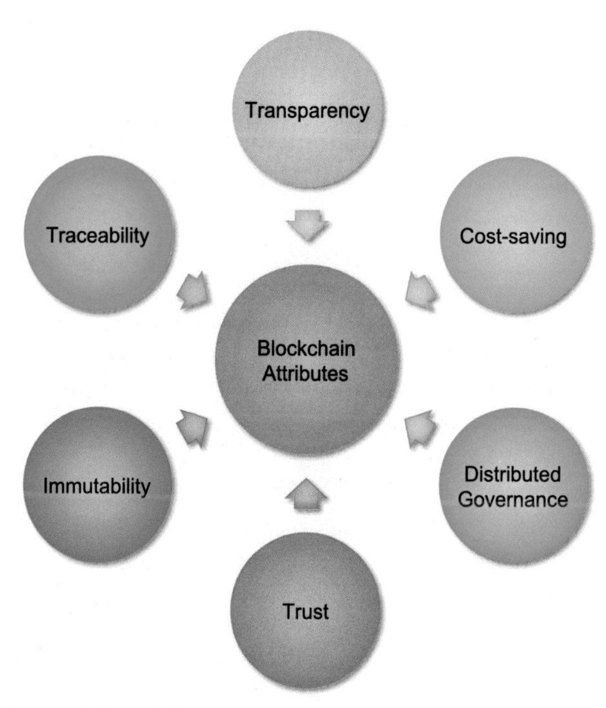

Fig. 3 Attributes of blockchain.

- *Transparency*: All the transactions done on the network are visible to all nodes and each node has a copy of the record of the transactions.
- *Traceability*: The blockchain network creates an immutable and verifiable record of the history of the transactions.
- *Immutability*: Data once stored on the blockchain network cannot be changed or altered.
- *Trust*: The blockchain technology is based on the cryptographic primitives that provide data integrity and data availability.
- *Distributed governance*: The validation and verification process of the transaction requires the agreement of all nodes, which is based on the consensus mechanism.
- *Cost-saving*: It does not involve any central authority to manage and control the user's relationships.

The above-mentioned attributes can be used in the sectors like mobility service and supply chain management of the transportation system, which are described in the following subsections (see Fig. 4).

3.2 Mobility as a service

Mobility as a service (MaaS) is defined on the concept of integrating different modes of transportation in one platform in which users can plan end-to-end trips and pay for private and public transportation services in one single platform. Any user can use this platform for using a taxi, bus, or any other mode of transportation. As a part of MaaS, all involved partners such as sharing companies, insurance enterprises, public transport organizations,

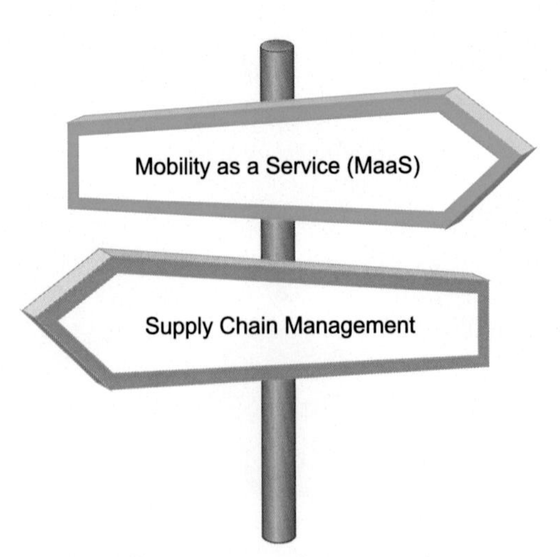

Fig. 4 Blockchain used in different sectors of transportation system.

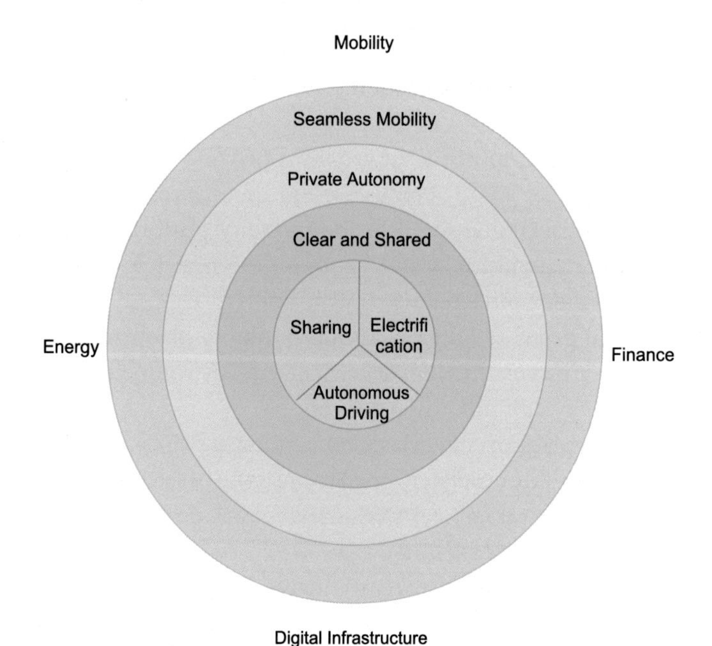

Fig. 5 Trust infrastructure for the cooperative ecosystem.

etc. are directly coordinated into mobility processes and interact digitally with each other.

Fig. 5 shows the trust infrastructure for the cooperative ecosystem represents the MaaS. It includes three main entities, which are described as follows.

- *Clear and shared transportation*: This entity means a clear transport, in the form of electric vehicles (bidirectional energy trading) with increasing shared mobility and increasing public transportation.
- *Private autonomy*: This entity is reached by using new vehicle technologies like self-driving cars, electric vehicles, and increased usage of car-sharing and ride-hailing concepts.
- *Seamless mobility*: This entity means the transportation system is from door to door and on-demand. The users have many ways to go from source to destination by using private, shared, and public transport. Mobility is delivered with a combination of self-driving shared vehicles and high-quality public transport.

3.2.1 Requirements for mobility Networks

The basic requirements of seamless mobility solutions for the future are described as follows.

- In the future, the integration of machines and vehicles must be possible and completely developed in mobility processes, which requires verified digital identities.
- The mobility platform of the future must provide trusted IoT connections to all vehicles and machines used in the network.
- Data security must be ensured in the mobility platform to protect the shared data and avoid hacks and frauds on the network.
- To enable efficient and effective new business models, all processes must be signed and processed digitally under mobility platform.
- To enable a trustworthy and decentralized environment for mobility transactions between unknown parties.
- The operating platform of a MaaS concept must be scalable and reliable. Therefore, it would completely open to any type of software and solution.
- A user can use several services from the network providers and solutions with his unique digital identity.
- There is no need for a third-party to the users to keep their primacy over their data.
- To ensure the best customer experience, all the members of the network must have equal opportunities to take participate in the network.

3.2.2 Decentralized mobility platform

A mobility network suitable for smart city solutions on a blockchain that includes four main components, which are described as follows.

1. *Digital twin standard*: A digital twin is the digital copy or replica of the vehicles that contain all status and current information of an available vehicle. It is a universal, interoperable, and open standard for vehicles of the transportation system. These enable vehicles to become full process members and make autonomous decisions. It is a part of a mobility concept, which allows all mobility providers to access information in the digital twin. The pictorial representation of the digital twin standard in mobility is as shown in Fig. 6.
2. *Distributed ledger and storage*: Blockchain technology has the potential to make a technical framework for neutral mobility networks. The best platform for a seamless mobility network is neither a private blockchain nor a public blockchain but it is a mixture of the properties of both blockchains, which is called public consortium blockchain. This platform is based on the real decentralization and data primacy with high performance during the exchange of information between untrusted parties. This type of cross-city neutral mobility concept will determine

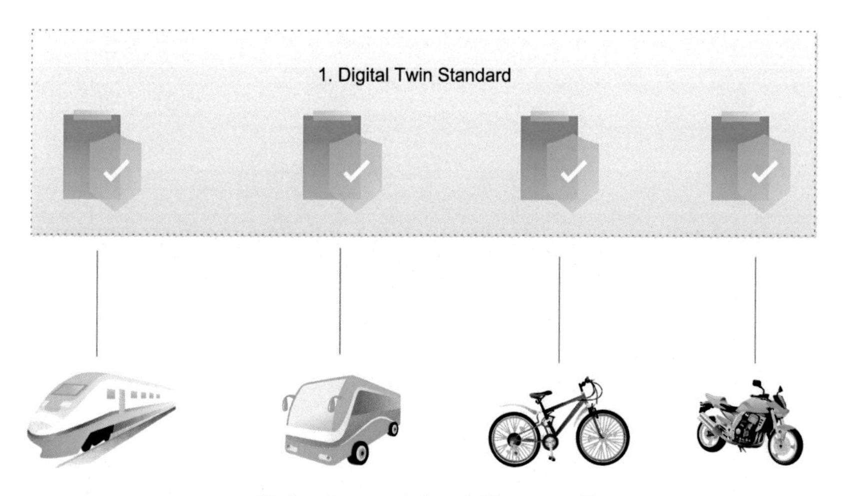

Fig. 6 Digital twins for all kind of vehicles.

the future of smart cities. The pictorial representation of mobility networks with distributed ledger is as shown in Fig. 7.

3. *Verified identities*: As the mobility networks operated with blockchain technology then, all the vehicles, users, service providers, cities, insurance companies, etc. get verified identities on the blockchain network. The verification process means all the entities have an immutable trust with their identity based on explicit verification by authorities (like driving license). So, these verified identities can be used by all mobility services within the network for preventing fraud and misuse. These digital identities are also the foundation for the trusted exchange of all data in a cross-company MaaS network. The pictorial representation of verified digital identities for all the involved parties is as shown in Fig. 8.
4. *Distributed applications*: The future of the neutral mobility network is based on acceptance in the real-world. The development of decentralized applications is up to everyone and to decide independently for whom and which areas they can be used. The combination of verified identities and digital twins on the blockchain can be used by any application vendor. This will further create an opportunity for a vendor to make a complete new multivendor mobility network across the globe. Based on such networks, companies can share their alliances and simplify the entire traveling for users. The pictorial representation of decentralized applications from different providers in different places is as shown in Fig. 9.

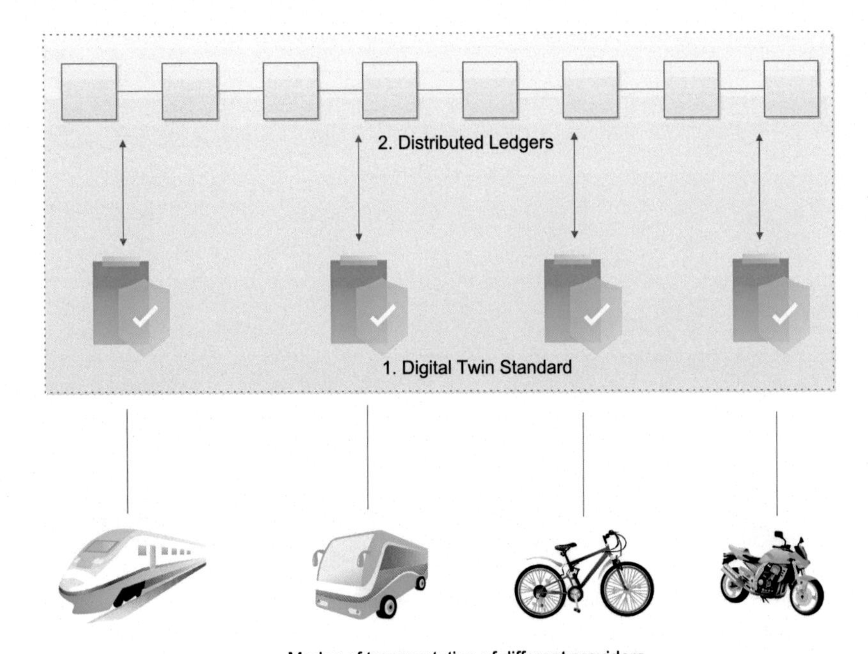

Fig. 7 Neutral operating mobility network.

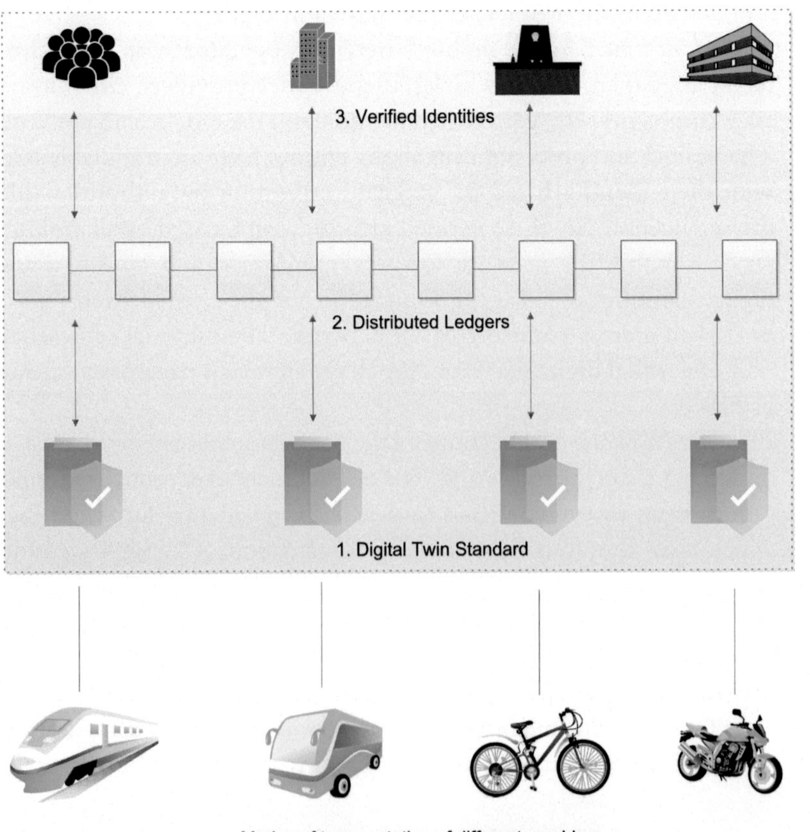

Fig. 8 Verified digital identities.

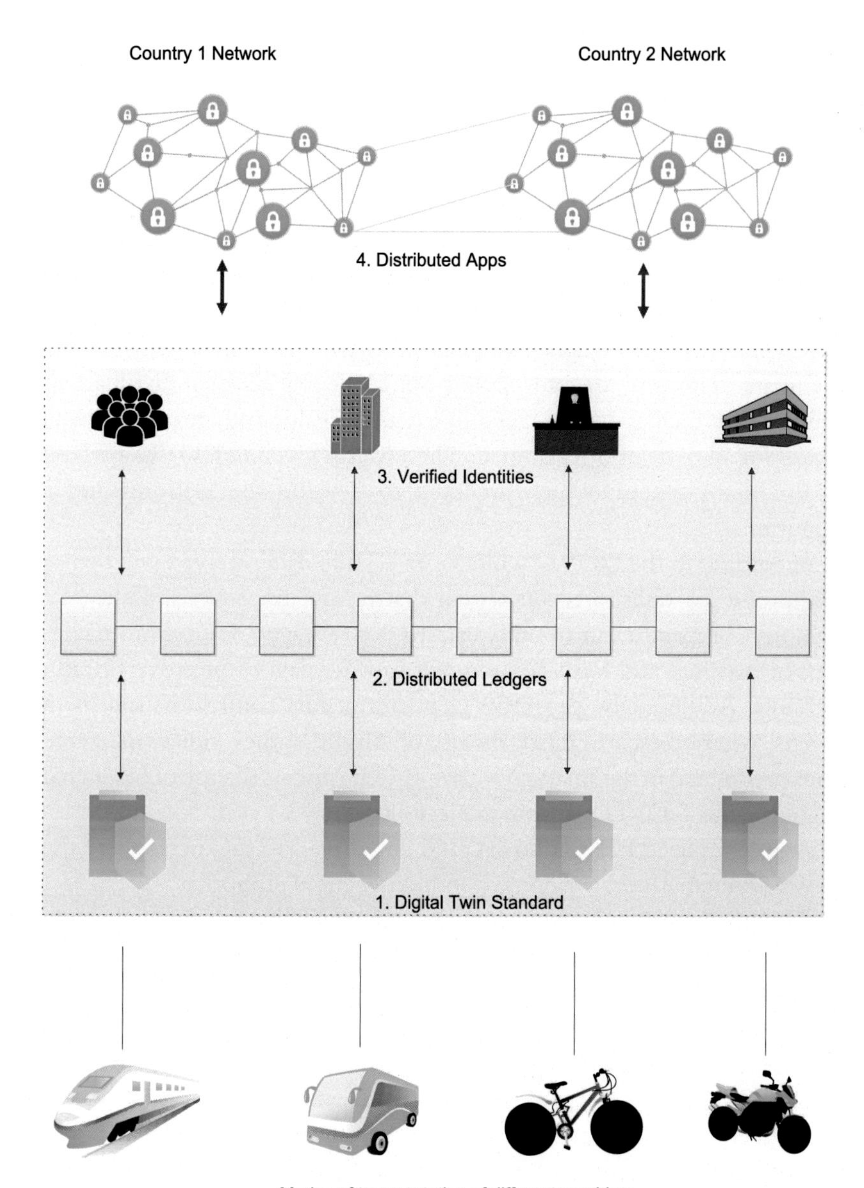

Fig. 9 Decentralized applications.

Blockchain allows P2P relationships that mitigate the effects of a centralized platform in a MaaS system. Therefore, this mitigates the dominance of a single platform regulating the entrance of service providers into a market. Also, the reward mechanisms allow a fair consensus between users and service providers, who can agree on what type of data can be shared [5, 6].

3.2.3 Use Case 1. DOVU: Create a DOVU in London using a blockchain

DOVU offers a public Ethereum-based blockchain platform where all members can exchange their vehicle information and services such as maintenance of vehicles. This platform can be used on any smart device, which connects to the Internet. Using this platform, users can use a shared vehicle and allow access to the car's data during driving. The customers and organizations that share their mobility-based data into this platform are rewarded. For example, sharing the type of vehicle they are driving, the infrastructure used such as roads or bridges, and the visionary information such as weather or climate. This type of information can be beneficial for individuals and companies, which are encouraged to exchange value and creating an equal reward for all users of the platform. The rewards given in DOVU are based on Ethereum tokens, which are used to develop the platform and its adoption.

According to the DOVU whitepaper [7], the data received will be used to improve the transportation system that further improves mobility. For example, companies can use this data to improve geolocalization, navigation, or parking, and local government can use data to improve planning decisions. Additionally, the DOVU platform puts control of data in the owners, where these can block the use of the data if they violate the agreed terms predefined in the smart contracts. Briefly, discuss the use of blockchain technology in DOVU platform are as follows.

- A decentralized P2P platform could link drivers to the users where they would establish their fares, without any central authority.
- The need for a manager who takes responsibility for the behavior of drivers and customers and supervising the exchange of values is minimized.
- The exchange of data is rewarded. All the users participated in the network can access data as agreed by the service providers and customers.
- Each user on the network has identified with a wallet to make secure transactions. Therefore, there is complete transparency of tokens flow and contractual terms while the identity of users remains the same.

DOVU is already working in London, and its critical challenge is to understand and use of this new technology in mobility.

3.3 Supply chain management

The distribution of products is essential for every industry. A supply chain is a network of individuals, organizations, resources, companies, as well as

Fig. 10 Working of supply chain.

technologies, which are combined for the manufacturing of a product or a service. The major five participants in the supply chain are suppliers, producers, distributors, retailers, and customers. The step-wise working of supply chain is as shown in Fig. 10 and consists of material flow, information flow, and financial capital flow. The main components of the supply chain are described as follows.

- *Natural resources*: In the production of any product, there could be a need for natural resources such as water, gas, electricity, etc.
- *Raw materials*: The raw materials are used for producing a certain material such as alloy, steel, farming goods, etc.
- *Finished products*: All the natural resources, components, and raw materials create the finished goods.
- *E-commerce and retail*: The dedicated shops and malls that will sell the finished products and goods to the consumers.
- *Recycling and returns*: These are the recyclable and reuse products that the customers gave back and got a refund.
- *Distribution*: These are several instructions within the supply chain, where many buyers and sellers would distribute their materials to respected manufacturing customers.

- *Transportation and storing*: All goods and services will require transportation and safe storing facilities to keep everything on track for production and distribution.

Supply chain management is the managing process of the production flow within the supply chain network. The management of the flow of goods, services, and information that involves storage and movement of products from source to destination is called supply chain management [8]. The proper implementation of supply chain management results in increased sales and revenues which will decrease fraud and overhead costs. Moreover, this will also lead to an increase in production and distribution. The different phases of traditional supply chain management is as shown in Fig. 11. But, the management of the supply chain is a difficult task as it interconnects with different

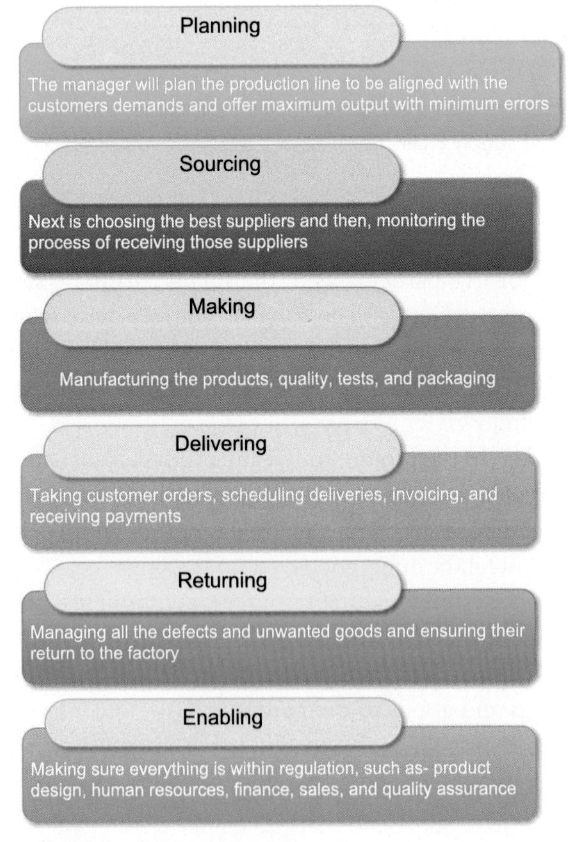

Fig. 11 Phases of supply chain management.

elements which become more inefficient when a business grows [9]. The key issues in supply chain management is described as follows.

- *Globalization*: Due to recent massive globalization, outsourcing production globally makes it difficult to coordinate within the supply chain.
- *Rapid changes in the market*: The technological revolution creates rapid changes in customer demand. Keeping up with them within the supply chain is quite difficult.
- *Compliance and quality*: With consumer expectation rising maintaining the same quality in every product becomes a burden. Also while handling thousands of products everyday, the enterprise have a hard time copying with all the regulations.
- *Lack of transparency*: Tracking every element in real-time is very difficult due to vast number of suppliers from different sources.
- *Corruption*: Bad personnels within the supply chain management harm the entire integrity of the chain. False transactions, no quality checks, lack of ethics are few corruptions.
- *High costs*: Multiple cost factors take up a lot of money over a proper supply chain such as inventory cost, quality cost, procurement cost, transportation cost, marketing cost, management cost, and many more.
- *Relatively slow*: Due to a lack in the management sector, the overall output of the supply chain is quite slow, which wastes a lot of time and cuts revenues.
- *Poor customer service*: The process of returning a product or refunding payment of a customer is complicated. Many companies and organizations does not offer best solutions for their customers in case of returns.

So, to resolve the aforementioned issues in supply chain management, blockchain technology is a viable solution [10]. The viability of blockchain consists of four main features, which make it highly applicable for the supply chain management.

1. *Transparent and controlled transactions*: Blockchain has no central authority that results in faster and more transparent settlements, as the ledger is updated automatically. When the smart contract runs, the processing of payment automatically starts that includes the visibility of a transaction to the authorized participants only.
2. *Preapproved transaction fees*: The commission for the cross-border payment transactions is deducted only after the completion of the transactions. But in the case of blockchain, there is a preapproved transaction fee that does not change during the processing of the transaction.

3. *Auditability*: All the transactions are immediately visible to the participated nodes and stored on the public ledgers where no one can change or tamper any information on the blockchain network.
4. *Reliable*: The distributed nature of blockchain does not include the central authority to validate the transactions. So, there is no single point of failure on the network. Also, the transactions processed on the blockchain are immutable and irreversible, which further eliminates the risks of fraud.

3.3.1 Benefits of decentralizing supply chain management

The following points described the benefits of decentralizing supply chain management.

- *Transparent ledger system*: Decentralized supply chain offers to trace every product movement using the transparent ledger system.
- *Tracking in real-time*: Blockchain-based supply chain supports real-time tracking of any product within the system along with their current condition for quality control.
- *Faster transactions*: The blockchain-based supply chain platform is capable of transacting thousands of transactions per second that saves a lot of time.
- *Trustless chain*: The smart contract integration will create a trustless medium among manufactures and vendors.
- *Product certification*: Everything will certified on the blockchain, promoting a platform without human errors and fraudulent activities.
- *Greater security*: Immutability of the blockchain for the supply chain network ensures greater security and also improves accuracy.
- *Lower carbon footprint*: This benefit of blockchain will decrease the need for returns, thus promoting a reduction in carbon footprint.
- *Reduced costs*: There is no middleman, which offers greater accuracy, no counterfeit products reduce the overall costs.

3.3.2 Blockchain use cases in supply chain

Several use cases of blockchain used in supply chain are described as follows.

- *P2P transaction settlement*: Direct P2P transactions with faster processing and minimum transaction fees.
- *Audit transparency*: Streamlining auditing process to offer sustainability and detect all inefficiencies.
- *Tracking consumer feedbacks*: Keeping track of all customer feedbacks and generating products based on customer demands.
- *Reducing counterfeit products*: Tracking every work flow and logs, blockchain reduces the possibility of counterfeit products.

- *Enforcing trade policies and tariffs*: Tariffs are trade policies that will automatically enforced if a product gets traded internationally.
- *Inventory management*: Distributed ledgers help the authorized users to manage the inventory with their exact location.
- *Accurate costing information*: Tracking all costs and timestamping them along the chain offers accurate costing information.
- *Better shipping data*: Real-time shipping data helps to forecast the arrival of products and raw materials.
- *Food safety*: Ensuring food safety by checking the quality of the production line within the supply chain network.
- *Automation*: It helps to reach a higher speed of data process and greater visibility.
- *Provenance*: Ensuring provenance for every product to satisfy the consumer of the authenticity of the product.
- *No human error*: Reduces the reliance on human tasks, which reduces any human errors that save money.
- *Identity verification*: Verification of every participant within the supply chain enforces safe play and get rid of fraud participants.
- *Avoiding compliance violation*: Helps to follow all compliances with proper authorization from the law that gets rid of any violation in the supply chain.

3.3.3 Applicability of blockchain for supply chain

The step-wise process of blockchain for supply chain is described as follows.

- *Step 1.*
  - Manufacturer and supplier find suitable parties using blockchain supply chain platform.
  - Initiates agreement with smart contracts.
- *Step 2.*
  - Suppliers uploads quality test documents and manufacturer checks authenticity.
  - Tags all material with RFID chip.
  - Ships the material and both parties track the process in real-time.
- *Step 3.*
  - The manufacturer gets the products and releases the payment.
  - Track all production line.
  - Validates final product quality test documents.
  - Add QR codes and the RFID chips to the packages.
  - Monitors storing in the warehouse.

  - Creates a smart contract agreement with verified distributor and ships the products.
- *Step 4.*
  - Distributor tracks the shipment and releases the payment upon product receipt.
  - Ships the products to retailers.
- *Step 5.*
  - Retailers tack the shipment.
  - Upon receipt, releases the payment to the distributors.
  - Forecast customer behaviors.
  - Offers applications to the end-users.
- *Step 6.*
  - Customer tracks the order.
  - Scans QR code for provenance.

3.3.4 Use Case 2. Walmart: To trace back mangoes and pork

Walmart's Hyperledger Fabric based on the private blockchain creates traceability for the entire supply chain [11]. It is open-source and vendor-neutral with two PoC projects that determine the traceability and authenticity of food from the farm in a quick and precise manner. The first project was based on tracing the origin of mangoes sold at Walmart US stores, and others to trace pork sold in China stores. For the mangoes, the experiment consisted of buying a bag of mangoes and asking the Walmart team to identify the farm they came from. This process is very time-consuming so Walmart developed an application based on blockchain, which focused on the supply chain system with the labeling authority that defines the attributes. All suppliers used the new labels and uploaded the data through a web-based interface, which means every stage of information is stored on the blockchain. For pork in China, it allowed uploading certificates of authenticity to the blockchain, bringing more trust to a system where that used to be a serious issue. This process allowed that even veterinary certificates were scanned and stored on the blockchain as an immutable digital copy that was accessible by any trusted user on the network. So, blockchain technology would not only help to trace back food but also to collect information about temperature, freshness, speed of the flow chain, and where the supply chain faces obstacles. It offers the following properties to the Walmart technology.

- *Traceability*: Stakeholders can track their goods from source to consumer delivery without any involvement of a third-party.

- *Transparency*: It allows the exchange of information on food attributes as it moves along the supply chain. This information can be used to prove origin, guarantee, quality, quantity of the product.
- *Trust*: The pork case allowed stakeholders to verify food authenticity and also verify animal vaccinations through the supply chain.
- *Decentralization*: Blockchain provides a decentralized platform that allows any stakeholder could verify the supply chain process without having to request information from third parties.

4. Challenges with blockchain in the supply chain

The transportation system will meet with a few challenges for implementing the blockchain technology successfully, which are described as follows.

- *Data standardization*: Data standardization is a data processing workflow that converts the structure of different datasets into one common format of data. It deals with the transformation of datasets after the data are collected from different sources and before it is loaded into target systems. It takes a lot of time and iteration to process and results in very precise, clear, and time-consuming integration and development work. So, the conventional industries trust a new blockchain network because data standardization is not easy. But in blockchain technology, each member must agree on how to characterize their data that includes detailed information of every document and bill. In this context, BiTA is a standard and advocacy organization to help members and establish standards for blockchain applications in the transportation system.
- *Data privacy*: As blockchain technology is immutable and tamper-proof, the data cannot be changed or altered. With a greater emphasis on data privacy, some provisions would need to be built in to protect consumer data, and even delete it if required. The immutability and transparency of blockchain also become a big concern in the transportation system as precise and appropriate data is rarely found in the real-world.
- *Technical challenges*: The blockchain technology is one of the most secure technologies but it takes a lot of time to get all of the verifications created and replicated. It must be improved as the process of verification and validation in the blockchain is not immediate and quick.

With this, there is a private and a public key to every blockchain transaction to control access to the information. If the private key of the blockchain transaction gets damaged then, that blockchain is lost and no longer verifiable.
In conclusion, blockchain has tremendous potential in the transportation system. But it still needs time to mature before becoming universal. At the same time, organizations need to start their work according to the blockchain strategy.

References

[1] U. Bodkhe, S. Tanwar, K. Parekh, P. Khanpara, S. Tyagi, N. Kumar, M. Alazab, Blockchain for industry 4.0: a comprehensive review, IEEE Access 8 (2020) 1–37.
[2] S. Tanwar, Q. Bhatia, P. Patel, A. Kumari, P.K. Singh, W.-C. Hong, Machine learning adoption in blockchain-based smart applications: the challenges, and a way forward, IEEE Access 8 (2019) 474–488.
[3] R. Chaudhary, A. Jindal, G.S. Aujla, S. Aggarwal, N. Kumar, K.-K.R. Choo, BEST: blockchain-based secure energy trading in SDN-enabled intelligent transportation system, Comput. Secur. 85 (2019) 288–299.
[4] R.S. Bali, N. Kumar, Secure clustering for efficient data dissemination in vehicular cyber-physical systems, Futur. Gener. Comput. Syst. 56 (2016) 476–492.
[5] C. Carter, L. Koh, Blockchain disruption in transport: are you decentralised yet? 2018.
[6] R. Rajbhandari, Exploring blockchain-technology behind bitcoin and implications for transforming transportation, final report, Texas A&M Transportation Institute, 2018. Tech. rep.
[7] DOVU, HOW the world's first mobility cryptocurrency will create a UBIQUITOUS reward ecosystem for the transport industry., Positioning document and Whitepaper, V2.0, WWW.DOVU.IO (2018).
[8] J.T. Mentzer, W. DeWitt, J.S. Keebler, S. Min, N.W. Nix, C.D. Smith, Z.G. Zacharia, Defining supply chain management, J. Bus. Logist. 22 (2) (2001) 1–25.
[9] D.M. Lambert, M.C. Cooper, Issues in supply chain management, Ind. Mark. Manag. 29 (1) (2000) 65–83.
[10] Z. Cekerevac, L. Prigoda, J. Maletic, Blockchain technology and industrial Internet of Things in the supply chains, Mest. J. 6 (2018) 39–47.
[11] How Walmart brought unprecedented transparency to the food supply chain with hyperledger fabric, Available: https://www.hyperledger.org/resources/publications/walmart-case-study (accessed 5 May 2020).

About the authors

Shubhani Aggarwal is pursuing PhD from Thapar Institute of Engineering & Technology (Deemed to be University), Patiala, Punjab, India. She received the BTech degree in Computer Science and Engineering from Punjabi University, Patiala, Punjab, India, in 2015, and the ME degree in Computer Science from Panjab University, Chandigarh, India, in 2017. She has many research interests in the area of Blockchain, cryptography, Internet of Drones, and information security.

Neeraj Kumar received his PhD in CSE from Shri Mata Vaishno Devi University, Katra (Jammu and Kashmir), India in 2009, and was a postdoctoral research fellow in Coventry University, Coventry, UK. He is working as a Professor in the Department of Computer Science and Engineering, Thapar Institute of Engineering & Technology (Deemed to be University), Patiala (Punjab), India. He has published more than 400 technical research papers in top-cited journals such as IEEE TKDE, IEEE TIE, IEEE TDSC, IEEE TITS, IEEE TCE, IEEE TII, IEEE TVT, IEEE ITS, IEEE SG, IEEE Netw., IEEE Comm., IEEE WC, IEEE IoTJ, IEEE SJ, Computer Networks, Information sciences, FGCS, JNCA, JPDC, and ComCom. He has guided many research scholars leading to PhD and ME/MTech. His research is supported by funding from UGC, DST, CSIR, and TCS. His research areas are Network Management, IoT, Big Data Analytics, Deep Learning, and Cybersecurity. He is serving as editor of the following journals of repute: ACM Computing Survey, ACM·IEEE Transactions on Sustainable Computing, IEEE·IEEE Systems Journal, IEEE·IEEE Network Magazine, IEE·IEEE Communication

Magazine, IEE·Journal of Networks and Computer Applications, Elsevier Computer Communication, Elsevier International Journal of Communication Systems, Wiley. Also, he has been a guest editor of various international journals of repute such as IEEE Access, IEEE ITS, Elsevier CEE, IEEE Communication Magazine, IEEE Network Magazine, Computer Networks, Elsevier, Future Generation Computer Systems, Elsevier, Journal of Medical Systems, Springer, Computer and Electrical Engineering, Elsevier, Mobile Information Systems, International Journal of Ad Hoc and Ubiquitous Computing, Telecommunication Systems, Springer, and Journal of Supercomputing, Springer. He has also edited/authored 10 books with international/national publishers like IET, Springer, Elsevier, CRC: Security and Privacy of Electronic Healthcare Records: Concepts, Paradigms and Solutions (ISBN-13: 978-1-78561-898-7), Machine Learning for Cognitive IoT, CRC Press, Blockchain, Big Data and IoT, Blockchain Technologies Across Industrial Vertical, Elsevier, Multimedia Big Data Computing for IoT Applications: Concepts, Paradigms and Solutions (ISBN: 978-981-13-8759-3), Proceedings of First International Conference on Computing, Communications, and Cyber-Security (IC4S 2019) (ISBN 978-981-15-3369-3). One of the edited text-book entitled, "Multimedia Big Data Computing for IoT Applications: Concepts, Paradigms, and Solutions" published in Springer in 2019 is having 3.5 million downloads till June 6, 2020. It attracts attention of the researchers across the globe (https://www.springer.com/in/book/9789811387586). He has been a workshop chair at IEEE Globecom 2018 and IEEE ICC 2019 and TPC Chair and member for various international conferences such as IEEE MASS 2020 and IEEE MSN 2020. He is a senior member of the IEEE. He has more than 12,321 citations to his credit with current h-index of 60 (September 2020). He has won the best papers award from IEEE Systems Journal and ICC 2018, Kansas City in 2018. He has been listed in the highly cited researcher of 2019 list of Web of Science (WoS). In India, he is listed in top 10 position among highly cited researchers list. He is an adjunct professor at Asia University, Taiwan, King Abdul Aziz University, Jeddah, Saudi Arabia, and Charles Darwin University, Australia.

CHAPTER TWENTY-THREE

Smart grid☆

Shubhani Aggarwal and Neeraj Kumar
Thapar Institute of Engineering & Technology, Patiala, Punjab, India

Contents

Abstract

The integration of blockchain technology and the electric grid makes the grid smart and decentralized. The adoption of blockchain technology allows the grid network to decentralize its operations. This means there is no need for a central authority in the smart grid to control, distribute, and manage the electricity. The decision making and transaction flow in the smart grid do not need to be transmitted through the centralized system. All the transactions and operations performed on the grid network are

☆ Applications.

Advances in Computers, Volume 121
ISSN 0065-2458
https://doi.org/10.1016/bs.adcom.2020.08.023

stored on the public ledgers of a blockchain. The buying and selling energy transactions across the network can be done through a computer program by validating and verifying the pre-determined clauses of the transaction. So, blockchain technology helps in setting up real-time energy markets and identity preserving transactions at much lower costs. In this chapter, we have described the role and benefits of blockchain technology in the smart grid in terms of energy management and demand–supply management.

Chapter points

- In this chapter, we discuss the role of blockchain in the smart grid.
- Here, we discuss the role of blockchain in the energy management, demand response management of the smart grid.

The amalgamation of ICT in power industry has led to a revolution known as smart grid (SG). Using these technologies, the entire working of SG ecosystem ranging from power generation, transmission, distribution, delivery to consumption and utilization is managed effectively with improved reliability, scalability, and efficiency [1, 2]. SG incorporates various type of smart devices and features such as smart meters, broadband and distributed applications, intelligent systems to provide low cost and efficient operations at all levels. It acts as a distributed platform for the smart devices connected to each other using advanced metering infrastructure (AMI). The energy consumers (smart homes, commercial development, industry) interact with the power utility using a bidirectional communication channel for energy trading in SG ecosystem.

1. Architecture of smart grid

SG is defined by an electricity grid, which integrates the behavior and actions of all entities connected. There are three types of entities; generation, which generates electricity; transmission and distribution, which distribute electricity; consumption, which consumes electricity. Together, these entities create a peer-to-peer network to ensure efficient distribution of the electricity; maintain low losses and a high level of quality, the security of electricity supply is as shown in Fig. 1.

SG technology is capable of managing the energy which is generated from various sources. Based upon the knowledge of demand and supply, utilities in SG system can predict energy requirement for future based upon the requirements of the users. Hence, forecasting the load and requirements generated from the consumer, the gap in demand and supply of energy can be minimized through optimal energy distribution in Smart Homes (SHs). If the SH knows its energy consumption, and future energy requirements

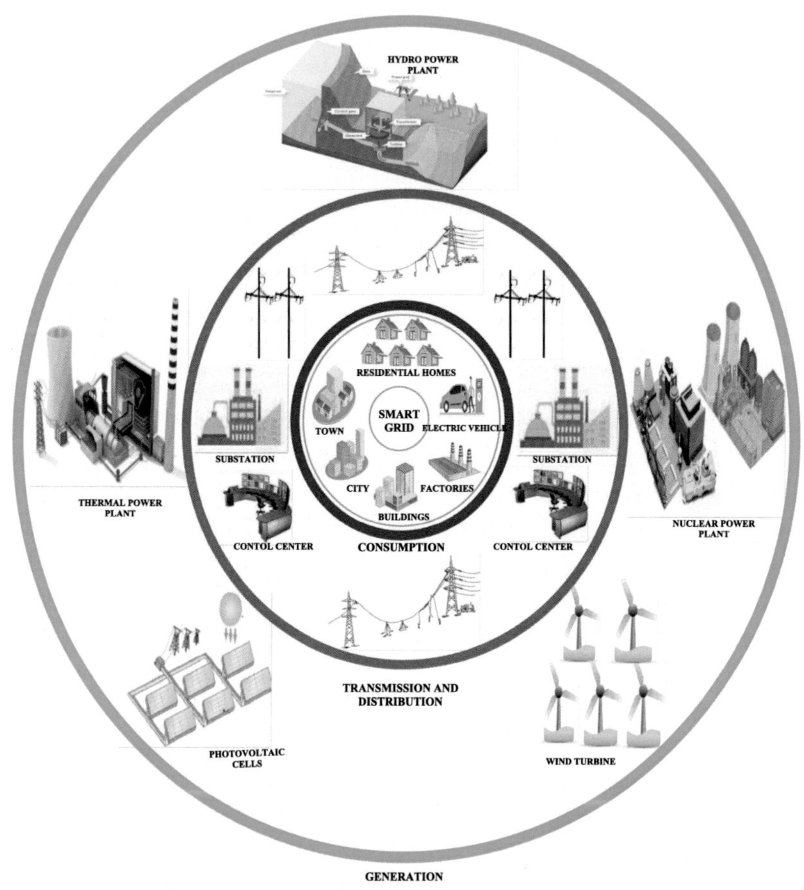

Fig. 1 Architecture of the smart grid.

then it can provide the energy demand to the grid, and accordingly grid can plan the distribution of available resources based upon the user's profile and utilizes its energy in a more efficient way as per the need of the user's requirement [3, 4].

A SH is a modern home in which all the devices or appliances are connected to a centralized system which monitors and controls their operational activities, thereby, making the appliances to work smartly as per the user demands. For example, users send short messaging service to the smart devices via application before entering the home and consequently switch on the lights and the heater in case of chilled weather. SH provides many benefits nowadays which includes ease of access control, enhance lifestyle, low energy, infotainment and security, and so on. As the energy resources are depleting gradually, so there is a requirement to use renewable resources for energy consumption, and generation, i.e., current resources should be managed effectively in order to last for longer duration as per the demands of the users. The flow of energy distribution is such that there is minimum loss and wastage of energy. The comparison of conventional/existing electrical grid with the SG is as presented in Table 1.

However, there exist various major security and privacy issues in the SG. The traditional SG mechanisms rely on trusted third parties which act as a single point of failure. Moreover, the anonymity and privacy leakage of energy consumers are one of the major concerns while depending on the third-party validator [5]. For this reason, it is very important to equip SG with a decentralized and secure system that can execute contracts and handle negotiations/transactions among several entities. To tackle these issues, a secure communication method is required, which can provide data security, integrity, authentication, non-repudiation, and confidentiality while transmitting the data generated by various SG devices [6]. The various security issues presented in the SG are described as follows.

- *Confidentiality*: The assurance that data will be disclosed only to authorized individual.
- *Availability*: The assurance that any network resource (data/ bandwidth/ equipment) will be available only for an authorized entity. Such resources are also protected against any incident that threatens their availability.
- *Integrity*: The assurance of accuracy and consistency of data will be maintained throughout the system. So, there are no unauthorized modifications, destruction, or losses of data that will go undetected.
- *Authenticity*: The validation that communicating parties are who they claim they are, and that messages supposedly sent by them are indeed sent by them. Authentication is basically validating your credentials like User Name/User ID and password to verify your identity. The system determines whether you are what you say you are using your credentials.
- *Authorization*: The assurance that the access rights of every entity in the system are defined for access control. Authorization determines your ability to access the system and up to what extent. Once your identity is verified by the system after successful authentication, you are then authorized to access the resources of the system.
- *Non-repudiation*: The assurance that undeniable proof will exist to verify the truthfulness of any claim of an entity.

Table 1 Comparison of conventional/existing electrical grid with the smart grid.

Features	Conventional/existing electrical grid	Smart grid
Two-way communication	Not in conventional grid	Involves efficient two-way communication through the advances in real-time devices
Complexity	Less complexity in comparison to SG	More complex as it continues generate and manage data at the same time through real-time devices
Self-healing capability	Does not have this capability. It corrects the faults only after the occurrence	SG has the capability as it automatically identifies the faults and make it correct immediately
Self-restoration capability	Does not have this capability	SG has the capability to control and self-restore the data
Plug-in hybrid vehicles	Cannot accommodate	Can accommodate Plug-in hybrid vehicles
Consumer involvement	Does not allow consumer involvement	Allows consumer involvement and helps the utilities to better manage and control energy demand every time
Dynamic pricing	Does not allow real-time pricing (RTP)	Allows RTP that helps to manage energy demand every time
Cyber attacks and security	Less prone to cyber attacks	High risks of cyber attacks and cyber security needs to be well built and tightened
Wide area control and measurement	Does not have this capability	SG has the capability through real-time and smart devices such as phasor measurement units (PMUs), supervisory control and data acquisition (SCADA), geographical information system (GIS)
Telecommunication devices	Very less telecommunication devices incorporated	SG is dependent on many telecommunication devices
Storage	Install conventional storage	New generation of storage techniques are incorporated

Continued

Table 1 Comparison of conventional/existing electrical grid with the smart grid.—cont'd

Features	Conventional/existing electrical grid	Smart grid
Penetration of renewable energy resources (RERs)	Conventional grids do not operate RERs efficiently in comparison to SG	Efficient operation and optimization of RERs
Distributed generation (DG)	Conventional grids do not operate DG efficiently in comparison to SG	Allows efficient and optimized operation of DG
Visualizations and monitoring capabilities	Conventional visualizations and monitoring capabilities gives alert to the operator	Advanced visualizations and monitoring capabilities will trigger and control itself in case of abnormal situation
AMI	Does not have AMI	Has AMI
Geographical information system (GIS)	Does not have GIS facilities	Incorporates GIS for better operation and control
Advanced control	Does not have advanced control	Has latest automation involved via distribution, analytical tools, and operational applications
Advanced sensors	Does not have advanced sensors It corrects the faults only after the occurrence	SG gives real-time information via advanced sensors and helps to maintain better power flow
Power quality	Poor power quality	SG has better power quality due to automation and control
Resiliency	Cannot resilient to natural disasters and fault conditions	Better resiliency in comparison to conventional grid

2. Salient features of smart grid

The important features of SG that make it different from the conventional/existing electrical grid in several manners. The salient features of SG are described as follows (see Fig. 2).

2.1 Integration of renewable energy resources (RERs) into the grid

Concern about climate change and carbon emissions have inserted pressure on the use of RERs as a primary source for energy generation and transmission. Renewable energy is generated through natural resources, i.e., wind, solar, hydropower, geothermal, tidal wave energy, bio-mass, etc. except fossil fuels. The generated energy from renewable resources is environment-friendly, sustainable, and economic sources of energy. Along with these advantages of renewable energy, they also have some disadvantages such as the stochastic and vague behavior of these resources. The major goal of the SG is to move toward 100% electricity generation from RERs, i.e., toward a 100% renewable grid. However, the disparate, intermittent, and typically widely geographically distributed nature of RERs complicates the integration of RERs into the SG. Moreover, individual RERs have generally lower capacity than conventional fossil fuel-based plants. Hence, the standards must be created and followed for the integration of RERs into the SG. A significant challenge associated with SGs is the integration of renewable power generation. The main challenge is to develop a sustainable solution

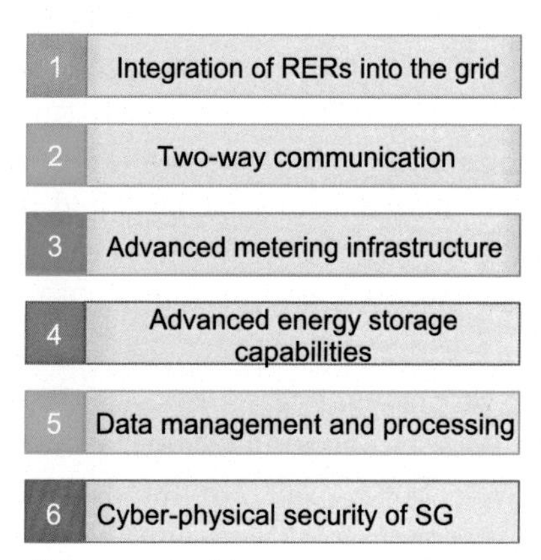

Fig. 2 Salient features of the smart grid.

space that can handle RERs, new pricing schemes, communication methodologies, optimal controls, real-time monitoring, and uncertainty issues [7].

In the SG system, sustainability requires the use of RERs, communication technologies require to improve environmental security as well as the ability to minimize the cost of the system's capacity and planning. For the modern SG, several computational technologies are required to design and study the long-term and short-term energy independence based on RERs [8, 9]. The current and conventional models are inadequate for handling the RERs based networks because of the randomness and the stochastic nature of the system. Several techniques like optimal allocation and control decisions minimize the losses and operating costs of renewable energy penetration into the grid and thus, increase the efficiency and sustainability of SGs. Hence, with the help of probability density functions, the RERs can be modeled, and the latest techniques such as stochastic programming, mixed integer programming, heuristic method, evolutionary programming, and adaptive dynamic programming [10–12]. The increasing use of RERs into the SG will be an important feature of future distribution systems.

2.2 Two-way communication

Communication is a process of transmitting, passing, and sharing ideas, opinions, facts, values, etc. from one person to another or one organization to another. It involves at least one sender, a message, a recipient, and the medium that transmits a message from the sender to the recipient. When the recipient gets the message and sends back a response to the sender. This model is called two-way communication, which is essential in the real-world. A smart grid uses two-way communication, cyber-secure communication technologies, and computational intelligence in an integrated manner across the entire spectrum of the energy system from the generation to the endpoints of consumption. So, communication infrastructure needs to be in place between the generating facilities and the system operator, electricity market and the transmission system [11]. Various smart devices like smart meters, smart sensors, and many other electronic appliances that collect information on various power system parameters and transmit it to control and monitor the equipment. These smart devices transmit the information over a two-way communication path. The communication medium may be power line carrier communications, radio frequency communications, Internet though WiFi networks, cellular communication, satellite, etc. The pictorial representation of two-way communication used in the SG is as shown in Fig. 3.

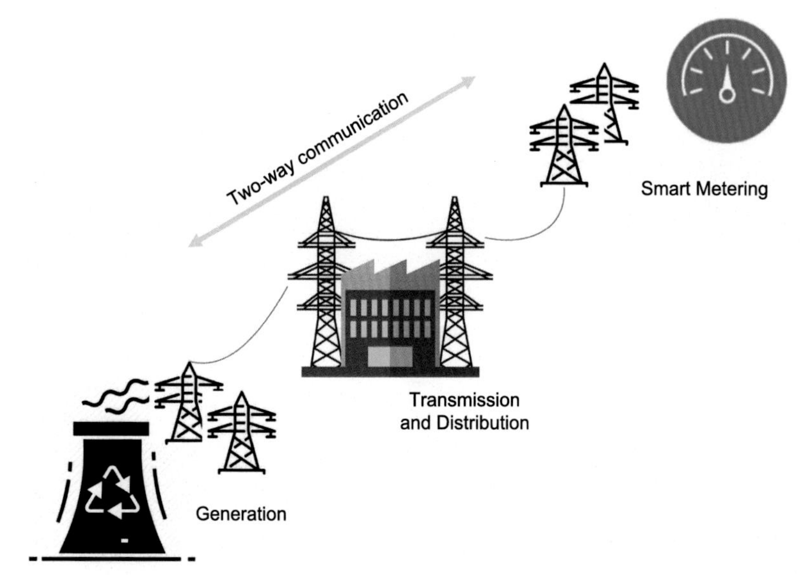

Fig. 3 Two-way communication.

The real-time information received from the smart devices is helpful for the system operator to predict, diagnose, and mitigate the issues that may damage the system. Moreover, with this new technology, the end-users will have more control over their energy consumption and cost. A strong and powerful communication network is very important for the reliable operation of a power grid.

2.3 Advanced metering infrastructure

Smart metering systems are considered as the next-generation power measurement systems, which are a revolutionary version of existing electric grids [13]. AMI is an integrated system of smart meters, data management systems and communication networks that enable two-way communication between the utilities and the customers. Hence, by using the AMI system, the issuance or communication of commands or the price signal from utility to a meter or load controlling devices are also possible. AMI system is a a platform that supports measurement, collection, and analysis of energy usage and communicates the same with the energy metering device [14]. The vital element of AMI system is smart meters that support various functions such as measuring electricity consumption of customers on different time-intervals, monitoring the on/off status of electricity, and measuring the voltage levels. The smart meters communicate these data to the utility companies for processing, analysis, and billing.

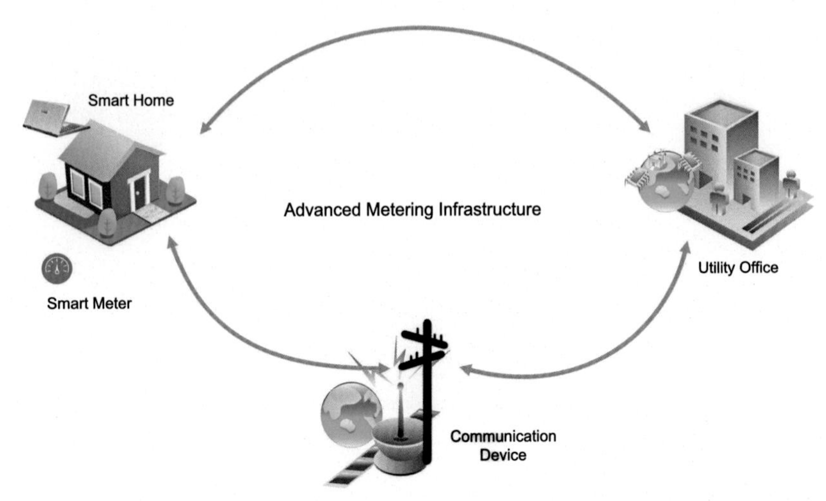

Fig. 4 Advanced metering infrastructure.

Various functionalities of smart meters include quantitative measurement, communication, control and calibration, power management, synchronization, and display. The pictorial representation of AMI is as shown in Fig. 4.

AMI system data integration with other information and management systems, including the geographic information and outage management systems enable the utilities to create detailed outage maps; this helps the end-users to know about the service restoration progress.

2.4 Advanced energy storage capabilities

Efficient and innovative storage technologies are required for the SG, especially when the RERs are integrated into the system. Most of the existing storage resources are hydro and pumped storage. However, the potential growth for these resources is slower than the need for the storage of growing net demand variability presented by RERs. Various storage technologies are emerging to fill this gap and several research projects are going on such as flywheels, supercapacitors, pumped hydro storage, storage batteries, etc. Storage integration also helps to improve the system reliability and removes the requirement to use electric power.

SG is one of the major challenges of the energy sector for both demand and supply in smart communities. Grid-connected energy storage systems play an important role in the development of SG. Energy storage can be useful in many applications such as voltage support, transmission and distribution, improved power quality, etc. Further, depending on their characteristics such as power rating and energy storage capacity, different storage technologies can be more effective. Hence, they are more suited for applications where the speed of response and power levels are favored like frequency regulation. So, to fully assess the operational value of energy storage service across technologies, there is a need for a operational value index of energy storage in electricity market operations [15].

2.5 Data management and processing

Data management and processing mainly involve four stages, which are creation, collection, management, and utilization of data. The description of the four stages is described as follows.

1. *Creation of data*: Real-time devices and sensors are used to capture and create the data for data management and processing.
2. *Collection of data*: Several smart devices and equipment are used to collect information from various sources and transmit this information to business applications via Gateways/routers using the Internet.
3. *Management of data*: Different types of software and analysis tools are used to monitor devices and collect data.
4. *Utilization of data*: Analysis, scrutinization, and software applications that add value to the information.

The amount of data is growing exponentially, and data management is an important necessity. So, to manage and control the bulk of receiving data, the following steps need to be followed.

- Recognize smart data classes and characteristics to manage and control a large amount of data.
- Consider distributed data and analytical architecture.
- Design data structure and architecture to match different types of data classes.
- Use of new tools and software that involve complex event processors for handling new classes of data.
- Develop new business processes into the transformation plans.

2.6 Cyber-physical security of SG

The term of cyber-physical systems (CPSs), coined in 2006 by the U.S. National Science Foundation, describes a connection between embedded systems and the physical world is as shown in Fig. 5. SG integrate the physical systems and cyber systems (sensors, ICT, and advanced technologies), and exhibit typical characteristics of cyber-physical system [16], which are described as follows.

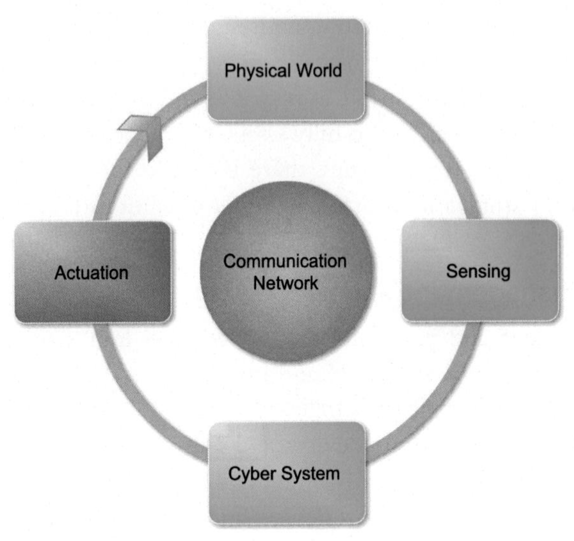

Fig. 5 Cyber-physical system.

- Integration of real and virtual worlds in a dynamic environment where situations are only controlled by a cyber-physical system that helps to adjust and control the simulation models perform in the future.
- A real-time parallel computation and distributed information processing of big data required to help deliver timely decisions for SG operations across transient, distribution, and scheduling layers through the cyber-physical system.
- Dynamic connections and interactions between components in both physical and cyber systems through communication networks where timely responses are important for dynamic cooperation.
- Self-adaption, self-organization, and self-learning by which the CPS can respond to faults, attacks, and emergencies that enable SG resilience and secure energy supply.

The cyber system is integrated with the physical power system for making the SG more energy-efficient and modernized. Although the adoption of the cyber system has made the grid more effective, it has introduced cyber-attack issues that are critical for communication infrastructure. Due to the cyber-attacks, SG may face operational failures and loss of synchronization. This operational failure may damage the critical power system that may interrupt the power supply and make the system unstable. So, the cybersecurity of power grid, encompassing attack prevention, detection, mitigation, and resilience, are among the most important research and development requirements for the emerging SG. The importance of physical and cybersecurity of SG have been highlighted by the recent intrusion incidents around the world. Various types of sensors, instrumentation, computational methodologies, and security controls have been developed to enhance the physical and cybersecurity of SG [17].

3. Applications of blockchain in smart grid

Fig. 6 shows the important applications of blockchain in the SG [18]. Here, we focus on five important application areas in SG where blockchain technology has been extensively researched. Each of these application areas is discussed below.

3.1 Peer-to-peer energy trading infrastructure

A blockchain-based trading infrastructure offers a decentralized platform that enables P2P secure energy trading between consumers and prosumers. The identity privacy and security of transactions is higher in the decentralized platform as compared to the traditional system [19]. Various aspects of P2P energy trading using blockchain are discussed below.

3.2 Blockchain architecture on P2P energy trading

Fig. 7 shows an overview of blockchain architecture on the P2P energy trading in SG environment comprising of a single cluster is presented. This cluster consists of a smart meter, electric appliances, buildings, SHs, and smart devices that communicate amongst themselves. To handle security in the SG ecosystem, the blockchain technology is used, which is based on the concept of cryptocurrency system. Blockchaining is responsible to provide secrecy and privacy to the users using a cryptography system, which authenticates the communication between the nodes.

In such a scenario, there are two types of nodes viz. normal and miner nodes. A miner node is used for authenticating and authorizing the transactions occurring in the network. On the contrary, the normal nodes work as an authority to coordinate and validate the transactions. The communication between these nodes inside a blockchain is secured using a cryptographic system, which provides integrity, security, and confidentiality. During the communication process in the blockchain technique, a consensus agreement is established for the transmission of information between the nodes. It is to be noted that the transactional data is added to the content of the block only if it is deemed secure and confidential by the miner nodes. Once the transactions are added into the block, there

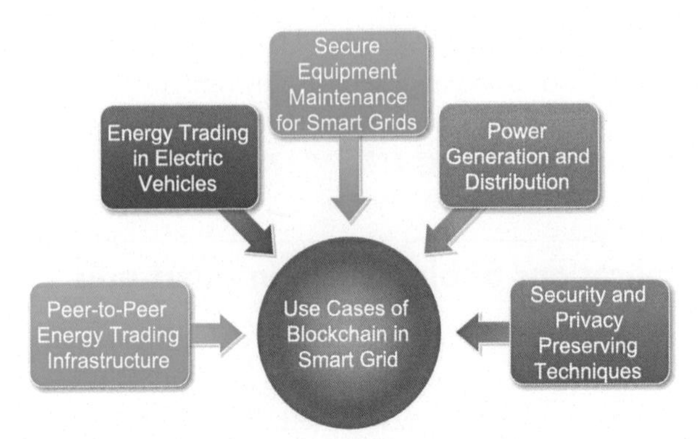

Fig. 6 Applications of Blockchain in smart grid.

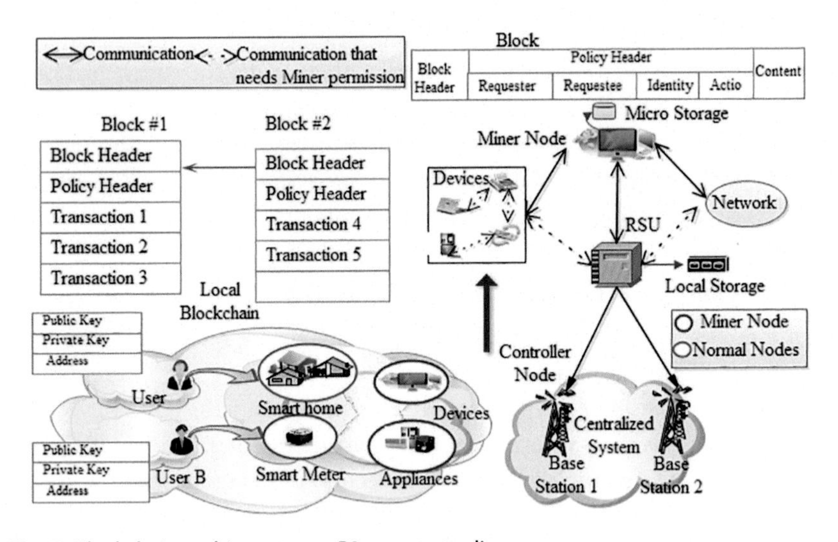

Fig. 7 Blockchain architecture on P2 energy trading.

is no mechanism to delete, update, or modify the data from the block, i.e., it is immutable. Once a block is full, it is pushed into the blockchain and a mining process is performed on it. The mining process is performed by the miner nodes, which provides authentication to the blockchain. The functionality and working of miner nodes well as normal nodes is described are detail as follows.

- *Miner node*: A miner node is a central entity between devices and the network. It is used for authenticating, authorizing, and auditing the transactions happening in the network. The miner node has its microstorage where devices can store their data temporarily. It communicates with the controller node through remote switching units (RSUs), which are used to pass all the transactions in the network. The transactional data which is passed through the controller node is stored in its local storage and it is authenticated by the miner node. Each miner node consists of various blocks and a ledger. A ledger is used to store the transaction information at any given instant of time. After the authentication of a transaction, the transactional data is appended into the blocks.

- *Normal node*: Normal nodes contain a full copy of the complete blockchain. They are used for verification, coordination, and validation of the transactions that are authenticated by the miner nodes.

3.2.1 Transaction handling

The requests raised by the smart devices deployed in the SHs are called transactions. There are two type of transactions in the SHs such as store and access transactions. These transactions are communicated through various nodes in the network to establish the consensus agreement for secure communication. There are two type of keys that are used in the blockchain technology, the requester key and the requestee key. The requester sends a request to the cloud storage for accessing the data through a *requester key*. The *requestee public key* is used by the cloud storage for granting the access permissions to the requesting SHs. The transactions that are passed through the network in the SHs are basically stored in a local blockchain. The detailed description about both types of transactions is provided as below.

- *Store transaction*: It is used to store data on the cloud storage. For example, storing the energy consumption data of smart meters on the utility server for billing. In this case, the data is stored on the cloud storage with different block IDs and a hash. No two data blocks can have the same block ID with same hash value. Thus, if the data is stored at cloud successfully, then the user is deemed authentic with the given block ID number and a hash value. Once the data is stored on the cloud storage, it can be accessed by the data owner using a public key as shown in Fig. 8.

  The procedure which is to be followed in order to store the data onto the cloud storage is discussed as follows.

  1. Firstly, the normal node sends a request to the miner node for storing the data on the cloud.

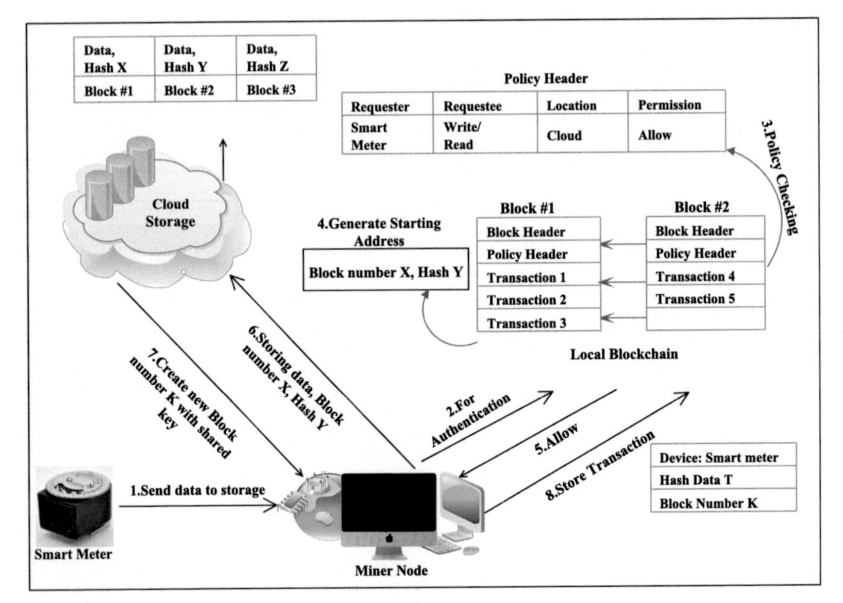

Fig. 8 Store transaction.

2. The miner node uses the local blockchain for authentication of the normal node. After the node authentication, the local blockchain generates a block ID number with a hash value, which gets updated at the miner node.
3. The miner node communicates with the cloud storage using the block ID number (with hash value) and the data which needs to be stored. The cloud storage computes a new hash value of received data and matches this value with the hash value of the data sent by the miner node. If this value is matched, then the cloud storage creates a new block ID number with shared key which is sent back to the miner node.
4. Miner node stores the received data in the local block and forwards the block ID along with the shared key to the normal node which initiated the storage request.
5. The normal node uses these entities to communicate its data to the cloud for storage purposes.

- *Access transaction*: The access transaction is used for accessing the information from the cloud storage by the data owner or the service provider (known as requestee node). The steps involved in access transaction are shown in Fig. 9 and are explained as follows.

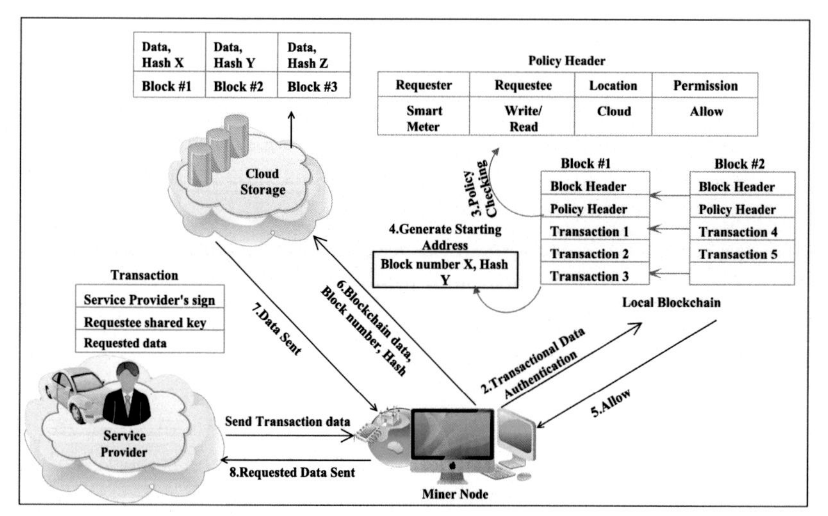

Fig. 9 Access transaction.

1. The requestee node sends the transaction data to the miner node, which consists of service provider's signature, requested data, and requestee's shared key.
2. Miner node sends the transaction data to the local blockchain for authentication, which generate a random block ID number and a hash value after authentication.
3. Miner node sends the requested data with the block ID and a hash value to the cloud storage, which matches the hash values. If they are same, then cloud storage allows the miner node to access the requested data.
4. The miner node gets the data from the cloud storage which is forwarded to the requestee node as an output after storing it in the local blockchain.

3.3 Energy trading in electric vehicles

Electric Vehicles (EVs) are eco-friendly vehicles with zero-emissions vehicles, which contribute to lower the level of population in a smart city. EVs are equipped with one or more batteries along with the communication infrastructure to support information sharing among various peers. They can be utilized to form a large network in which their battery can either store, transport, and discharge energy to the grid or it can charge energy from the grid. This bi-directional charging/discharging capability of EVs can provide many benefits such as energy management, demand response management, load balancing, stability, and several other services. To meet the demand of EVs energy, various Charging Stations are deployed at geolocated sites in a smart city. EVs have to physically connect to these Charging Stations and charge the required amount of energy after paying a specific price. Various energy service providers support such services by providing real-time energy trading platforms such as Plug Share [20]. However, real-time communication for handling payment handling has to face various security and privacy bottlenecks. In this context, blockchain is very popular for providing secure transaction management for energy trading between EVs and Charging Stations.

Fig. 10 shows a secure energy trading process between EVs and Charging Stations using blockchain technology. In this process, an EV sends a request to all the available Charging Stations for energy trading. The available Charging Stations announce the energy prices to the requesting EV according to its energy requirement. Once, the requesting EVs accepts the energy price, the P2P energy trading process between the EVs is validated by miner nodes using blockchain technology. For charging, the main task for EVs is to select the most suitable CS. In this direction, several researcher articles have been published. They have

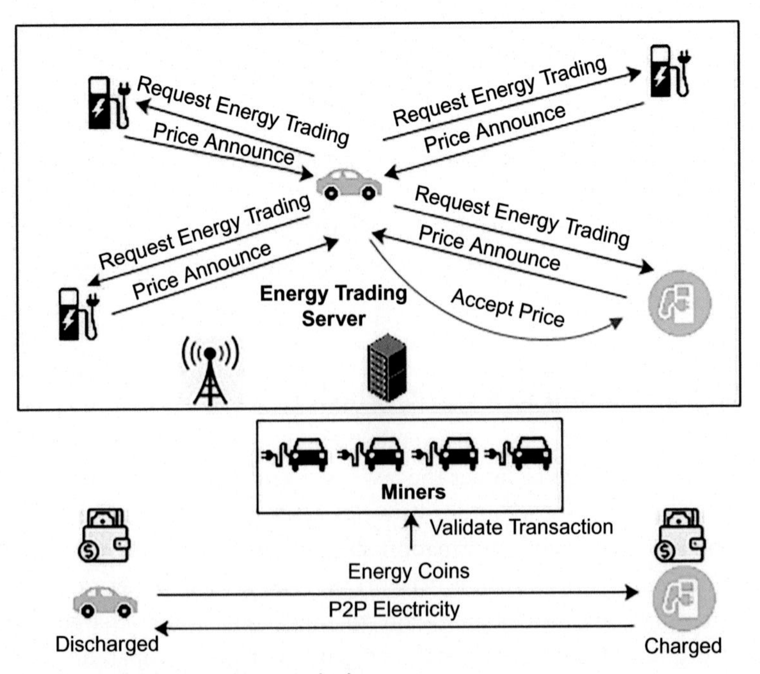

Fig. 10 Energy trading in electric vehicles.

designed an interconnection network between the smart vehicles which benefits the transportation authorities, owners of vehicles, service providers, and manufacturers. Also, communication between vehicles can be protected from hijackers, attacks, and privacy threats using blockchain technology [21, 22].

3.4 Secure equipment maintenance for smart grids

Equipment health monitoring, fault diagnosis, and maintenance are integral parts of SG. The traditional methods of diagnosis involve the necessity for technicians to visit the field for diagnosis and maintenance. It also involves an investment of manpower and other expenses at the risk of the client being unsatisfied with the services. This leads to a need for developing systems that can reduce the maintenance time and are unaffected by regional restrictions. SG encompasses several types of equipment ranging from substations to smart meters installed in homes. Such a complex smart system, in turn, requires smart equipment maintenance measures to ensure high efficiency and reliability [23]. A blockchain integrated framework, as shown in Fig. 11, can be utilized to create a platform for interaction among the vendors, diagnosis depots, and clients in a secure manner to decide upon the required maintenance measures for a mutually agreeable price

3.5 Power generation and distribution

Numerous cyber attacks on SG have been undertaken in the past where the malicious attackers have used various methods such as DoS, Data Injection Attacks, etc. to manipulate data and gain control in the grid [24]. This has resulted in complications such as regional power outages and even complete

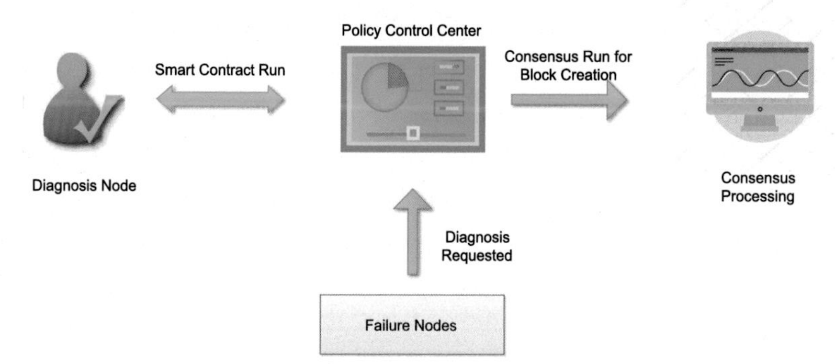

Fig. 11 Architecture for secure equipment maintenance.

blackouts [25]. Incorporating blockchain into the power generation and distribution systems help in the prevention of data manipulation that ensures data immutability.

Blockchain in the SG environment can not only secure the transactions but also protect the privacy of the consumer domain. In SG, power is generated from various types of renewable sources such as thermal power plants, solar panels, wind farms, etc. After generation, the electricity is transmitted through the transmission lines to the substations where a high amount of power voltage is step-up and step-down by the transformers. The substations transfer energy to the smart home through distribution lines. The data related to energy trading in smart homes are monitored at a centralized DC. Moreover, edge DC is connected in between the electricity consumers and centralized DC to reduce overall network latency. At the consumer's end, a miner node is located in each cluster of smart homes to provide security and privacy. The scenario of SG using blockchain technology is as shown in Fig. 12.

3.6 Security and privacy-preserving techniques

Smart meters in the SG are placed at every home to get information about electricity consumption in real-time, which is used by the utilities for various purposes [26]. By analyzing the electricity consumption profile of the users, malicious entities can track the electricity usage pattern, thereby disclosing the users' private information [27]. Although blockchain technology does not directly ensure privacy preservation, advanced cryptographic mechanisms can be incorporated for enabling data privacy. Zero-knowledge proof, ECDSA, and linkable ring signatures are some of the techniques which can protect the privacy of the devices involved in the SG.

4. Challenges for blockchain into the smart grid

The following are the list of challenges that can be occurred by blockchain incorporation into the SG.

4.1 Scalability issues

The number of transactions increases exponentially in a blockchain that needs large storage to accommodate these transactions. Currently, the Bitcoin storage has exceeded 200 GB while that for Ethereum has 1 TB. Even though a considerably high number of transactions are being carried

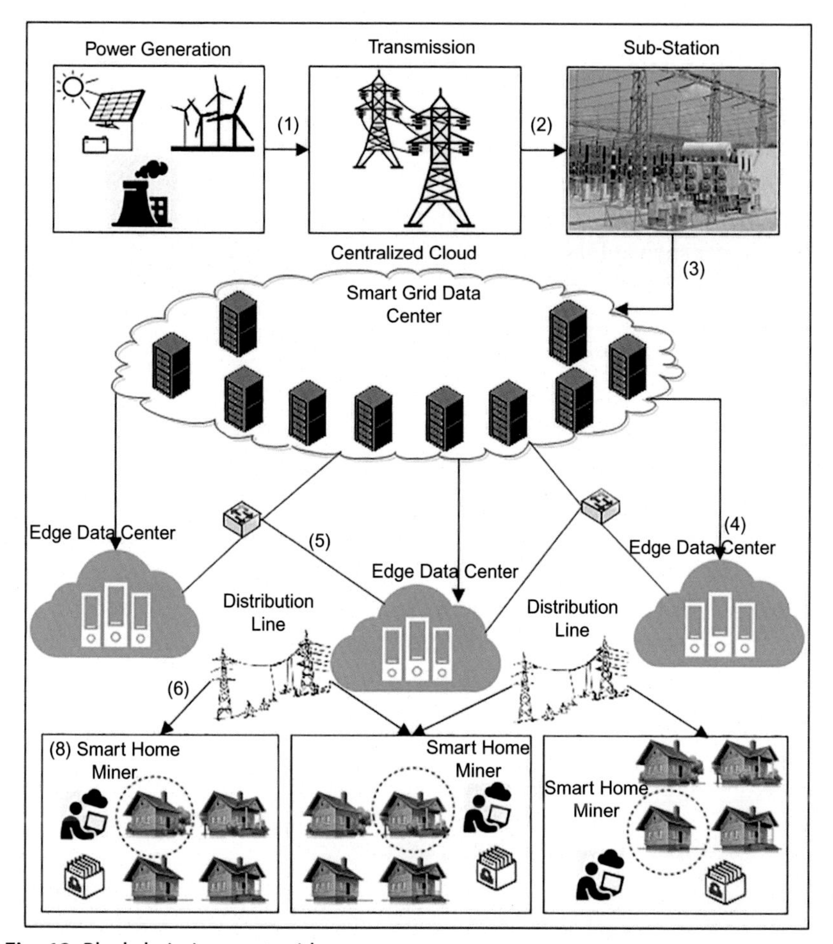

Fig. 12 Blockchain in smart grid.

out using Bitcoin, the processing rate of data into blocks is approximate 7 per second in a blockchain. Meanwhile, the average number of transactions in Ethereum is up to 15 per second. Also, Bitcoin technology requires high processing power and time for making a block in a blockchain by using the PoW consensus mechanism. According to the report in [28], Bitcoin technology spent 30 billion kWh electricity to process 30 million transactions, which accounted for about 0.13% of global electricity consumption. In the energy sector, the number of transactions per second is very high for large scale operations. Because thousands of users are simultaneously involved in the process of buying and selling energy, which creates a large overhead upon the nodes of the blockchain. This problem can be solved by replacing the PoW with the PoS or PoA algorithms. These algorithms require much less computing capacity and support much higher rates of transactions.

Further research and innovations have to be carried out to find solutions to scale up the requirements of the SG without compromising on the security aspects [29]. Research is also leading to advancement in enabling technologies such as DHT [30], IPFS [31] to potentially address the scalability and throughput challenges.

4.2 Chances of centralization

Currently, blockchain in the SG is still a growing technology and is prone to attacks from the energy assorted who might exploit it for financial advantages. The first reason for the centralization is the clustering of mining nodes for better computational capacity. By clustering the mining nodes into mining pools, there exists a risk of acquiring enough resources to plot a 51% attack. The other reason for centralization is that most of the architecture in SG is based on consortium or private blockchains to reduce the wastage of power and latency, which is associated with public blockchain architectures. Hence, a limited number of nodes are responsible to verify and validate in the public blockchains that chances of mismanagement. Therefore, strict supervision under governmental laws should be enforced before the start of the blockchain project that ensures privacy and security.

4.3 Development and infrastructure costs

Implementing blockchain architecture in the SG requires high infrastructural costs and re-framing the traditional/existing frameworks in the current electric grid. Also, re-architecting the simple meters into smart meters to upgrade the energy transactions through smart contracts, ICT-based infrastructures for blockchain operations, and other AMI and software for the development of the whole platform. Such high infrastructure costs may divert the grid operators from the incorporation of blockchain into the grid structure. The current infrastructure of the grid has been adopted after years of research and development, which yields optimal results with much less overall expenditure. For example, the current electric grid communication system employs telemetry technology, which is more mature as well as much less expensive in comparison to blockchain technology.

4.4 Legal and regulatory support

The regulatory bodies do support the active participation of users in the energy market, and the formation of community energy structures.

However, when it comes to natural changes in the main power grid architecture, the existing grid system does not support energy trading from prosumers to consumers. Also, it does not approve the distributed ledger technology into the architectural framework. So, new types of contracts and new changes have to be developed for the P2P energy trading that supports such type of services. These matters are heavily regulated in an existing grid system. For these reasons, it is very challenging to adopt the blockchain technology into the main grid and has proven its worth in the formation of microgrids without the amendment of the legal structure's framework.

5. Initiatives on smart grid in India

India is developing various SG projects, such as "National Smart Grid Mission," "Smart Grid Vision and Roadmap for India," "Nehru National Solar Mission," etc. to enhance the reliability and performance of the grid. These projects support economical, technical, and social development of the country. During the last decade, the installation of RERs has grown at an annual rate of 25%. POWERGRID Corporation of India has taken an initiative to develop a SG pilot project at Puducherry. The objective of this project is to determine the technology effectiveness, policy advancement and regulatory framework for cost effective and energy-efficient design. It provides input for standardization and interoperability plan of various methods and technologies like net metering, EVs deployment with charging through renewables, etc. [32].

References

[1] J. Gao, Y. Xiao, J. Liu, W. Liang, C.L.P. Chen, A survey of communication/networking in smart grids, Futur. Gener. Comput. Syst. 28 (2) (2012) 391–404.

[2] G.S. Aujla, M. Singh, A. Bose, N. Kumar, G. Han, R. Buyya, BlockSDN: blockchain-as-a-service for software defined networking in smart city applications, IEEE Network 34 (2) (2020) 83–91.

[3] C. Lin, D. He, N. Kumar, X. Huang, P. Vijaykumar, K.-K.R. Choo, HomeChain: a blockchain-based secure mutual authentication system for smart homes, IEEE Internet Things J. 2 (2019) 818–829.

[4] P.K. Sharma, N. Kumar, J.H. Park, Blockchain-based distributed framework for automotive industry in a smart city, IEEE Trans. Ind. Inf. 15 (7) (2018) 4197–4205.

[5] N. Kumar, A.V. Vasilakos, J.J.P.C. Rodrigues, A multi-tenant cloud-based DC nano grid for self-sustained smart buildings in smart cities, IEEE Commun. Mag. 55 (3) (2017) 14–21.

[6] N. Komninos, E. Philippou, A. Pitsillides, Survey in smart grid and smart home security: issues, challenges and countermeasures, IEEE Commun. Surv. Tutorials 16 (4) (2014) 1933–1954.

[7] G. Bansal, A. Dua, G.S. Aujla, M. Singh, N. Kumar, SmartChain: a smart and scalable blockchain consortium for smart grid systems, in: 2019 IEEE International Conference on Communications Workshops (ICC Workshops), IEEE, 2019, pp. 1–6.
[8] C.P. Vineetha, C.A. Babu, Smart grid challenges, issues and solutions, in: 2014 International Conference on Intelligent Green Building and Smart Grid (IGBSG), IEEE, 2014, pp. 1–4.
[9] A. Bari, J. Jiang, W. Saad, A. Jaekel, Challenges in the smart grid applications: an overview, Int. J. Distrib. Sens. Netw. 10 (2) (2014) 974682.
[10] S.S. Reddy, J.A. Momoh, Realistic and transparent optimum scheduling strategy for hybrid power system, IEEE Trans. Smart Grid 6 (6) (2015) 3114–3125.
[11] J. Zheng, K.O.U. Yanni, L.I. Mengshi, W.U. Qinghua, Stochastic optimization of cost-risk for integrated energy system considering wind and solar power correlated, J. Modern Power Syst. Clean Energy 7 (6) (2019) 1472–1483.
[12] S.S. Reddy, V. Sandeep, C.-M. Jung, Review of stochastic optimization methods for smart grid, Front. Energy 11 (2) (2017) 197–209.
[13] K. Sharma, L.M. Saini, Performance analysis of smart metering for smart grid: an overview, Renew. Sustain. Energy Rev. 49 (2015) 720–735.
[14] R.R. Mohassel, A. Fung, F. Mohammadi, K. Raahemifar, A survey on advanced metering infrastructure, Int. J. Electr. Power Energy Syst. 63 (2014) 473–484.
[15] S.S. Reddy, C.-M. Jung, Overview of energy storage technologies: a techno-economic comparison, Int. J. Appl. Eng. Res. 12 (22) (2017) 12872–12879.
[16] J. Zhao, F. Wen, Y. Xue, X. Li, Z. Dong, Cyber physical power systems: architecture, implementation techniques and challenges, Dianli Xitong Zidonghua(Automation of Electric Power Systems) 34 (16) (2010) 1–7.
[17] J. Xie, A. Stefanov, C.-C. Liu, Physical and cyber security in a smart grid environment, Wiley Interdiscip. Rev. Energy Environ. 5 (5) (2016) 519–542.
[18] S. Aggarwal, R. Chaudhary, G.S. Aujla, N. Kumar, K.-K.R. Choo, A.Y. Zomaya, Blockchain for smart communities: applications, challenges and opportunities, J. Netw. Comput. Appl. 144 (2019) 13–48.
[19] S. Aggarwal, R. Chaudhary, G.S. Aujla, A. Jindal, A. Dua, N. Kumar, Energychain: Enabling energy trading for smart homes using blockchains in smart grid ecosystem, in: Proceedings of the 1st ACM MobiHoc Workshop on Networking and Cybersecurity for Smart Cities, 2018, pp. 1–6.
[20] W. Yuan, J. Huang, Y.J.A. Zhang, Competitive charging station pricing for plug-in electric vehicles, IEEE Trans. Smart Grid 8 (2) (2015) 627–639.
[21] M. Pustišek, A. Kos, U. Sedlar, Blockchain based autonomous selection of electric vehicle charging station, in: 2016 International Conference on Identification, Information and Knowledge in the Internet of Things (IIKI), IEEE, 2016, pp. 217–222.
[22] A. Dorri, M. Steger, S.S. Kanhere, R. Jurdak, Blockchain: a distributed solution to automotive security and privacy, IEEE Commun. Mag. 55 (12) (2017) 119–125.
[23] S. Huh, S. Cho, S. Kim, Managing IoT devices using blockchain platform, in: 2017 19th International Conference on Advanced Communication Technology (ICACT), IEEE, 2017, pp. 464–467.
[24] V. Hassija, V. Chamola, V. Saxena, D. Jain, P. Goyal, B. Sikdar, A survey on IoT security: application areas, security threats, and solution architectures, IEEE Access 7 (2019) 82721–82743.
[25] K. Singh, S.C. Choube, Using blockchain against cyber attacks on smart grids, in: 2018 IEEE International Students' Conference on Electrical, Electronics and Computer Science (SCEECS), IEEE, 2018, pp. 1–4.
[26] A. Ipakchi, F. Albuyeh, Grid of the future, IEEE Power Energy Mag. 7 (2) (2009) 52–62.
[27] T. Alladi, V. Chamola, B. Sikdar, K.-K.R. Choo, Consumer iot: security vulnerability case studies and solutions, IEEE Consumer Electron. Mag. 9 (2) (2020) 17–25.
[28] Index, Bitcoin Energy Consumption, Digiconomist.(2018), 2019.

[29] M. Andoni, V. Robu, D. Flynn, S. Abram, D. Geach, D. Jenkins, P. McCallum, A. Peacock, Blockchain technology in the energy sector: a systematic review of challenges and opportunities, Renew. Sustain. Energy Rev. 100 (2019) 143–174.
[30] G. Zyskind, O. Nathan, Decentralizing privacy: using blockchain to protect personal data, in: 2015 IEEE Security and Privacy Workshops, IEEE, 2015, pp. 180–184.
[31] M. Klems, J. Eberhardt, S. Tai, S. Härtlein, S. Buchholz, A. Tidjani, Trustless intermediation in blockchain-based decentralized service marketplaces, in: International Conference on Service-Oriented Computing, Springer, 2017, pp. 731–739.
[32] I.S. Jha, S. Sen, R. Kumar, Smart grid development in India— a case study, in: 2014 Eighteenth National Power Systems Conference (NPSC), IEEE, 2014, pp. 1–6.

About the authors

Shubhani Aggarwal is pursuing a PhD from Thapar Institute of Engineering and Technology (Deemed to be University), Patiala, Punjab, India. She received the BTech degree in Computer Science and Engineering from Punjabi University, Patiala, Punjab, India, in 2015, and the ME degree in Computer Science from Panjab University, Chandigarh, India, in 2017. She has many research interests in the area of Blockchain, cryptography, Internet of Drones, and information security.

Neeraj Kumar received his PhD in CSE from Shri Mata Vaishno Devi University, Katra (Jammu and Kashmir), India in 2009, and was a postdoctoral research fellow in Coventry University, Coventry, United Kingdom. He is working as a professor in the Department of Computer Science and Engineering, Thapar Institute of Engineering and Technology (Deemed to be University), Patiala, Punjab, India. He has published more than 400 technical research papers in top-cited journals such as IEEE TKDE, IEEE TIE, IEEE TDSC, IEEE TITS, IEEE TCE, IEEE TII, IEEE TVT, IEEE ITS, IEEE SG, IEEE Network, IEEE Communications, IEEE WC, IEEE IoTJ, IEEE SJ, Computer Networks, Information Sciences, FGCS, JNCA, JPDC, and ComCom. He has guided many research scholars leading to PhD and ME/

MTech. His research is supported by funding from UGC, DST, CSIR, and TCS. His research areas are Network Management, IoT, Big Data Analytics, Deep learning, and cyber-security. He is serving as an editor of the following journals of repute: ACM Computing Survey, ACM·IEEE Transactions on Sustainable Computing, IEEE·IEEE Systems Journal, IEEE·IEEE Network Magazine, IEE·IEEE Communication Magazine, IEE·Journal of Networks and Computer Applications, Elsevier · Computer Communication, Elsevier·International Journal of Communication Systems, Wiley. Also, he has been a guest editor of various International Journals of repute such as IEEE Access, IEEE ITS, Elsevier CEE, IEEE Communication Magazine, IEEE Network Magazine, Computer Networks, Elsevier, Future Generation Computer Systems, Elsevier, Journal of Medical Systems, Springer, Computer and Electrical Engineering, Elsevier, Mobile Information Systems, International Journal of Ad hoc, and Ubiquitous Computing, Telecommunication Systems, Springer, and Journal of Supercomputing, Springer. He has also edited/authored 10 books with International/National Publishers like IET, Springer, Elsevier, CRC, Security and Privacy of Electronic Healthcare Records: Concepts, paradigms and solutions (ISBN-13: 978-1-78561-898-7), Machine Learning for Cognitive IoT, CRC Press, Blockchain, Big Data and IoT, Blockchain Technologies across industrial vertical, Elsevier, Multimedia Big Data Computing for IoT Applications: Concepts, Paradigms and Solutions (ISBN: 978-981-13-8759-3), Proceedings of First International Conference on Computing, Communications, and Cyber-Security (IC4S 2019) (ISBN 978-981-15-3369-3). One of the edited text-book entitled "Multimedia Big Data Computing for IoT Applications: Concepts, Paradigms, and Solutions" published in Springer in 2019 is having 3.5 million downloads till June 06, 2020. It attracts the attention of the researchers across the globe (https://www.springer.com/in/book/9789811387586). He has been a workshop chair at IEEE Globecom 2018 and IEEE ICC 2019 and TPC Chair and member for various International conferences such as IEEE MASS 2020 and IEEE MSN 2020. He is a senior member of the IEEE. He has more than 12,321 citations to his credit with current h-index of 60 (September 2020). He has won the best papers award from IEEE Systems Journal and ICC 2018, Kansas-city in 2018. He has been listed in the highly cited researcher of 2019 list of web of science (WoS). In India, he is listed in top 10 position among highly cited researchers list. He is an adjunct professor at Asia University, Taiwan, King Abdul Aziz University, Jeddah, Saudi Arabia and Charles Darwin University, Australia.

Healthcare system ☆

Shubhani Aggarwal and Neeraj Kumar
Thapar Institute of Engineering & Technology, Patiala, Punjab, India

Contents

Abstract

In this chapter, we will be going through the blockchain healthcare use cases and blockchain healthcare examples. These use cases will help us better understand the impact of blockchain on healthcare. The blockchain healthcare applications will come equipped with the latest technology and solves the challenges that the healthcare industry is going through. The focus should be on improving the quality of healthcare and ensuring that it takes a patient-centric approach rather than maximizing profit. There is no denying that the blockchain impact on healthcare will be substantial in the future. Here, we are trying to learn blockchain healthcare use cases by sharing blockchain healthcare examples and applications.

☆ Applications.

Advances in Computers, Volume 121
ISSN 0065-2458
https://doi.org/10.1016/bs.adcom.2020.08.024

Chapter points

- In this chapter, we discuss the role of blockchain in the healthcare system.
- Here, we discuss the changes made in the traditional healthcare system after using blockchain technology.

The demand for a revolution in technology or any sector is necessary. The healthcare industry needs to make changes and provide itself with the best possible services and the solution. The technology, which will revolutionize the healthcare system, is *Blockchain* [1]. This technology can change how physicians access patient's data, how clinical research will take place, and other aspects of the healthcare system. Before, we go deep into the blockchain healthcare uses cases and applications, we first need to understand the existing healthcare system [2].

1. Current healthcare system

The current healthcare system relies on the interaction between patient and physician and works on limited data. The limited aspect of the healthcare system keeps failing to take advantage of the data. Also, the current process of healthcare is time-consuming and tiring, which results in the noneffective handling of the patient. The current healthcare system faces some challenges, which are described as follows.

1.1 Supply chain and drug counterfeit

The adverse effect of the supply chain leads to manipulated prices, delays in the supply of drugs, and much more. Also, drug counterfeiting is a big problem as it leads to massive losses for the healthcare industry. The current/existing supply chain systems are not capable of keeping the counterfeit drugs at bay, which leads to a massive $200 million loss for the healthcare industry.

1.2 Data segmentation and no proper management

Data maintenance in the current healthcare system is another aspect where this system works not well. Critical information and data of the patients are distributed all over the systems and departments, which leads to issues and delays to get the right information at the right time. So, due to the lack of critical data availability, many healthcare systems fail to provide the necessary treatment to the patients. Even the patients are not in full control as they have too many reports from different physicians which are hard to manage in a single place. The management system is also impacted as the system is not equipped with the right tools for a smooth process.

1.3 Healthcare security and data storage

The major problem in the healthcare system is to misuse healthcare data. The data is mostly sold to third party companies. There is more to the data misuse issue that can be understood by the following points.

- Not all clinical reports are reported in the government agency that means data loss.
- The healthcare reports are filled with misleading information and errors.
- Healthcare organizations have lost approx. $380 per record in data discontinuity.

The reason for poor data maintenance in healthcare is the use of outdated systems, which are not connected with most of the healthcare systems and applications.

2. Blockchain for healthcare system

The healthcare system consists of a group of hospitals that are sponsored, managed, owned by a central authority. It provides a medical care center where all types of medical services are provided to the end users [3, 4]. But, till now the healthcare system is managed and controlled by a central authority, which can lead to a single point of failure. So, to provide decentralization in the healthcare system for security, blockchain technology is the best solution [5].

Blockchain has been used in the healthcare system in various domains like information management, data sharing, data storing, and access control systems. It helps in providing security to the patient's personal information in a decentralized manner. The steps used to secure the healthcare system using blockchain technology are shown in Fig. 1 and are discussed below.

- Initially, IoT sensors monitor and collect the patient's health information such as temperature, blood pressure, respiratory rate, heart rate, blood sugar, pulse rate, etc.
- Data collected from sensors is monitored by an administrator and reports are generated by the system.
- After receiving the patient's report, doctors analyze and recommend the required treatment.

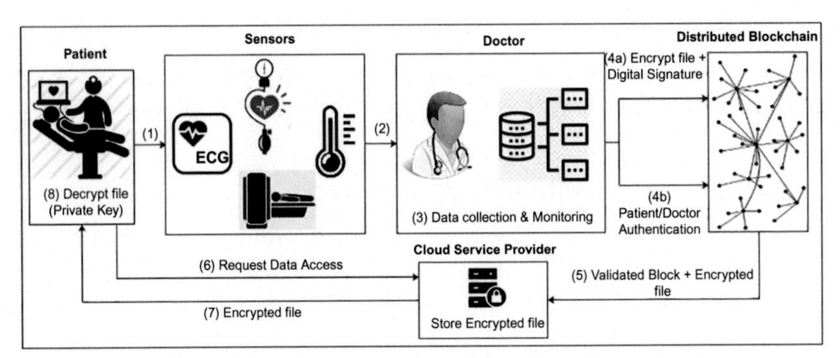

Fig. 1 Blockchain in healthcare system.

- Doctors even share the treatment reports with the distributed database in an encrypted format, which is then validated using the blockchain technology.
- The patient submits the request to the cloud service provider (CSP) for accessing the treatment record.
- After the validation at CSP, the patient receives the encrypted file from CSP.
- The encrypted file is decrypted by the patient's private key for accessing the treatment records.

3. Blockchain can solve the healthcare issues

Blockchain technology can completely change the healthcare system. It can help the healthcare industry to overcome the aforementioned challenges. It can help improve universal access, data integrity, security, traceability, and interoperability. Blockchain applications hold the key to improve the current healthcare system. With this technology, multiple healthcare systems can come together and exchange data with each other using a distributed framework [6]. The following challenges in the healthcare system, which can be solved by blockchain.

3.1 Interoperability

The biggest advantage of using blockchain in the healthcare system is interoperability. The current multiple healthcare systems use different types of protocols and standards to ensure data that can be accessed across hospitals. However, this can lead to complexity when it comes to integrating into new systems or platforms. So, blockchain can solve all the interoperability issues as it acts as a decentralized database. The data can be accessed to the APIs with a focus on the standard data format. This technology can also work well with the current platforms and protocols, which are used to access and store data in the healthcare system.

3.2 Security

The current healthcare industry suffers from data leaks that can cost millions of dollars. Data tampering and theft is becoming a big concern and should be properly dealt with, especially in the healthcare system. Blockchain utilizes data encryption using private keys, and only the receiver can decrypt the content using his key. So, we can secure the medical data using blockchain technology rather than the currently available solutions out there [7].

3.3 Maintenance cost

Maintenance cost is also an integral problem with the current healthcare systems. The current system requires maintenance across different operations and needs a specialized team to ensure that all the functions and operations are run in synchronization. But, blockchain does not suffer from this problem as it is a distributed decentralized network. The data is distributed across the network which means that there is no single point of failure. If a node goes down, data can be fetched from other nodes as there are multiple copies of the data on the network. Each node has its copy of the database. The public ledger concept of blockchain will help hospitals to cope with the emergencies better. Another benefit of having a public ledger on each node means that there is less transaction cost when it comes to storing or retrieving information.

3.4 Data integrity

Blockchain provides integrity, which solves a lot of problems that exist in the current healthcare industry. With this technology, data integrity can be maintained at all levels all the time by storing all the data in public ledgers. Once the data is uploaded on the blockchain, it cannot be changed or altered by malicious actors, preserving the integrity of the data. Only the patients can change the details when working with a physician or doctor. There are many case studies for blockchain in healthcare that this technology can bring the necessary integrity to protect data from being stolen or misused.

3.5 Universal access

Public blockchain provides universal access to all its users. It is not dependent on a central authority which makes universal access possible. Authorized entities can easily access data whenever they needed. The whole process can be automated with different mechanisms similar to smart contracts. In short, blockchain for medical is promising and should also be encouraged in the healthcare system.

4. Use cases of blockchain in healthcare system

Blockchain technology can solve many issues faced by banks, finance, and healthcare. Healthcare is a big industry with a lot of problems that still need to be solved. Blockchain, on the other hand, can solve the healthcare problems. There are multiple use cases for blockchain technology in the healthcare system are described as follows.

4.1 Billing and claims management

Solve.Care is a healthcare benefits administration and payments that aims to solve administrative issues for healthcare [8]. It will solve processing issues as well as billing and claims management. It is only because blockchain technology provides transparency and data integrity. It will also process payments and ensure that extra billing by entities can be eliminated. The scenario of billing and claims management using blockchain in the healthcare system is as shown in Fig. 2.

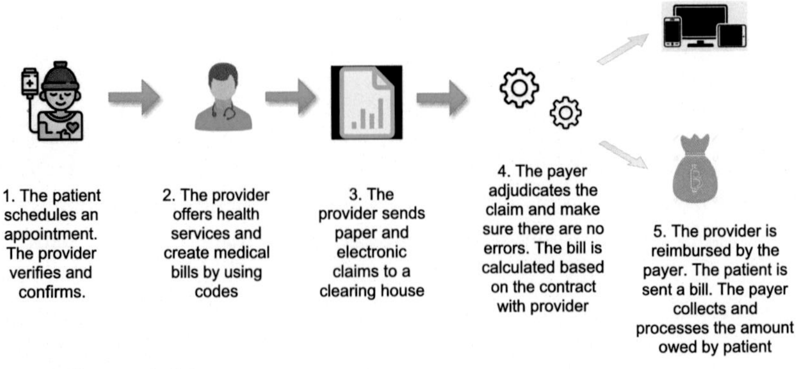

Fig. 2 Billing and claims management.

4.2 Clinical data exchange and interoperability

Another critical use of blockchain is to exchange clinical data. Medicalchain is a smart medical system, which uses blockchain technology to securely manage health records for a collaborative, smart approach to healthcare [9]. It will solve the health data exchange problems. It will create a network of data where anyone including laboratories, hospitals, doctors, nurses, etc. can access records. For accessing the data, there is a need for permission to become a part of the blockchain network. It will also address privacy issues. The scenario of clinical data exchange using blockchain in the healthcare system is as shown in Fig. 3.

4.3 Cyber security and healthcare IoT

The healthcare system is also a target for cybercriminals, especially when they can lock up the whole database for money exchanges [10]. Data theft is also a severe issue where cybercriminals steal the medical data and sell it in the market. In this context, patientory is a distributed application that builds a blockchain-powered health information exchange, which is the Health Insurance Portability and Accountability Act-based (HIPPA) compliant. It will enhance cybersecurity protocols and electronic medical records interoperability to secure important medical data. The scenario of cyber security and healthcare IoT using blockchain is as shown in Fig. 4.

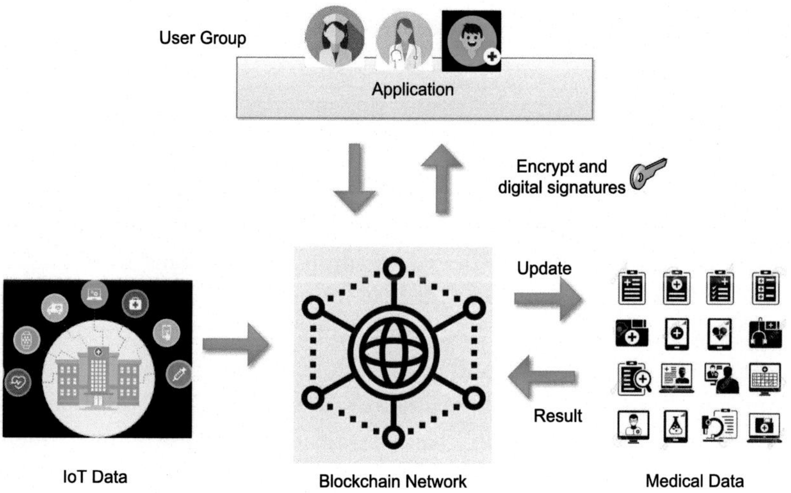

Fig. 3 Clinical data exchange using blockchain.

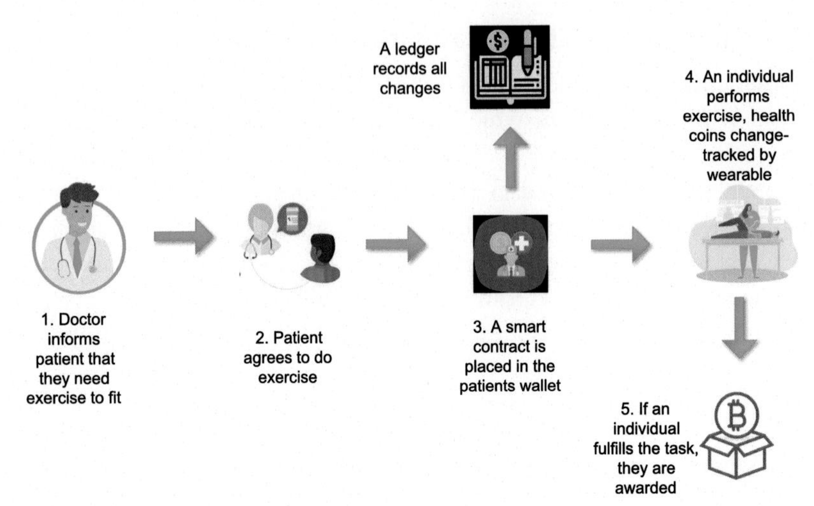

Fig. 4 Healthcare IoT using blockchain.

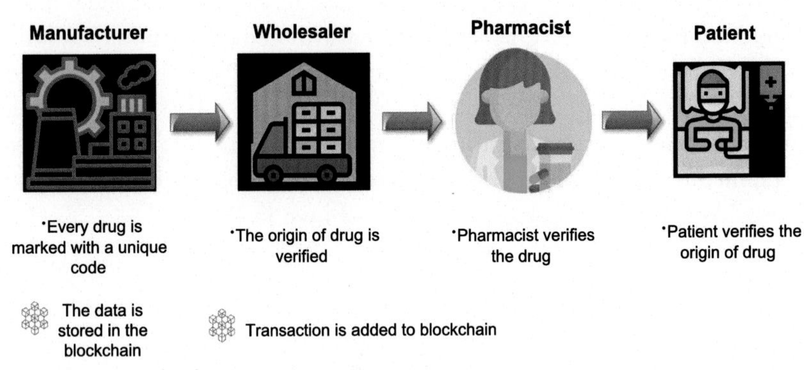

Fig. 5 Drug supply chain provenance.

4.4 Drug supply chain provenance and integrity

The supply chain is always a big challenge for the healthcare industry. With the blockchain technology, the whole drug supply chain issue can be improved. For example, Farmatrust is working for the drug supply chain in the healthcare system [11]. The project aims to eliminate the use of abusive drugs and hence save lives. It will also control and manage the transformation of counterfeit drugs in the market. They will also provide transparency for the drug supply chain. To achieve all of this, a unique identity will be combined with a company's digital supply chain. The scenario of supply chain using blockchain in the healthcare system is as shown in Fig. 5.

References

[1] P. Bhattacharya, S. Tanwar, U. Bodke, S. Tyagi, N. Kumar, BinDaaS: Blockchain-based deep-learning as-a-service in Healthcare 4.0 applications, IEEE Trans. Netw. Sci. Eng. (2019).
[2] R. Gupta, S. Tanwar, S. Tyagi, N. Kumar, M.S. Obaidat, B. Sadoun, Habits: Blockchain-based telesurgery framework for Healthcare 4.0, in: 2019 International Conference on Computer, Information and Telecommunication Systems (CITS), IEEE, 2019, pp. 1–5.
[3] G.S. Aujla, R. Chaudhary, K. Kaur, S. Garg, N. Kumar, R. Ranjan, SAFE: SDN-assisted framework for edge-cloud interplay in secure healthcare ecosystem, IEEE Trans. Ind. Inf. 15 (1) (2018) 469–480.
[4] R. Chaudhary, A. Jindal, G.S. Aujla, N. Kumar, A.K. Das, N. Saxena, Lscsh: lattice-based secure cryptosystem for smart healthcare in smart cities environment, IEEE Commun. Mag. 56 (4) (2018) 24–32.
[5] J. Vora, A. Nayyar, S. Tanwar, S. Tyagi, N. Kumar, M.S. Obaidat, J.J.P.C. Rodrigues, Bheem: a blockchain-based framework for securing electronic health records, in: 2018 IEEE Globecom Workshops (GC Wkshps), IEEE, 2018, pp. 1–6.
[6] J. Hathaliya, P. Sharma, S. Tanwar, R. Gupta, Blockchain-based remote patient monitoring in healthcare 4.0, in: 2019 IEEE 9th International Conference on Advanced Computing (IACC), IEEE, 2019, pp. 87–91.
[7] S. Challa, A.K. Das, V. Odelu, N. Kumar, S. Kumari, M.K. Khan, A.V. Vasilakos, An efficient ECC-based provably secure three-factor user authentication and key agreement protocol for wireless healthcare sensor networks, Comput. Electr. Eng. 69 (2018) 534–554.
[8] Solve.Care, Available: https://solve.care/ (accessed 11 May 2020).
[9] Medicalchain, Available: https://medicalchain.com/en/ (accessed 11 May 2020).
[10] Patientory Inc, Available: https://patientory.com/ (accessed 11 May 2020).
[11] Farmatrust, Available: https://farmatrust.io/ (accessed 11 May 2020).

About the authors

Shubhani Aggarwal is pursuing PhD from Thapar Institute of Engineering & Technology (Deemed to be University), Patiala, Punjab, India. She received the BTech degree in Computer Science and Engineering from Punjabi University, Patiala, Punjab, India, in 2015, and the ME degree in Computer Science from Panjab University, Chandigarh, India, in 2017. She has many research interests in the area of Blockchain, cryptography, Internet of Drones, and information security.

Neeraj Kumar received his PhD in CSE from Shri Mata Vaishno Devi University, Katra (Jammu and Kashmir), India in 2009, and was a postdoctoral research fellow in Coventry University, Coventry, UK. He is working as a Professor in the Department of Computer Science and Engineering, Thapar Institute of Engineering & Technology (Deemed to be University), Patiala (Punjab), India. He has published more than 400 technical research papers in top-cited journals such as IEEE TKDE, IEEE TIE, IEEE TDSC, IEEE TITS, IEEE TCE, IEEE TII, IEEE TVT, IEEE ITS, IEEE SG, IEEE Netw., IEEE Comm., IEEE WC, IEEE IoTJ, IEEE SJ, Computer Networks, Information sciences, FGCS, JNCA, JPDC, and ComCom. He has guided many research scholars leading to PhD and ME/MTech. His research is supported by funding from UGC, DST, CSIR, and TCS. His research areas are Network Management, IoT, Big Data Analytics, Deep Learning, and Cybersecurity. He is serving as editor of the following journals of repute: ACM Computing Survey, ACM·IEEE Transactions on Sustainable Computing, IEEE·IEEE Systems Journal, IEEE·IEEE Network Magazine, IEE·IEEE Communication Magazine, IEE·Journal of Networks and Computer Applications, Elsevier Computer Communication, Elsevier International Journal of Communication Systems, Wiley. Also, he has been a guest editor of various international journals of repute such as IEEE Access, IEEE ITS, Elsevier CEE, IEEE Communication Magazine, IEEE Network Magazine, Computer Networks, Elsevier, Future Generation Computer Systems, Elsevier, Journal of Medical Systems, Springer, Computer and Electrical Engineering, Elsevier, Mobile Information Systems, International Journal of Ad Hoc and Ubiquitous Computing, Telecommunication Systems, Springer, and Journal of Supercomputing, Springer. He has also edited/authored 10 books with international/national publishers like IET, Springer, Elsevier, CRC: Security and Privacy of Electronic Healthcare Records: Concepts, Paradigms and Solutions (ISBN-13: 978-1-78561-898-7), Machine Learning for Cognitive IoT, CRC Press, Blockchain, Big Data and IoT, Blockchain Technologies Across Industrial Vertical, Elsevier, Multimedia Big Data Computing for IoT Applications: Concepts, Paradigms and Solutions (ISBN: 978-981-13-8759-3), Proceedings of First

International Conference on Computing, Communications, and Cyber-Security (IC4S 2019) (ISBN 978-981-15-3369-3). One of the edited text-book entitled, "Multimedia Big Data Computing for IoT Applications: Concepts, Paradigms, and Solutions" published in Springer in 2019 is having 3.5 million downloads till June 6, 2020. It attracts attention of the researchers across the globe (https://www.springer.com/in/book/9789811387586). He has been a workshop chair at IEEE Globecom 2018 and IEEE ICC 2019 and TPC Chair and member for various international conferences such as IEEE MASS 2020 and IEEE MSN 2020. He is a senior member of the IEEE. He has more than 12,321 citations to his credit with current h-index of 60 (September 2020). He has won the best papers award from IEEE Systems Journal and ICC 2018, Kansas City in 2018. He has been listed in the highly cited researcher of 2019 list of Web of Science (WoS). In India, he is listed in top 10 position among highly cited researchers list. He is an adjunct professor at Asia University, Taiwan, King Abdul Aziz University, Jeddah, Saudi Arabia, and Charles Darwin University, Australia.

Voting system☆

Shubhani Aggarwal and Neeraj Kumar
Thapar Institute of Engineering & Technology, Patiala, Punjab, India

Contents

Abstract

This chapter aims to solve the issues of the digital voting process using blockchain technology. Firstly, introducing the problems in the current voting system then goes into a brief explanation of blockchain technology that how it is currently being used in the digital voting process. This chapter represents the digital voting process in detail and the issues they face. Here, we represent the blockchain-based platform to the voting system that ensures the privacy of the voter's vote. This platform also provides data immutability and transparency to the digital voting process.

Chapter points

- In this chapter, we discuss the role of blockchain in the voting system.
- Here, we discuss the changes made in the traditional voting system after using blockchain technology.

Democratic voting is a very crucial and serious event in any country. The most common way of voting is through a paper-based system, but this is not time to vote using a paper into the 21st century of this modern world. The digital voting process is the use of electronic devices, such as voting machines, Internet browsers, to cast voters to vote. This is also known as e-voting system or an online voting system.

The major concern of the digital voting process is the security and privacy of the votes. With such monumental decisions at stake, there can be no doubt about the system's ability to secure data and defend against

☆ Applications.

Advances in Computers, Volume 121
ISSN 0065-2458
https://doi.org/10.1016/bs.adcom.2020.08.025

potential attacks. The only solution is the blockchain technology that can be potentially solved the issues of the digital voting system. It is a form of a distributed database where records stored in the form of transactions in a block. With the use of blockchain technology, a secure and robust platform for the digital voting system can be provided.

1. Current digital voting system

Several countries are currently using digital voting systems around the world. For example, Estonia has had the electronic voting system since 2005 and in 2007 was the first country in the world, which allows the online voting system to cast the votes at their respective places. The bases of this system are the national ID card that all Estonian citizens are given. These cards contain encrypted files that identify the owner and allows the owner to carry out online and electronic activities including online banking services, digitally signing documents, access their information on government databases, and e-voting [1].

To cast a vote in an online system, the voter should enter their card number into a card reader and then access the voting website on the computer. Then, the voter enters the PIN number and an online system checks that the voter is eligible to vote or not. Once confirmed, the voter can cast the vote on election day. The voter can also use a mobile phone to cast a vote in an online voting system. When the voter submits its vote, a vote is passed through the publicly accessible server to the vote storage server where it is encrypted and stored until the online voting process is over. Then, the vote has all identifying information cleaned from it and is transferred by DVD to a vote counting server which is disconnected from all networks. This server decrypts and counts the votes and then outputs the results. Each stage of this process is logged and audited [2]. In this process, there is the possibility of risks via malware on the client-side machine that monitors the user placing their vote and then later changing the votes to a different candidate. The other possible risk is for an attacker to directly infect the servers though malware being placed on the DVDs used to set up the servers and transfer the votes [3].

The scenario of online voting system in Estonia [1] is as shown in Fig. 1.

2. Blockchain-based online voting system

The traditional and online electoral systems have to handle many issues and problems related to security, privacy, and transparency in its

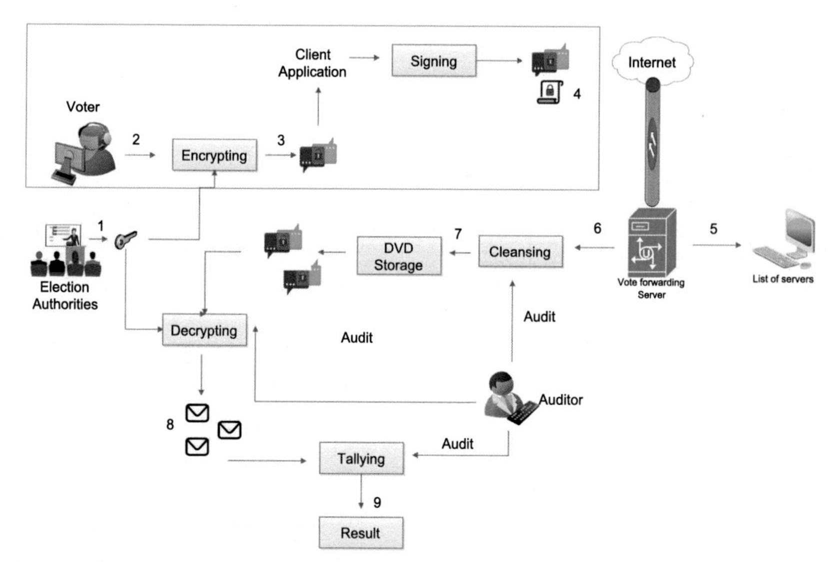

Fig. 1 Online voting system in Estonia [1].

existing manner of execution. Electoral systems use a centralized network wherein all the processing is handled by a trusted third party. However, this leads to some other major concerns in the form of security and transparency in these systems. To solve these problems in the online voting system, blockchain technology is a promising solution. It not only authenticates the voters but also provides confidentiality of the votes in online voting system. The architecture of the blockchain-based voting process system is as shown in Fig. 2. In this architecture, blockchain safeguards and secures the entire voter registration, voting, and counting processes in a distributed manner.

The usage of blockchain in the online voting system in two aspects, which is described as follows.

2.1 Record-keeping database or database manipulation

In an online voting process, the prevention of voting database manipulation is of utmost importance. Moreover, the manipulation in the result voting data is also an important concern. The use of blockchain technology with permission protocol provides a distributed record-keeping framework to an online voting system. It uses SHA-256 to compute the hash values for recording the voting results of different stations that are linked to each other. The use of digital signatures having blockchain in the voting system provides

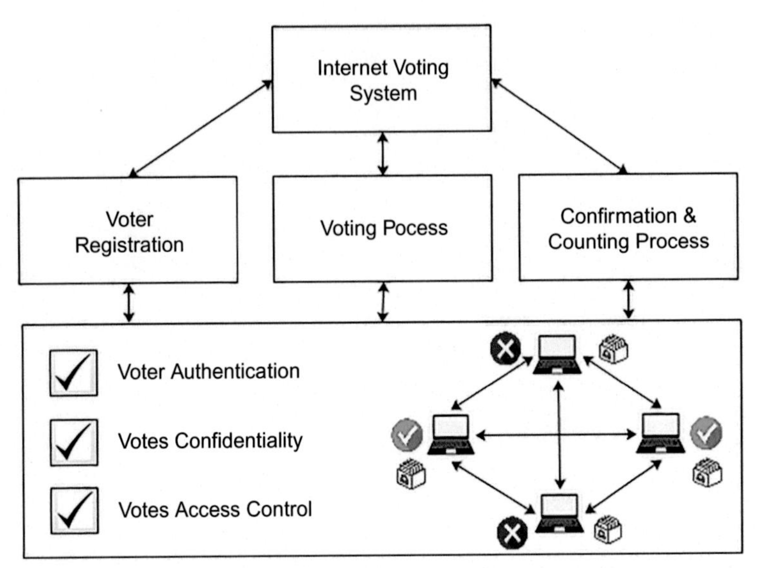

Fig. 2 Blockchain in online voting system.

reliability to the system. Hence, by integrating the blockchain in the online voting system ensures decentralized, non-interactive, self-managed, and free receipt platform for a large scale electoral process [4,5]

2.2 Tamper-proof vote counting

Vote counting in the online voting process must be tamper-proof and secure in every manner. An attacker can gain control of the counting process and manipulate the votes and alter the winning candidate, which may lead to serious obligations. The verifiability process of blockchain in the online voting system provides a reliable, verifiable, transparent, secure, and efficient platform [6].

In the end, it has been observed that the blockchain technology provides transparent, tamper-proof, and reliable results in the online voting process. To ensure a secure voting process, the private blockchain platform can be used in which nodes will behave in a restricted manner. In this network, every node cannot participate in the blockchain and there is strict authority management on data access. It would establish more trust and thereby encourage more people participation and vote in the democratic process. But in addition to this, blockchain technology is the biggest challenge for those countries which have low Internet connectivity and mobile penetration.

References

[1] Valimised, Available: https://www.valimised.ee/et (accessed 10 May 2020).
[2] E-voting is (too) secure, Available: https://www.ria.ee/en/news/e-voting-too-secure.html (accessed 10 May 2020).
[3] D. Springall, T. Finkenauer, Z. Durumeric, J. Kitcat, H. Hursti, M. MacAlpine, J.A. Halderman, Security analysis of the Estonian internet voting system, in: Proceedings of the 2014 ACM SIGSAC Conference on Computer and Communications Security, 2014, pp. 703–715.
[4] R. Hanifatunnisa, B. Rahardjo, Blockchain based e-voting recording system design, in: 2017 11th International Conference on Telecommunication Systems Services and Applications (TSSA), IEEE, 2017, pp. 1–6.
[5] B. Wang, J. Sun, Y. He, D. Pang, N. Lu, Large-scale election based on blockchain, Procedia Comput. Sci. 129 (2018) 234–237.
[6] S.H. Shaheen, M. Yousaf, M. Jalil, Temper proof data distribution for universal verifiability and accuracy in electoral process using blockchain, in: 2017 13th International Conference on Emerging Technologies (ICET), IEEE, 2017, pp. 1–6.

About the authors

Shubhani Aggarwal is pursuing a PhD from Thapar Institute of Engineering and Technology (Deemed to be University), Patiala, Punjab, India. She received the BTech degree in Computer Science and Engineering from Punjabi University, Patiala, Punjab, India, in 2015, and the ME degree in Computer Science from Panjab University, Chandigarh, India, in 2017. She has many research interests in the area of Blockchain, cryptography, Internet of Drones, and information security.

Neeraj Kumar received his PhD in CSE from Shri Mata Vaishno Devi University, Katra (Jammu and Kashmir), India in 2009, and was a postdoctoral research fellow in Coventry University, Coventry, UK. He is working as a professor in the Department of Computer Science and Engineering, Thapar Institute of Engineering and Technology (Deemed to be University), Patiala (Pb.), India. He has published more than 400 technical research papers in top-cited journals such as IEEE TKDE, IEEE TIE, IEEE TDSC, IEEE TITS, IEEE TCE, IEEE TII, IEEE TVT, IEEE ITS, IEEE SG, IEEE Netw., IEEE Comm., IEEE WC, IEEE IoTJ, IEEE

SJ, Computer Networks, Information sciences, FGCS, JNCA, JPDC, and ComCom. He has guided many research scholars leading to PhD and ME/MTech. His research is supported by funding from UGC, DST, CSIR, and TCS. His research areas are Network Management, IoT, Big Data Analytics, Deep Learning, and Cybersecurity. He is serving as editors of the following journals of repute: ACM Computing Survey, ACM·IEEE Transactions on Sustainable Computing, IEEE·IEEE Systems Journal, IEEE·IEEE Network Magazine, IEE·IEEE Communication Magazine, IEE·Journal of Networks and Computer Applications, Elsevier·Computer Communication, Elsevier·International Journal of Communication Systems, Wiley. Also, he has been a guest editor of various International Journals of repute such as—IEEE Access, IEEE ITS, Elsevier CEE, IEEE Communication Magazine, IEEE Network Magazine, Computer Networks, Elsevier, Future Generation Computer Systems, Elsevier, Journal of Medical Systems, Springer, Computer and Electrical Engineering, Elsevier, Mobile Information Systems, International Journal of Ad hoc and Ubiquitous Computing, Telecommunication Systems, Springer and Journal of Supercomputing, Springer. He has also edited/authored 10 books with International/National Publishers like IET, Springer, Elsevier, CRC. Security and Privacy of Electronic Healthcare Records: Concepts, paradigms and solutions (ISBN-13: 978-1-78561-898-7), Machine Learning for cognitive IoT, CRC Press, Blockchain, Big Data and IoT, Blockchain Technologies across industrial vertical, Elsevier; Multimedia Big Data Computing for IoT Applications: Concepts, Paradigms and Solutions (ISBN: 978-981-13-8759-3), Proceedings of First International Conference on Computing, Communications, and Cyber-Security (IC4S 2019) (ISBN 978-981-15-3369-3). One of the edited text-book entitled, "Multimedia Big Data Computing for IoT Applications: Concepts, Paradigms, and Solutions" published in Springer in 2019 is having 3.5 million downloads till 6 June, 2020. It attracts attention of the researchers across the globe (https://www.springer.com/in/book/9789811387586). He has been a workshop chair at IEEE Globecom 2018 and IEEE ICC 2019 and TPC Chair and member for various International conferences such as IEEE MASS 2020, IEEE MSN2020. He is senior member of the IEEE. He has more than 12,321 citations to his credit with current h-index of 60 (September 2020). He has won the best papers award from IEEE Systems Journal and ICC 2018, Kansas-city in 2018. He has been listed in the highly cited researcher of 2019 list of web of science (WoS). In India, he is listed in top 10 position among highly cited researchers list. He is adjunct professor at Asia University, Taiwan, King Abdul Aziz University, Jeddah, Saudi Arabia and Charles Darwin University, Australia.

Printed in the United States
By Bookmasters